Results and Problems in Cell Differentiation

W0245774

A Series of Topical Volumes in Developmental Biology

18

Editors

W. Hennig, L. Nover, and U. Scheer

Results and Problems in Cell Differentiation

W. Hennig (Ed.)

Early Embryonic Development
of Animals

With 38 Figures and 9 Tables

Springer-Verlag Berlin Heidelberg GmbH

Professor Dr. Wolfgang Hennig
Dept. of Molecular and Developmental Genetics
University of Nijmegen
Toernooiveld, NL-6525 ED Nijmegen
The Netherlands

ISBN 978-3-662-21836-5

Library of Congress in Publication Data
Early embryonic development of animals / W. Hennig, ed. p. cm.
(Results and problems in cell differentiation; 18)
Includes bibliographical references and index.
ISBN 978-3-662-21836-5   ISBN 978-3-540-47191-2 (eBook)
DOI 10.1007/978-3-540-47191-2
1. Developmental genetics. 2. Embryology.
I. Hennig, Wolfgang, 1941- . II. Series.
QH607.R4 □ vol. 18 □ [QH453] □ 574.87´612 s--dc20 □ [591.3´3] □ 92-20781 CIP

© Springer-Verlag Berlin Heidelberg 1992
Originally published by Springer-Verlag Berlin Heidelberg New York in 1992
Softcover reprint of the hardcover 1st edition 1992

Typesetting Best-set, Hong Kong
31/3145 - 5 4 3 2 1 0 - Printed on acid-free paper

Introduction

There has been extremely rapid progress in developmental biology over the past 10 years. Thus. few conceptual publications are available which may serve as an introduction to a particular field and give a suffiently broad range of insight without remaining too superficial. Such fundamental reviews, however, are needed, not only for advanced university teaching but also for graduate and postgraduate students entering a new field of interest.

This situation has induced me to prepare this volume of the Results and Problems in Cell Differentiation Series which might fulfill some of these requirements. Obviously, it is impossible to cover the wide range of topics in a volume of this kind. This requires not only a selection focusing on certain representative topics, but also presenting a variety of approaches and the different facets of developmental processes studied in different organisms. The most extensively studied organisms are *Drosophila*, *Caenorhabditis* and the mouse. Therefore these are the systems chosen for this volume. In addition, the zebrafish is being introduced in the field of developmental biology research so quickly that I considered it important to include an evaluation of its advantages and possibilities.

I believe that the different contributions cover their topics extremely well. I have used the reviews for teaching and found them very suitable for this purpose. Certainly, I myself have learned a great deal from this lecture and I enjoyed reading the different articles. Hopefully also other readers will share this opinion and will be encouraged to further extend their knowledge not only in a particular system but in the entire extent of the field.

It is with much gratitude that I thank the authors. Two of the articles were completed very early and had to be revised to keep them updated because of the retraction of other contributions. I feel very much indebted to these authors for their patience and understanding and I am grateful to the other for doing their best to submit their manuscripts quickly.

Nijmegen, Juni 1992   *Wolfgang Hennig*

Contents

2 Embryonic Development
in *Caenorhabditis elegans* (With 7 Figures)
PAUL E. MAINS

3 Analysis of Early Development
in the Zebrafish Embryo (With 3 Figures)
ERIC S. WEINBERG

4 Early Mouse Development (With 9 Figures)
ACHIM GOSSLER

5 Genomic Imprinting in Mammals (With 8 Figures)
WOLF REIK

1 The Molecular Genetic Basis of Positional Information in Insect Segments

Joan E. Hooper[1] and Matthew P. Scott[2]

1 Introduction

Pattern formation is the process by which a group of initially unspecified cells organizes itself to generate a precise array of structures. Pattern forming systems as diverse as the vertebrate limb, the insect segment, and the hydra body column, are remarkably resistant to environmental effects and to malicious manipulations perpetrated by researchers. Large variations in the overall size of the organism, the number of cells, and the growth conditions can be tolerated. Injuries result in attempts by the surviving cells to regenerate the pattern. For example, if part of an insect segment is damaged or surgically removed then some or all of the missing structures will be regenerated during subsequent molts. Such regenerative responses to injury are termed "pattern regulation". Pattern regulation almost certainly reflects the same mechanisms that accommodate variations in cell number and arrangement during normal development. Thus, studies of pattern regulation in response to injury should reveal the processes that form pattern de novo during development.

In the earliest stages of embryogenesis, pattern regulation involves the whole embryo, as in twinning. Later, pattern regulation is independent in discrete parts of the embryo such as a limb or a segment. The unit within which pattern regulation occurs has been termed a "morphogenetic field" (Weiss 1939). The relative position within the morphogenetic field determines the ultimate fate of each cell. Thus, "positional information", the information a cell possesses on its own position within the morphogenetic field (Wolpert 1969), is the key to pattern formation. The concepts of morphogenetic fields and pattern regulation are based upon observation of a small number of experimentally favorable systems. However, the similarities between the regulative responses in diverse organisms and tissues suggests a common mechanism in many pattern forming systems.

Polarity refers to the orientation of an asymmetric pattern element relative to the morphogenetic field. It might refer to the orientation of

[1] Department of Cellular and Structural Biology, University of Colorado Health Sciences Center Denver, Colorado 80262, USA
[2] Departments of Developmental Biology and Genetics, Stanford University School of Medicine, Stanford, California 94305-5427, USA

Results and Problems in Cell Differentiation 18
W. Hennig (Ed.)
Early Embryonic Development of Animals
© Springer-Verlag Berlin Heidelberg 1992

a bristle or of an ordered groups of cells such as the rhabdomere of a compound eye. In cellular terms, polarity is an intrinsic property of each cell, reflected in internal asymmetries. As a cell or group of cells forms a structure, it must make the pattern element appropriate to its position and must orient that pattern element within the field. Thus, two decisions are involved in the formation of a bristle: whether to make a bristle and which way to point it.

Two experimental approaches are converging to reveal the mechanisms that underlie positional information within the insect segment. First, transplantation and ablation experiments reveal how cells respond to abnormal cell-cell contacts or abnormal positions. They suggest that local cell interactions are critically important in pattern regulation. Second, genetic screens using the fruit fly *Drosophila* have identified genes whose products mediate pattern formation. Studies of these genes and their products will elucidate the molecular and cellular basis for positional information within the insect segment. Here we review "segment polarity" genes, those genes responsible for pattern formation within the segment, and relate their functions to classical studies (also reviewed in Martinez-Arias 1989). Homologues of many *Drosophila* pattern-forming genes play critical roles in vertebrate development. What we learn from insects apparently applies to the development of a broad spectrum of animals.

2 The Experimental Basis for Positional Information and Its Local Character

In 1940, Wigglesworth initiated a new era in studies of pattern formation. He observed that larval epidermal cells of the blood-sucking bug, *Rhodnius*, are determined with respect to polarity (e.g. bristles oriented in the anteroposterior axis) and to general region (e.g. dorsal, ventral, lateral), but not with respect to specific cell type (e.g. bristle or smooth cuticle). His experimental paradigm, grafting or ablating and then monitoring the ensuing pattern perturbations, has established the following properties of the determined state. First, cells have a memory both of their polarity and of their position within the segment. Second, cells respond to the polarity and the positional state of neighboring cells. Third, incompatible neighboring cells trigger responses that ultimately stabilize with compatible neighborly relationships.

2.1 Ablation and Grafting Experiments Reveal the Importance of Local Cell-Cell Interactions

Piepho (1955) rotated by 180° bits of integument from the caterpillar, *Galleria*. After pupation he noted that scales that formed in the middle of

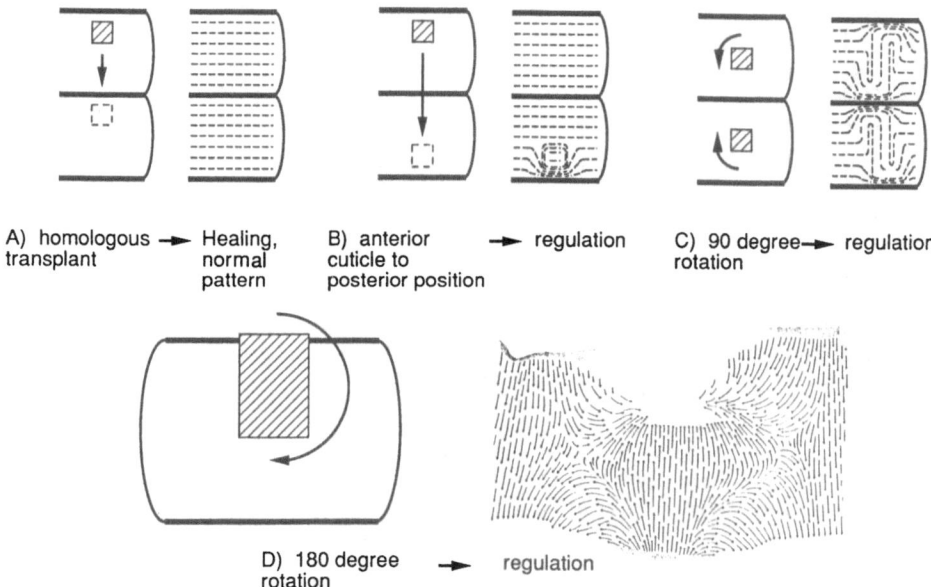

Fig. 1. Summary of cuticle transplantation experiments in the abdomen of *Rhodnius* and of *Galleria*. **A** When a piece of larval epidermal epithelium is transplanted to a homologous position in another segment, the cells reproduce the normal cuticle pattern. **B** When a piece of larval epidermal epithelium is transplanted to a nonhomologous position in a different segment, the cells around and within the graft contribute to new cuticular patterns. **C** When a piece of larval epidermal epithelium is rotated 90°, the cells around and within the graft contribute to new cuticular patterns. Note the handedness of the new patterns. (**A–C**; *Rhodnius*, redrawn from Locke 1959). **D** When a piece of larval epidermal epithelium is rotated through 180°, the new pattern is globally disorganized, but locally the continuity between polarities of the pattern elements is reconstituted. (*Galleria*, redrawn from Piepho 1955)

the graft had the expected reversed polarity while scales near the graft junction reoriented to directions intermediate between graft and host (Fig. 1D). Intrinsic cell polarity, seen in the center of the graft, was subject to influences by neighboring host cells near the edge of the graft. Those influences respecified polarity to minimize the difference in orientation between each cell and its neighbor.

Locke (1959) investigated the factors controlling the transverse ripples of the dorsal abdomen of *Rhodnius*. Grafts to the homologous position within the same segment, to homologous positions of different segments, or across the dorsal midline formed patterns indistinguishable from unoperated controls (Fig. 1A). Ninety degree or 180° rotations produced results similar to the *Galleria* experiments; ripples in the center of the graft reflected the original orientation of the graft; ripples far from the graft presented a normal pattern. Near the graft-host junction, the ripples deviated from both graft and host patterns, restoring continuity of individual ripples (Fig. 1C). Surprisingly, grafts to nonhomologous positions generated whorls of ripples even when they were not rotated (Fig. 1B). Apposition of cells with the

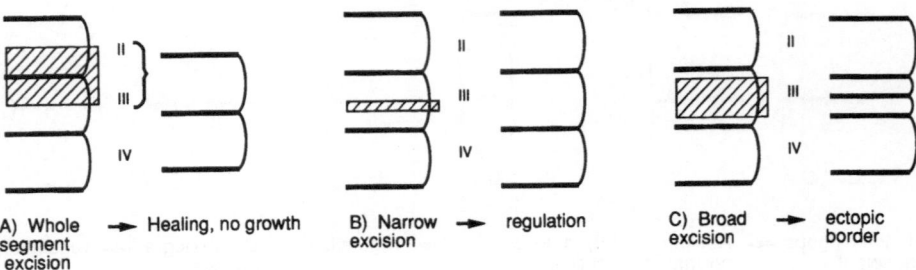

Fig. 2. Summary of cuticle ablation experiments in the abdomen of *Oncopeltus*. The consequences of surgical removal of different amounts of the segment repeat from the abdominal cuticle of the third-stage larva is assayed in the fifth-stage larva, two molts later. **A** If a whole segment equivalent is excised, the wound heals generating a normally sized segment of hybrid character. **B** If only a narrow strip of a segment is removed, the excised part of the pattern is regenerated by regulation. **C** If more than half of a segment is removed, the regulative response regenerates an ectopic segment border in reversed polarity rather than regenerating the excised part of the segment. (Redrawn from Wright and Lawrence 1981)

same polarity, but from different anteroposterior levels within a segment triggered respecification just like apposition of cells of different polarity. He characterized the cellular attribute revealed by these experiments as a "segmentally repeating gradient of incompatibility" (Locke 1966). Cells from similar positions within the segment were "compatible" so graft junctions simply healed. However, cells from different positions within the segment were "incompatible" and triggered responses that smoothed the discontinuities to restore compatibility. In other words, cells have records of their position within the segment and of their polarity, information that is repeated and equivalent in every segment. Cells also sense those properties in neighboring cells, determine whether they are "compatible", and respond accordingly.

Nübler-Jung (1977) demonstrated respecification using the cotton bug, *Dysdercus*, where anterior cells are heavily pigmented and posterior cells are unpigmented. Cells at the edge of unpigmented posterior grafts acquired the pigmented character of anterior cells when transplanted to anterior locations. This was clearly respecification of graft cells since graft and host cells were genetically marked. Reciprocal respecification was seen at the edge of anterior-to posterior grafts.

Wright and Lawrence (1981) cauterized transverse strips in the abdomen of *Oncopeltus* larvae to appose cells from different levels within the segment. Wounds that excised a full segment healed with no regeneration (Fig. 2A). Small wounds healed to reconstitute the complete pattern, even if that involved regenerating a segment border (Fig. 2B). But when wound healing apposed cells whose positions differed by more than half a segment, the regenerative response produced ectopic segment borders in mirror-image polarity (Fig. 2C). Despite the global pattern disruption, the new tissue restores local continuity to the pattern so that each pattern element

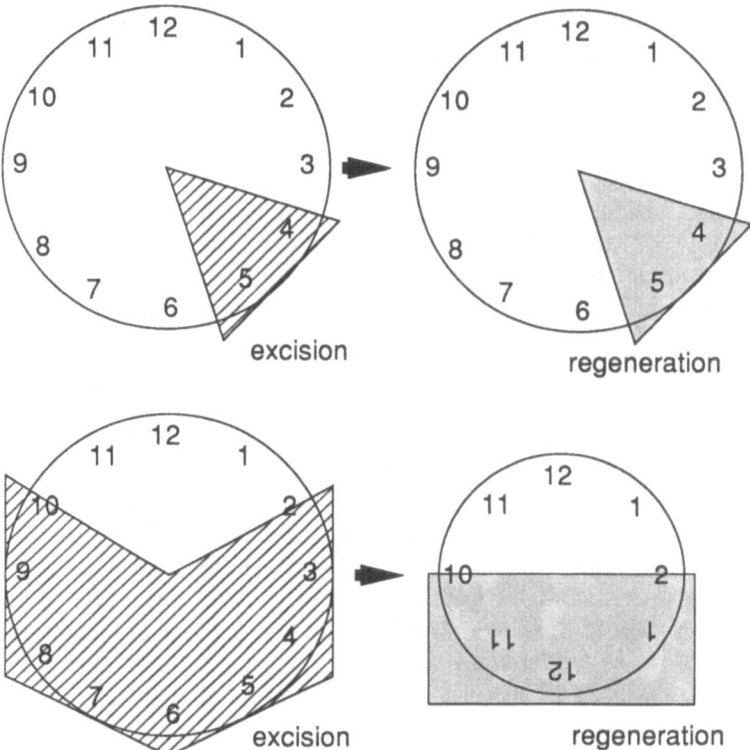

Fig. 3. Polar coordinate representation of pattern regulation. Different cell "states" along the anteroposterior axis of each segment or the circumference of a limb are formally represented by a clock face numbered *1–12*. *Upper panel* If a small part of the pattern is removed (e.g. *4* and *5*), creating a small discontinuity as the wound edges meet, it will be regenerated by the cells along the wound edges (i.e., the juxtaposed 3 and 6 cells), recreating continuity with appropriate polarity. *Lower panel* If a sufficient fraction of the cells is removed (e.g. deletion of *3–9* to juxtapose 2 and *10*), the regenerated cells form pattern elements 11, 12, and 1 rather than 3–9. The duplication of 11, 12, and 1 is oriented opposite to the original tissue. The discontinuity at the wound edges is resolved by intercalation of a mirror-image duplication of the remaining pattern. In both cases, it is the "shortest distance" or the "smallest part of the pattern" which is formed, regardless of polarity reversals that may be generated. Local continuity is retained at the expense of the overall pattern. (French et al. 1976)

is apposed only to normal neighbors. They explain the mirror-image duplications based on polar coordinates (French et al. 1976; Bryant et al. 1981), with the following rules. Cells along the anteroposterior axis of each segment have different "states", formally represented by a clock face numbered 1–12 (Fig. 3). Continuity between neighboring cells (i.e. contiguous numbered states) is imperative. Discontinuity induces regeneration via intercalation of the smallest number of states. If a small part of the pattern is removed (e.g. 4 and 5), it will be regenerated by the cells along the wound edges (i.e. the juxtaposed 3 and 6 cells; Fig. 3, upper panel).

However, if a sufficient fraction of the cells is removed (e.g. deletion of 3–9 to juxtapose 2 and 10), the regenerated cells form pattern elements 11, 12, and 1 rather than 3–9 (Fig. 3, lower panel). The duplication of 11, 12, and 1 is oriented opposite to the original tissue. Within the mirror-image duplication each cell has a polarity 180° out of phase with the body axis; local interactions between cells in different states determines the polarity of each cell. Local continuity of the pattern is maintained so that each pattern element is apposed only to normal neighbors. Because local continuity is maintained at the expense of global pattern, we conclude that local rather than long range cues dominate the interactions that underlie pattern regulation.

The clock, or polar coordinate, model (French et al. 1976; Bryant et al. 1981) presented above in the context of the insect segment, accurately describes pattern regulation in *Drosophila* imaginal discs (the precursor tissues of much of the adult body), cockroach legs, and amphibian limbs. In each case, normal neighbors are restored by intercalation of the smallest possible part of the pattern, regardless of orientation relative to the body axis. Apparently local interactions, local compatibility, and local continuity guide regeneration, and by implication also pattern formation. The long range organization of the pattern might simply be the sum of local processes. The difficulty is to understand local continuity in cellular and molecular terms and to understand how it is produced and maintained.

2.2 Mechanisms of Regulative Responses

The regeneration phenomena suggest that in many instances cells do not sense position or polarity with respect to the whole animal. Rather, they respond to local signals which could be secreted molecules that move over a short range or surface components of neighboring cells. Cells are normally arranged with one type of neighbor on one side and another type on the other. If either neighbor is replaced by a third type, the cells sense a problem and respond. The repertoire of responses includes respecification (Marcus 1962; Abbott et al. 1981; Bryant and Fraser 1988), movement (Nübler-Jung 1974; Nardi and Kafatos 1976), mitosis (Nübler-Jung 1977; Dale and Bownes 1980; Campbell and Shelton 1987; Campbell and Caveney 1989), and possibly also cell death.

Is it the lack of a normal signal, the presence of an abnormal one, or both that induces a regulative response? Bryant (1987) proposed that the growth of duplicated structures could be triggered by interfering with the cells' ability to recognize the appropriateness of their neighbors. This was based partly on the observation that mutations that cause random cell death in *Drosophila* imaginal discs also produce random mirror-image duplications, albeit at low frequencies. A dying cell might no longer transmit the appropriate signal, and thereby trigger a regenerative response in a neighboring cell.

Locke (1966) attempted to show that regulative responses are induced by the interruption of cell communication as well as by wounding or grafting. He monitored cell migration, an early response to wounding, when filter strips were implanted in a slit in the insect epidermis to form a barrier between cells on either side of the slit. Permeable (nitrocellulose) filters triggered no response, while impermeable filters triggered cell migration. This was interpreted as evidence that blocking a signal is sufficient to trigger at least the initial stages of a regulative response. However, it is unclear whether the diffusion of factors through the filters, cell contact through the filter pores, or differences in the surface properties of the filters caused the different responses to the different filters.

It is unclear whether direct cell contact is required for regenerative responses, or whether diffusion of a chemical substance is sufficient. In the caterpillar stage of *Galleria*, the polarity of some of the host insect's cells was changed by a fragment of donor epithelium implanted in reverse orientation underneath the host epithelium (Piepho and Hintze-Podufal 1971). The respecification occurred in the absence of obvious cell migration between donor and host. Direct cell contact between the parallel epithelial sheets was not excluded, thus presenting a problem since epithelial cells extend processes as long as several cell diameters (Locke and Huie 1981; Nardi and Magee-Adams 1986).

Whatever the mechanisms, cells must learn what to do themselves, what to expect from their neighbors, and how to respond if the neighbor is not the expected one. What types of information allow the cell to develop such biases?

2.3 Models of Positional Information

Three classes of models have been advanced to explain how the insect segment is organized. In the first (Fig. 4A), positional information is embodied by a graded distribution of some factor across the developing or regenerating segmental field (e.g. Lawrence 1966; Stumpf 1968). Cells detect the level of this "morphogen" and respond accordingly. Pattern elements are specified by the level of the morphogen, polarity by the direction of the gradient, and size/growth control by the steepness of the gradient. In various models, the gradient (or two opposing gradients) is generated by some combination of local sources, local or general sinks, and diffusion of the morphogen (e.g. Gierer and Meinhardt 1972). One problem with gradient models is that two adjacent cells often need to respond differently. It is difficult to imagine how a quantitative difference in morphogen as small as 1% could consistently be interpreted to direct the formation of two different cell types. Some sort of local cross-checking system should be necessary, in addition to the gradient, to accomplish such precision. A second difficulty is raised by repeating patterns such as insect

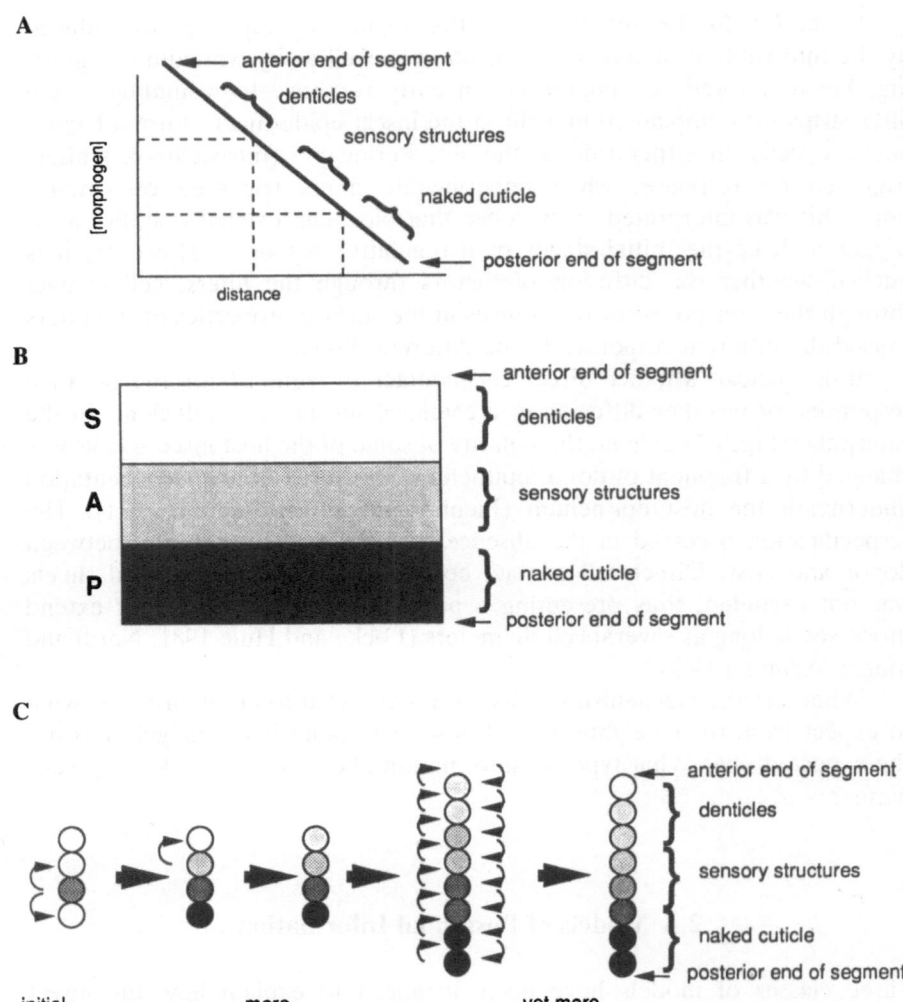

Fig. 4. Models for generation of diversity within a developing field. **A** In gradient models, the quantity of a differentiation-determining substance, a morphogen, is present in a concentration gradient across the segment, with high and low points at the segment (or parasegment) border. Cells read the local concentration and then maintain that level in their local environment by acting as a sources and/or sinks for the morphogen. As cells proliferate, they continue to maintain the morphogen gradient. At some later stage in development, they read the morphogen level and differentiate accordingly. **B** Some small number of states (*S, A, P*) are established when or soon after the segment primordia are established. After they are established, these states are independent of the states of neighboring cells and are transferred to progeny with high fidelity. The control of morphogenesis is not addressed. **C** One or more cells within the segment primordium is initially specified. That cell(s) induces near neighbors to take on different states. As cells proliferate, a chain of near-neighbor interactions achieves a stable array of states within the segment primordium. Different fates then follow from the different states. (Redrawn from Martinez-Arias 1989; see text for references)

segments which require discontinuities at segment boundaries. The high point of a morphogen gradient in one segment would be adjacent to a low point in the next segment. Experimental evidence speaks against a discontinuity anywhere in the segment repeat (Wright and Lawrence 1981) or in the circumference of a limb (Bryant et al. 1981). The product of the *wingless* (*wg*) gene may act as a morphogen to organize the posterior half of each segment (Sects. 5 and 6).

The second class of models proposes that permanent cell states or "compartments" underlie the segmental repeat (Fig. 4B). Each state is marked by the heritable expression of a "selector" gene, analogous to the specification of segmental differences by homeotic "selector" genes. Lineally related cells are in the same state and there are lineage restrictions between cells in different states. At least two discrete, lineally distinct states clearly exist in *Drosophila* imaginal discs, defining an anterior and a posterior compartment (Garcia-Bellido 1975; Lawrence and Morata 1983; Lawrence 1989). Activity of the *engrailed* gene specifies cells as members of the posterior compartment (reviewed in Lawrence 1981). With only two cell states, polarity must be an intrinsic property of all cells, established by the primary embryonic axis (Lawrence 1989). It cannot be a product of neighboring cells since the neighboring cells in *both* directions from any compartment are in the opposite state. But intrinsic polarity is at odds with the experimentally induced polarity reversals that accompany mirror-image pattern duplications. This relatively simple picture also fails to explain growth control and the mirror-image duplications themselves.

If three (Meinhardt 1986) or more (Martinez-Arias et al. 1988) cell states occupy each segment, then polarity, growth control, mirror image duplications can be explained. Three or more states (e.g. SAP/SAP/SAP for three segment primordia) create an inherent polarity. To cells of the A state, S neighbors mean anterior and P neighbors mean posterior and so forth. Upon cell ablation, patterns are regenerated if the sequence remains intact, but mirror-image duplications occur if the sequence is reversed. For instance, deletion of A cells to give SAP/SXP/SAP would lead to regeneration of the normal pattern, but deletion of SA to give SAP/XXP/SAP could lead to the generation of SAP/S\P/SAP, with an inverted polarity at the new S. This accounts for mirror-image duplications. If the different states are mutually activating (Meinhardt and Gierer 1980), then size and/or proliferation is governed by the distance over which the activation is effective. This accounts for the cell proliferation that usually accompanies regeneration or mirror-image duplication (Nübler-Jung 1977; Dale and Bownes 1980), and the growth regulation that accompanies normal development (Bryant 1987).

Multiple cell states within each segment provide a crude basis for pattern, but do not explain how the final, continuous, differentiated pattern is generated. A refinement of the cell state model (Meinhardt 1983; Lawrence et al. 1987) postulates that the boundaries between states are the real organizing centers from which patterns are specified. Pattern elements are arrayed around segment and parasegment borders. Experimental evidence that these borders are organizing centers for the pattern, rather than elements of the pattern, is lacking. However, molecular analysis of segment polarity genes suggests that borders are established early in development, and their integrity is essential for pattern formation within the segment (Sect. 5).

Mittenthal (1981) restated the polar coordinate description in cellular terms as "the rule of normal neighbors" (interpreted in Fig. 4C; Martinez-Arias 1989). The organization of cells within the segment is mediated by local cell interactions which establish positional states. Positional state is not the permanent state of compartment models; nor is it the quantity of gradient models. It is a quality which expresses the current state of determination of the cell. Certain neighborly relationships are mutually enforcing, while others are unstable. Inappropriate neighbors trigger a chain of events which ultimately results in only appropriate neighborly relationships. The field is gradually organized through a sequence of local cell interactions. This model might be thought of as a refinement of cell state models; distinct states at early stages interact as development progresses, to intercalate additional states and refine the pattern. An example of mutually reinforcing interactions, between *wg* and *engrailed* (*en*)-expressing cells in the *Drosophila* embryo is discussed in Section 5.

These models are consistent with the experimental results and provide stimulating ideas, but they provide little guidance for further experiments. Another approach is to identify and study the molecules responsible for pattern regulation and positional information. Mutations in genes for the information-carrying molecules might prevent a cell or group of cells from sensing that the neighbors are as they should be. Such mis-sensing would mimic physical ablations, leading to respecification or regeneration and thence, aberrant or duplicated patterns. During the past decade or so, *Drosophila* mutants have been found that have at least some of the properties expected for defects in the patterning molecules mediating cell-cell communication. These mutants are the main subject of this chapter.

3 *Drosophila* Development and Segmentation Genes

3.1 How a Syncytium Becomes a Metameric Embryo

The stages of *Drosophila* development are summarized in Table 1. After fertilization, 13 extremely rapid nuclear division cycles generate a syncytial blastoderm of some 6000 nuclei in a monolayer at the cortex of the embryo (Foe and Alberts 1983). The anterior-posterior axis of the egg is established by maternally encoded mRNAs. For example, mRNA from the gene *bicoid* (*bcd*), is deposited at the anterior pole of the egg during oogenesis. After fertilization, translation of the *bcd* mRNA leads to a gradient of the *bcd* protein (BCD)[3], with its highest concentration at the anterior pole of the syncytial embryo. BCD is a transcription factor (Driever and Nüsslein-Volhard 1989; Struhl et al. 1989; Hoch et al. 1991). The gradient of BCD initiates a cascade of transcriptional activation (summarized in Fig. 5) which ultimately organizes the anterior embryo.

The zygotically active segmentation genes act sequentially to control finer and finer aspects of segmental patterning (Nüsslein-Volhard and Wieschaus 1980). After nuclear cycle nine (stage 4, Table 1), zygotic nuclei respond to the BCD gradient by the local activation or repression of region-specifying "gap" genes. Different gap genes specify different blocks of embryonic segments. Four gap genes have been cloned. *Krüppel*, *knirps*, *hunchback*, and *tailless*, all encode zinc-finger proteins (Rosenberg et al. 1986; Tautz et al. 1987; Nauber et al. 1988; Pignoni et al. 1990) that probably act as transcription factors (Stanojevic et al. 1989; Treisman and Desplan 1989; Pankratz et al. 1990; Zuo et al. 1991). Transcriptional cross-regulation between the gap genes generates overlapping domains of gap gene expression that subdivide the anterior-posterior axis of the embryo (Hülskamp et al. 1990; Mohler et al. 1989; Eldon and Pirotta 1991; Kraut and Levine 1991; reviewed in Akam 1987; Scott and Carroll 1987; Ingham 1988).

Early in nuclear division cycle 14 (stage 5, Table 1), gap gene activities modulate the ubiquitous transcription of the pair-rule genes *hairy*, *runt*, and *even-skipped* into repeating transverse stripes. Each stripe plus interstripe corresponds to a two-segment repeat along the body axis. Mutations in pair-rule genes cause defects at two segment intervals, the defects primarily occurring in cells that would normally have expressed the gene. Each stripe of *hairy* or *even-skipped* is controlled by a different "stripe response element", a cis-acting DNA sequence with clustered binding sites for gap gene products (Howard et al. 1988; Carroll and Vavra 1989; Hooper et al. 1989). The stripes are formed as specific ratios of gap gene activities repress and activate transcription of *hairy* or *even-skipped* in each interstripe and stripe (Carroll and Vavra 1989; Stanojevic et al. 1989; Harding et al. 1989; Carroll 1990; Pankratz et al. 1990; Warrior and Levine 1990; Small et al. 1991; Stanojevic et al. 1991).

The pair-rule genes *hairy*, *runt*, and *even-skipped* control the striped expression patterns of other pair-rule genes such as *fushi tarazu* and *paired* (Howard and Ingham 1986; Carroll and Scott 1986; Baumgartner and Noll 1991). During cycle 13 the latter genes are uniformly transcribed throughout the primordia of the body segments. As the stripes of *hairy*, *runt*, and *even-skipped* form, the uniform transcription of *fushi tarazu* and *paired* resolves into stripes. Cross-regulatory and autoregulatory interactions among pair-rule genes during cycle 14 stabilize and refine the stripes (Carroll and Scott

---

[3] Genetic notations: italics (e.g. *en*) refers to the gene or its transcription, uppercase (e.g. EN) refers to the protein product.

Table 1. Stages of *Drosophila* embryogenesis (Campos-Ortega and Hartenstein 1985)

Stage	Time (AEL[a])	Features
1	0–0:25 h	Cleavage divisions 1–2
2	0:25–1:05 h	Cleavage divisions 3–8; nuclei begin to migrate towards cortex
3	1:05–1:20 h	Cleavage division 9; blastoderm nuclei reach cortex; pole cell formation; midblastula transition (zygotic transcription activated)
4	1:20–2:10 h	Syncytial blastoderm: last four cleavage divisions; "gap" genes interacting to delineate broad domains
5	2:10–2:50 h	Cellularization of the blastoderm; interphase of cycle 14; pair-rule genes activated in stripes in alternate segment primordia; *en*, *wg*, *hh* and *gsb* activated in stripes in part of each segment primordium
6	2:50–3:00 h	Early gastrulation: mesoderm invaginates: cephalic, anterior, and posterior transverse furrows form; posterior midgut primordium forms
7	3:00–3:10 h	Gastrulation completed; anterior midgut invagination begins
8	3:10–3:40 h	Germ-band elongation; two mitoses in mesoderm: one asynchronous round of mitosis in ectoderm
9	3:40–4:20 h	Segmentation of the mesoderm; neuroblasts begin to segregate; *ptc* and *ciD* resolve into stripes
10	4:20–5:20 h	Parasegment grooves form; tracheal placodes form; stomodeum (foregut primordium) invaginates; second asynchronous round of mitosis in the ectoderm; *wg* stripes modulated
11	5:20–7:20 h	Segmental furrows form; final mitosis in the ectoderm; cell death in the ectoderm (second half of stage 11); final mitosis in mesoderm, visceral and somatic mesoderm split; *ptc* stripes split
12	7:20–9:20 h	Germ-band retraction; nerve cord separates from epidermis, axon outgrowth begins; anterior and posterior midgut primordia meet and fuse

Table 1. *Continued*

Stage	Time (AEL[a])	Features
13	9:20–10:20 h	Peripheral nervous system forms; somatic muscles begin to form
14	10:20–11:20 h	Head involution.
15	11:20–13 h	Dorsal closure.
16	13–16 h	Midgut constrictions appear; cuticle secretion begins
17	16 h-hatching	Further organogenesis; movements and circulation begin

[a] After egg laying.

1986; Frasch and Levine 1987; Hiromi and Gehring 1987; Jiang et al. 1991). The stripes formed by the different pair-rule gene products are out of phase so that each cell in the two-segment repeat expresses a unique combination of pair-rule gene products. Therefore, a combinatorial code of pair-rule genes could control the determination of each cell within the segment (Gergen et al. 1986; Ingham et al. 1988; Martinez-Arias et al. 1988). *hairy, odd-skipped, even-skipped, fushi tarazu,* and *paired* all encode proteins that have been shown to be, or are likely to be, transcription factors (Laughon and Scott 1984; Frigerio et al. 1986; Macdonald et al. 1986; Rushlow et al. 1989). Thus the initial specification of the segmented body plan of *Drosophila* is a cascade of combinatorial actions of transcription factors integrated on complex promoters within the syncytial embryo (Fig. 5; reviewed in Akam 1987; Scott and Carroll 1987; Ingham 1988; Nüsslein-Volhard and Roth 1989).

Segment polarity genes control the formation of pattern within each segment. Mutations in these genes cause parts of the pattern to be absent from each segment, as described further below. Transcription of the segment polarity genes *en* (DiNardo et al. 1985; Fjose et al. 1985; Kornberg et al. 1985), *gooseberry* (*gsb*; Baumgartner et al. 1987; Côté et al. 1987), *wg* (Baker 1987; van den Heuvel et al. 1989), and *hedgehog* (*hh*; Mohler and Vani, 1992; P. Beachy, pers. comm.) is activated late in cycle 14 in single-cell wide stripes, one per segment, girdling the blastoderm. At the same time, expression of the pair-rule genes *even-skipped* and *paired* changes to segment-periodicity stripes. Segment polarity genes are probably activated by specific combinations of positive and negative signals from the pair-rule gene products (Gergen et al. 1986; Harding et al. 1986; Howard and Ingham 1986; Macdonald et al. 1986; Scott and O'Farrell 1986; DiNardo and O'Farrell 1987; Ingham et al. 1988). This is the final step in the transcriptional cascade that generates segmental divisions of the embryo. At the end of nuclear cycle 14, cell membranes extend inward from the surface of the

14   Joan E. Hooper and Matthew P. Scott

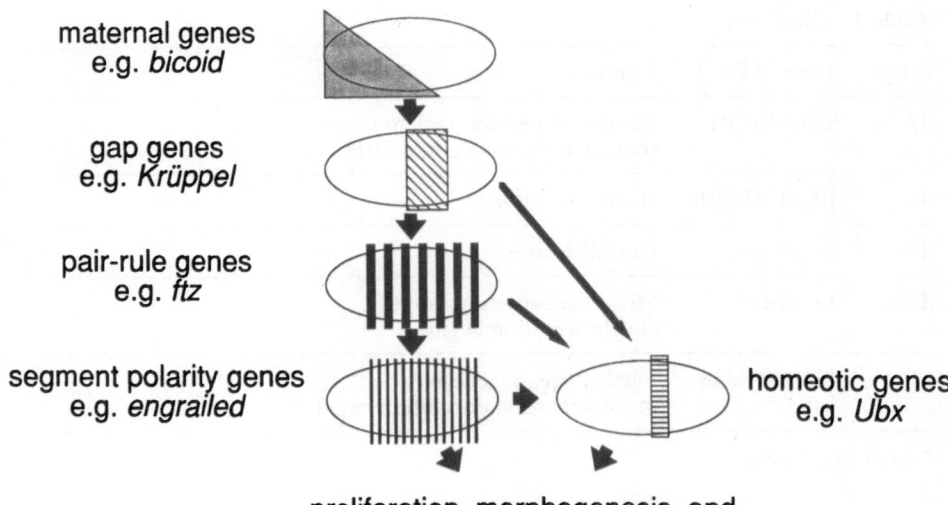

Fig. 5. The progressive refinement model of segmentation. Maternally encoded mRNAs deposited or activated asymmetrically in the egg establish the polarity of the embryo. "Gap" genes interpret graded maternal information to specify regions of the embryo, and interact with each other to refine their patterns. "Pair-rule" genes respond to the gap genes to establish repeating patterns of transcripts and protein within the anlage of the segmented embryo, at two-segment periodicity. "Segment polarity" genes respond to combinations of pair-rule gene products to become expressed in a single stripe in each segment primordium. Interactions among the segment polarity genes refine or maintain their patterns and activities. Homeotic genes respond to at least some of the gap genes, pair-rule genes, and segment polarity genes to establish the differences between segments within the embryo. Homeotic and segment polarity genes continue to be expressed through embryogenesis and direct cell proliferation, morphogenesis, and differentiation. This model is a general summary and does not include all the details and exceptions. (See text for references)

embryo to enclose the blastoderm nuclei (Foe and Alberts 1983). This transition from syncytial to cellular development adds a second level of complexity to patterning the embryo, and cellular dialogues become important during the ensuing proliferation and differentiation events. Segment polarity genes mediate and respond to cell communication events which establish the intrasegmental pattern.

Cells are allocated to specific segmental fates during cycle 14 (Simcox and Sang 1983), by localized activation of homeotic selector genes including *labial, Deformed, Sex combs reduced, Antennapedia (Antp), Ultrabithorax (Ubx), abdominalA,* and *AbdominalB* (Garcia-Bellido 1977; Lewis 1978; Kaufman et al. 1980; reviewed in Mahaffey and Kaufman 1988). All of these genes contain homeoboxes and their products can regulate transcription (Thali et al. 1988; Jaynes and O'Farrell 1988; Krasnow et al. 1989; Samson et al. 1989; Winslow et al. 1989). The initial activation of homeotic genes is controlled by the levels and/or ratios of gap gene-encoded transcription factors (White and Lehmann 1986; Riley et al. 1987; Harding and Levine 1988; Irish et al. 1989; Reinitz and Levine 1990) and also by some segmentation gene-encoded transcription factors (Ingham and Martinez-Arias 1986; Martinez-Arias et al. 1988).

Specification of cells along the dorsal-ventral body axis is independent of specification by the anterior-posterior gene system (reviewed in Anderson 1987; Nüsslein-Volhard and Roth 1989). Maternally active dorsal-ventral genes set up the axis and zygotically active genes become locally functional under the initial control of the maternal genes. The dorsal-ventral gene products include serine proteases (DeLotto and Spierer 1986; Chasan and Anderson 1989), a transmembrane protein (Hashimoto et al. 1988), an epidermal growth factor receptor-like protein (Price et al. 1989), and a homologue of the *rel* oncogene (Steward 1987). Loss-of-function mutations in most dorsal-group genes lead to the ventral side of the embryo developing with dorsal characteristics. The development of dorsal-ventral pattern is beyond the scope of this chapter.

The ultimate fate of an ectodermal cell is specified by these three systems: position within the segment by pain-rule and segment polarity genes, dorsal-ventral position by dorsal-ventral genes, and segment identity by homeotic genes. Homeotic transformations demonstrate the independence of segment identity, dorsal-ventral, and intrasegmental positional information. Loss of *Ubx* function in the third thoracic segment causes it to develop as a second thoracic segment (Lewis 1978). The activation of *Antp* function in the head, where the gene is normally silent, causes legs to develop in lieu of antennae (Denell et al. 1981; Struhl 1981; Hazelrigg and Kaufman 1983; Frischer et al. 1986; Schneuwly et al. 1987). In imaginal discs mosaic for mutations in *Antp* or *Ubx*, mosaic appendages result. In these monsters, distal structures always differentiate in distal positions and anterior structures in anterior positions, even when the appendage is part antenna and part leg or part wing and part haltere (Postlethwait and Schneiderman 1969). Thus, cells know their relative positions and differentiate accordingly, regardless of the actual structures they are making.

3.2 Parasegments and Segments

Gastrulation (stages 6 and 7; Table 1) begins at the end of cycle 14, as soon as the cell membranes are completed. The mesoderm precursors invaginate along the ventral midline and furrows form near the boundary of the head and thoracic primordia. The cells along the ventral midline, in the "germ band" which will form the segmented part of the embryo, move around the posterior pole and then anteriorly along the dorsal surface of the embryo toward the back of the head (stage 8). At the end of germ band extension, the abdominal primordia lie dorsally, behind the head, and the embryonic axis is folded back on itself. The central nervous system precursors segregate from the ectoderm near the ventral midline while the germ band is extended (stages 9 and 10). Neuroblasts delaminate from the epidermal precursors and move to the interior of the embryo. While the germ band is extended there are three rounds of mitosis in the ectoderm. The first morphological metamerization of the ectoderm is shallow grooves that appear during stage 10. These metameres are parasegments rather than segments. Each parasegment consists of the posterior part of one segment primordium and the anterior part of the next segment primordium (Martinez-Arias and Lawrence 1985). The early domains of most homeotic genes are para-segments rather than segments, so parasegments are thought to represent the primary metameric divisions.

Near the end of germ band extension, deeper grooves form in the ectoderm, demarcating the segment borders (stage 11). The germ band retracts (stage 12) so that the posterior end of the embryo is again at the posterior end of the egg, and organogenesis begins. After about 22 h the first instar larva hatches and begins a life of continuous eating. Externally, the hatching larva has three thoracic and eight abdominal segments, with repeating patterns of setae, denticles, and sensory structures punctuated by segment borders. While there are segment-specific variations to the pattern, its iterated nature is clear (Lohs-Schardin et al. 1979).

3.3 The Developmental Origins of Adult Structures

Primordial cells for imaginal, i.e., adult, tissues are segregated in the embryo soon after cellularization of the blastoderm (Wieschaus and Gehring 1976; Bate and Martinez-Arias 1991). The primordium for the head capsule, the genitalia, and for each dorsal and ventral thoracic segment consists of a separate bag of epithelial cells, the imaginal discs. The disc cells proliferate and become organized during the larval stages. As pupation begins, most larval tissues break down and the disc epithelia evaginate and differentiate to form adult structures such as legs, antennae, or eyes. While the hair, bristle, sense organ, and coloration patterns in the adult are more complex than in the larva, similar regulative phenomena are seen in larval development and in imaginal discs. The same homeotic genes that specify larval segment differences also specify the segment differences in the imaginal discs. Many segment polarity genes required for patterning larval segments are also required to pattern imaginal structures. Thus, the mechanisms for patterning larval and adult structures may be similar, despite the apparently different modes of development.

Imaginal discs are subdivided into anterior and posterior compartments (Garcia-Bellido et al. 1973; Crick and Lawrence 1975; Lawrence and Morata 1976; Morata and Lawrence 1977; Garcia-Bellido et al. 1979; Martinez-Arias and Lawrence 1985). Compartments are defined by lineage restrictions; each compartment is a polyclone consisting of the descendants of a founder group of cells that is established during early development. The border between compartments corresponds to the edge of the expression domain of a number of homeotic or segment polarity genes (e.g. *Ubx* in the wing disc, *en* in all imaginal discs). Yet there is no morphological feature that distinguishes the boundary, either in the developing imaginal disc or in the adult cuticle. While this lineage restriction/compartment border apparently has great significance for the expression of regulators such as segmentation and homeotic genes, it has no obvious morphological correlate.

4 Segment Polarity Genes, the Genetically Defined Components of the Intrasegmental Patterning System

Segment polarity genes are defined as the group of genes required to organize the segmentally repeating pattern of the integument. By studying the products of these genes, we hope to learn what information is passed from cell to cell and to understand the cellular and molecular mechanisms employed in pattern formation. Our working hypothesis is that the activity of segment polarity genes is the molecular basis for positional information. The combination of segment polarity genes active in a cell determines its "positional state". The state of each cell is dependent on the states of nearby cells and is regulated by cues from its neighbors. These cues are the basis for the "compatible neighbors" phenomenon of the grafting experiments (Sect. 2). The outstanding challenges are to determine how many different states there are, to identify which genes mediate which cell interactions, to understand how the segment becomes increasingly complex as development proceeds, and to understand how localized activities of segment polarity genes drive morphogenesis.

4.1 The Phenotypes of Mutations in Segment Polarity Genes

The ventral cuticle of a hatching-stage wild-type embryo is decorated with a segmentally repeating pattern of fine bristles (denticles) in the anterior part of each segment and naked cuticle in the posterior part of each segment (Fig. 6). Embryos deficient for any of the segment polarity genes produce cuticle missing a contiguous set of pattern elements in every segment. The missing pattern elements are often replaced by mirror-image duplications of the remaining pattern elements (Nüsslein-Volhard and Wieschaus 1980). Sixteen segment polarity genes have been identified. They form four loose groupings, based on their recessive embryonic-lethal mutant phenotypes (Fig. 6; Table 2). Mutations in segment polarity genes of the first class delete ventral denticles, that is anterior pattern elements, and replace them with naked cuticle. *naked* (*nkd*) and *shaggy* (*sgg*, also known as *zeste-white 3*) are in this class (Jürgens et al. 1984; Perrimon and Smouse 1989).

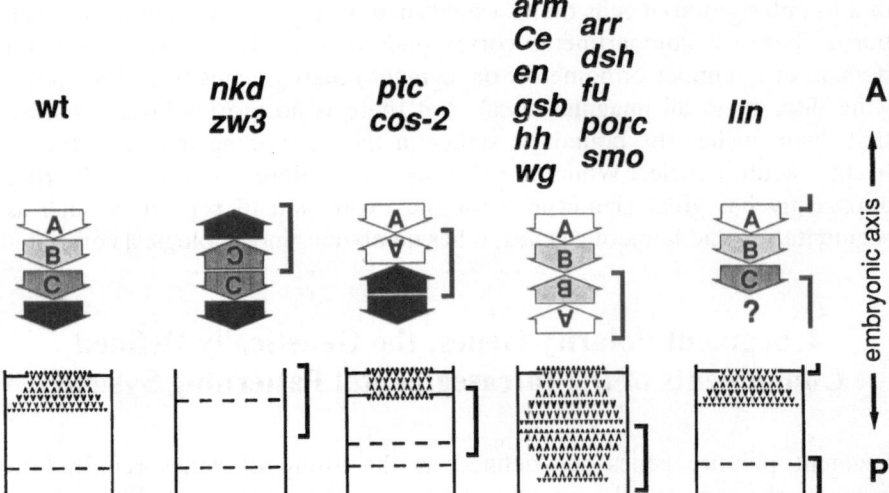

Fig. 6. Phenotypic classes of segment polarity genes. Schematic drawings of the ventral cuticle pattern of one segment, including denticles (*arrowheads*), segment borders (*horizontal solid lines*), and parasegment borders (*horizontal dashed lines*) are shown beneath each mutant class. The *brackets* indicate the affected region of the segment. The *broad arrows* (*A*, *B*, *C*, and *D*) represent abstractions of the mutant patterns. *naked (nkd), shaggy (sgg), patched (ptc), costal-2 (cos-2), armadillo (arm), arrow (arr) cubitus interruptus Dominant (ciD), dishevelled, (dsh), engrailed (en), fused (fu), gooseberry (gsb), hedgehog (hh), porcupine (porc), smoothened (smo), wingless (wg), lines (lin)*. **A** and **P** indicate anterior and posterior. (See text for references)

Mutations in genes of the second class result in the deletion of central pattern elements and their replacement by segment border pattern elements. *patched (ptc), costal-2 (cos-2)*, and perhaps *costal-1 (cos-1)* form this class (Nüsslein-Volhard and Wieschaus 1980; Grau and Simpson 1987). The largest group of segment polarity genes includes those that delete ventral naked cuticle, that is, posterior pattern elements, and replace it with denticles. This group includes *arrow (arr), armadillo (arm), cubitus interruptus Dominant (ciD; Cell is the same gene[4]), dishevelled (dsh), engrailed (en), fused (fu), gooseberry (gsb), hedgehog (hh), porcupine (porc), smoothened (smo)*, and *wingless (wg)* (Nüsslein-Volhard et al. 1980, 1984; Kornberg 1981; Orenic et al. 1987; Perrimon and Mahowald 1987; Perrimon et al. 1989). *lines (lin)* represents a fourth class, in which mutations cause the absence of the anterior part of each denticle band, including the segment border (Nüsslein-Volhard and Wieschaus 1980; Nüsslein-Volhard et al. 1984).

---

[4] *Cell* and *ciD* mutations identify the same gene. *ciD* is probably a gain-of-function mutation (Lindsley and Grell 1968). *Cell* represents the null phenotype of the gene (Orenic et al. 1987).

Table 2. The segment polarity genes

Gene and cytology	Embryonic phenotype	Expression pattern; protein product	Maternal effect?	Adult phenotypes	Clonal analysis	References[a]
naked (*nkd*) 75D-76B	Naked cuticle replaces denticles: cell death					1, 2
shaggy (*sgg*) = *zw3* 3B1	Like *nkd*: cell death	Homogeneous in egg and embryo; serine/threonine kinase	Yes		Required throughout the wing, cell autonomous	3, 4, 5, 6, 7
patched (*ptc*) 44D3-4	Ectopic segment and parasegment borders replace central pattern elements: little cell death	Homogeneous, stages 5–8; anterior 3/4 of segment, stages 9–10; thin stripes flanking *en* after stage 11; stripe anterior to *en* in imaginal discs; transmembrane protein	No	Anterior wing hypertrophy (*ptc^tuf*)	Required in anterior wing; domineering local non-autonomy	2, 8, 9, 10, 11, 12
costal-2 (*cos-2*) 43B3-43C3	Like *ptc*, but without ectopic segment borders		Yes	Anterior wing hypertrophy	Not cell autonomous in wings	13, 14, 15
armadillo (*arm*) 2B15–17	Denticles replace naked cuticle: cell death	Homogeneous in egg and embryo; plakoglobin homologue	Yes		Required for oogenesis; cell autonomous in embryos, wings	3, 16, 17, 18, 19, 20, 21, 22, 23, 24

Table 2. *Continued*

Gene and cytology	Embryonic phenotype	Expression pattern; protein product	Maternal effect?	Adult phenotypes	Clonal analysis	References[a]
cubitus interruptus Dominant (*ciD*) = *Cell* 101F2-102A5	Like *arm*: cell death	Homogeneous, st. 5–8; anterior 3/4 of segment after st. 9; zinc-finger protein	No	L4 wing vein interrupted (*ciD*); L3–L4 wing vein fusions, ocellar, and scutellar defects in hemizygotes		8, 16, 25, 26, 27
dishevelled (*dsh*) 10B6-7	Like *arm*: cell death		Yes	Thoracic, eye, and wing defects (*dsh¹*)	Cell autonomous in embryos, and wing	3, 28
engrailed (*en*) 48A2	Like *arm*: cell death; duplicated gene (*invected*)	Posterior 1/4 of segment after stage 5; posterior compartment in imaginal discs; homeodomain protein	No	Anterior cross-vein defect (*en^{CX1}* heterozygotes); posterior wing hypertrophy, notched scutellum (*en¹*)	Required in post. compartments; cell autonomous in imaginal discs	3, 8, 9, 29, 30, 31, 32, 33, 34
fused (*fu*) 17C4-6	Like *arm*, weak: cell death	Homogeneous in egg and embryo; serine/threonine kinase	Yes	L3, L4 wing veins fused, scutellar and ocellar defects	Cell autonomous in embryos	3, 8, 16, 17, 35, 36, 37, 38
gooseberry (*gsb*) 60E9-F1	Like weak *arm*, ventral only: little cell death: duplicated gene:	Posterior 1/2 of, segment stages 5–7; coincident with *wg* after stage 9; homeodomain protein			Cell autonomous in embryos	3, 8, 9, 41, 42, 43, 44
arrow (*arr*) 48B-49B (?)	Like very weak *arm*					9

				Small eye (hh^{bar}); anterior wing hypertrophy (Mrt)	Required in ocelli and scutellum; domineering local non-autonomy	1, 3, 8, 45, 46
hedgehog (hh) 94E	Like arm: cell death	Posterior 1/4 of segment after stage 5				
porcupine (porc) 17A-B	Like arm		Yes		Not cell autonomous in embryos	28, 47
smoothened (smo) 22B-?	Like arm: more severe at 18°C				Not cell autonomous in embryos	9
wingless (wg) 28A1-3	Like arm: cell death	1–2-cell-wide stripe anterior to en after stage 5; extracellular matrix-associated protein	No	Wing transformed to notum (wg^l); short wing (spd)	Required for distal appendages; small clones rescued	3, 8, 9, 48, 49, 50, 51, 52, 53, 54, 55
lines (lin) 44F-46A	Segment border deleted					9

a References: (1) Jürgens et al. 1984; (2) Martinez-Arias et al. 1988; (3) Perrimon and Mahowald 1987; (4) Simpson et al. 1988; (5) Perrimon and Smouse 1989; (6) Bourouis et al. 1990; (7) Siegfried et al. 1990; (8) Nüsslein-Volhard and Wieschaus 1980; (9) Nüsslein-Volhard et al. 1984; (10) Hooper and Scott 1989; (11) Nakano et al. 1989; (12) Phillips et al. 1990; (13) Whittle 1976; (14) Grau and Simpson 1987; (15) Simpson and Grau 1987; (16) Wieschaus et al. 1984; (17) Gergen and Wieschaus 1986; (18) Wieschaus and Noell 1986; (19) Wieschaus and Riggleman 1987; (20) Klingensmith et al. 1989; (21) Riggleman et al. 1989; (22) Peifer and Wieschaus 1990; (23) Riggleman et al. 1990; (24) Peifer et al. 1991; (25) Orenic et al. 1987; (26) Eaton and Kornberg 1990; (27) Orenic et al. 1990; (28) J. Klingensmith and N. Perrimon, pers. comm; (29) Lawrence and Morata 1976; (30) Kornberg 1981; (31) DiNardo et al. 1985; (32) Fjose et al. 1985; (33) Kornberg et al. 1985; (34) Poole et al. 1985; (35) Fausto-Sterling 1971; (36) Fausto-Sterling 1978; (37) Martinez-Arias 1985; (38) Mariol et al. 1987; (39) Busson et al. 1988; (40) Préat et al. 1990; (41) Bopp et al. 1986; (42) Baumgartner et al. 1987; (43) Côté et al. 1987; (44) Hidalgo 1991; (45) Mohler 1988; (46) Mohler and Vani, 1992; (47) Perrimon et al. 1989; (48) Sharma 1973; (49) Morata and Lawrence 1977a; (50) Baker 1987; (51) Rijsewijk et al. 1987; (52) Baker 1988a; (53) Baker 1988b; (54) van den Heuvel et al. 1989; (55) Tiong and Nash 1990.

Most segment polarity genes are required for the proper development of imaginal discs (reviewed in Phillips and Whittle 1990) as well as for embryonic segments. Mosaic analysis indicates that *en* and *hh* are required only in the posterior compartment (Lawrence and Morata 1976; Mohler 1988). Mosaic analysis and adult viable alleles indicate that *ptc* and *cos-2* are required in the anterior compartment, particularly its anterior part (Whittle 1976; Phillips et al. 1990). *ciD* and *fu* are primarily required near the compartment border, especially around wing vein L4, as judged by adult viable phenotypes (Lindsley and Grell 1968). *arm*, *dsh*, *porc*, and *wg* are required in different parts of different imaginal discs (Sharma 1973; Baker 1988; Peifer et al. 1991; J. Klingensmith and N. Perrimon, pers. comm.). *sgg* is required throughout the whole imaginal disc (Simpson et al. 1988). The requirements for *arr*, *gsb*, *nkd*, *smo*, and *lin* in the adult are not known.

4.2 Cell Fates Are Transformed by Segment Polarity Gene Mutations

Cell death and regeneration can cause mirror-image pattern duplications in imaginal discs (Russell 1974; Bryant 1975), but the pattern defects in segment polarity mutants are not a consequence of cell death followed by regeneration. In *ptc* mutant embryos the deletion/duplication phenotype results from respecification of cell fates (Martinez-Arias et al. 1988; Hooper and Scott 1989). Mis-programming of cells is also evident in *en*, *nkd*, *sgg*, *smo*, *hh*, *wg*, and *dsh* mutants by stage 9/10, several hours before any cell death is apparent (Perrimon and Mahowald 1987; Martinez-Arias et al. 1988; Perrimon et al. 1989; Hidalgo and Ingham 1990; Ingham et al. 1991; J. Hooper, unpub. observ.). Cell death in the epidermis during germ band shortening (stage 11–12) does contribute to the final phenotype of *arm*, *dsh*, *en*, *fu*, *hh*, *nkd*, *wg*, and *sgg* mutants, causing short embryos and segment fusions (Martinez-Arias and Ingham 1985; Perrimon and Mahowald 1987; Klingensmith et al. 1989). However, the cell death is a secondary consequence of disorganization and/or transformation within the segment rather than the primary cause of the confusion.

4.3 Segment Polarity Genes Mediate Cell Communication

Analyses of embryos and adults mosaic for wild-type and *fu*, *porc*, *wg*, *gsb*, *hh*, or *ptc* cells show that cell communication is intimately involved in the function of each of these genes. *fu*, *porc*, *wg*, and maybe also *gsb* and *hh*, small mutant clones are phenotypically rescued by neighboring wild-type cells (Morata and Lawrence 1977; Fausto-Sterling 1978; Wieschaus and Riggleman 1987; Baker 1988; J. Klingensmith and N. Perrimon, pers. comm.). Mutant cells can respond to a wild-type signal from neighboring cells. Such genes might be involved in generating or transmitting a signal. *hh* or *ptc* mutant clones can confer a mutant phenotype on neighboring

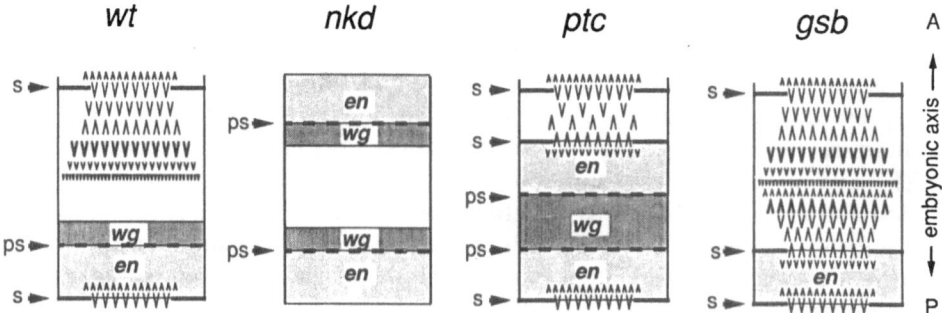

Fig. 7. The expression patterns of *wg* and *en* position borders and bristles. *en* (*light shading*) and *wg* (*dark shading*) expression is indicated superimposed on morphological features including denticles (*arrowheads*), parasegment grooves (*ps* and *dashed lines*), and segment borders (*s* and *solid lines*), in wild-type (wt), *nkd*, *ptc*, and *gsb* homozygous mutant embryos. A parasegment border forms at interfaces between *wg*- and *en*-expressing cells. A segment border forms at interfaces between *en*- and non-*wg*-expressing cells. Naked cuticle forms in the vicinity of *wg*-expressing cells

genotypically wild-type cells (Mohler 1988; Phillips et al. 1990). These genes might be involved in relaying or suppressing the relay of a signal: mutant cells might transmit an inappropriate signal to their neighbors causing those wild-type neighbors to exhibit a mutant phenotype, or the mutant cells might fail to send a signal that maintains the wild-type cells on the proper pathway. In embryos or adults mosaic for *arm*, *dsh*, *en*, or *sgg*, mutant cells always show the mutant phenotype (Lawrence and Morata 1976; Wieschaus and Riggleman 1987; El Messal et al. 1988; Peifer et al. 1991; J. Klingensmith and N. Perrimon, pers. comm.). These genes might be involved in receiving or implementing a signal. Nonautonomy, the lack of correspondence between genotype and phenotype, is expected in genetically mosaic flies if nearest neighbor interactions are necessary to generate pattern.

4.4 Segment Polarity Genes Specify Morphology

Specific morphological features are predicted by the expression of particular segment polarity genes (Fig. 7). For instance, *en* is expressed in the primordium of the posterior part of the segment (DiNardo et al. 1985; Fjose et al. 1985; Kornberg et al. 1985). *wg* is expressed in the cells immediately anterior to the cells that contain *en* products (Baker 1988). Parasegmental grooves form at the junction between *en*- and *wg*-expressing cells, and segment borders form at the edge of *en*-expressing cells where *wg* is not expressed. The same relationship between *en* and *wg* expression and the morphology of the segment is maintained when these genes are misexpressed in mutants. *ptc* mutant embryos have ectopic segment and parasegment borders in every segment (Hooper and Scott 1989) which correlate with ectopic *wg* and *en* expression (Martinez-Arias et al. 1988; Figure 7). Ectopic *en* expression produces an additional junction between *en*- and *wg*-expressing cells, and also, an additional edge of *en*-expressing cells where *wg* is not expressed. An ectopic parasegment border forms at the former ectopic interface and an ectopic segment border forms at the latter ectopic interface. The *nkd* phenotype is similarly predicted by the ectopic *en* and *wg* expression observed in *nkd* mutants (Martinez-Arias et al. 1988). Thus, the *en* and *wg* genes, acting in adjacent cells, may direct the formation of specific structures by those cells. Other morphological correlates of segment polarity gene expression are presented in Section 6.

4.5 Segment Polarity Gene Functions in Internal Tissues

Components of the epidermal patterning system are also required for development of internal tissues. *ptc*, *en*, *gsb*, *Ce*, *wg*, and *sgg* participate in the development of the nervous system (Simpson et al. 1988; Bourouis et al. 1989; Patel et al. 1989; Simpson 1990; Siegfried et al. 1990). *en* is not required in the mesoderm (Lawrence and Johnston 1984), which is segmented but not divided into anterior and posterior compartments (Lawrence 1982). *wg* is required for proper morphogenesis in the visceral mesoderm (van den Heuvel et al. 1989; Immerglück et al. 1990; Reuter et al. 1990). *arm* and *fu* are required for oogenesis, the former in the germ line (Wieschaus and Noell 1986), and the latter in some other cells of the mother (Busson et al. 1988). The maternal component of *cos-2* may contribute to the overall anterior-posterior polarity of the embryo (Grau and Simpson 1987).

4.6 Maternal and Zygotic Requirements for Segment Polarity Genes

The maternal genome substantially affects the embryonic phenotype of *sgg*, *cos-2*, *arm*, *fu*, *dsh*, and *porc* (Fausto-Sterling 1971; Wieschaus and Noell 1986; Grau and Simpson 1987; Perrimon and Mahowald 1987; Busson et al. 1988; Perrimon et al. 1989; Perrimon and Smouse 1989). For instance, the segment polarity phenotype of *dsh* is only seen when the embryo is homozygous for the mutation *and* the eggs are derived from homozygous mutant germ lines. A single functional copy of the gene, either in the germ line during oogenesis or in the embryo, is sufficient to prevent the mutant phenotype. mRNAs deposited in the egg during oogenesis provide a physical basis for the genetically defined maternal contribution (Martinez-Arias 1985; Mariol et al. 1987; Riggleman et al. 1989). Mosaic analysis of *fu* indicates that in spite of the maternally deposited mRNA, its function is not activated until after blastoderm cellularization (Fausto-Sterling 1971). Why should a process that takes place significantly after activation of the zygotic genome require components contributed during oogenesis? Is it simply a matter of dosage – the zygotic genome cannot make gene product fast enough? In the case of *arm*, the maternal and zygotic contributions to the phenotype are simply additive (Perrimon and Mahowald 1987). Apparently the embryo requires some threshold level of *arm* activity by the mid-germ band extension stage. It is immaterial whether the activity is contributed maternally or zygotically, as long as the threshold is reached. Whether this is also true for the other genes remains to be determined.

4.7 How Many Segment Polarity Genes Are There?

Most of the segment polarity genes whose phenotypes are strictly dependent on zygotic genotype have probably been identified in screens of the zygotic genome (Jürgens et al. 1984; Nüsslein-Volhard et al. 1984; Wieschaus et al. 1984). Of the 16 segment polarity genes currently known, 5 require that both the maternal and zygotic contributions be removed before the embryonic phenotype is evident. Four of these maternal-effect genes are on the X-chromosome, and two of those four were identified in a screen specifically designed to detect maternally masked, zygotically required genes (Perrimon et al. 1989). Since the X-chromosome comprises approximately 20% of the *Drosophila* genome, simple arithmetic predicts that there are about 16 more maternally masked segment polarity loci on the autosomes, of which only *cos-2* is currently identified. Thus the current tally of 16 segment polarity genes may only represent about half of the total genes whose null state produces a segment polarity phenotype.

5 Segment Polarity Genes Maintain the Integrity of the Parasegment Border

The early function of many segment polarity genes seems to be to maintain the integrity of the parasegment border. The parasegment border is the interface between *wg*-expressing cells anteriorly and *en/hh*-expressing cells posteriorly (e.g. Fig. 7). Pair-rule genes initiate expression of *wg* and of *en/hh* in adjacent cells (Fig. 8). The *wg* and *en/hh* cells then produce short-range signals that maintain their "identities"; *en/hh* expression requires a signal from *wg* cells (the Wg signal) and *wg* expression requires both a signal from *wg* cells and a signal from *en/hh* cells (the E/H signal; diagrammed in Figs. 9, 10, 11). *wg* transcription is repressed by *ptc* if cells do not receive the E/H signal. *en/hh* transcription is repressed by *nkd* if cells do not receive the Wg signal. The net effect is to retain a precise and unique interface in every segment despite the cell movements and proliferation that accompany development.

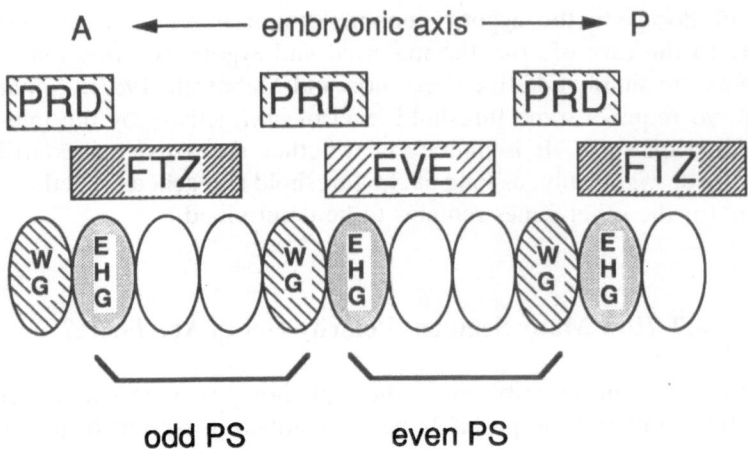

Fig. 8. The initiation of segment polarity gene expression by pair-rule genes. Schematic representation of *en*, *wg*, *hh*, and *gsb* (**E, W, H, G**, respectively) expression in the cellular blastoderm. The expression patterns of the pair-rule genes *paired* (*prd*), *even-skipped* (*eve*), and *fushi tarazu* (*ftz*) are indicated by the *boxes* over the cells (after Ingham et al. 1988). **EVE** and **FTZ** negatively regulate *wg*. In even parasegments (**even PS**) *en* is specified by the overlap of **PRD** and **EVE**. In odd parasegments (**odd PS**) *en* coincides with the overlap of **PRD** and **FTZ**. The even stripes of both *wg* and *en* require *paired* function; the odd stripes of both *wg* and *en* require *odd-paired*. The pattern of *odd-paired* expression is unknown. *hh* is initiated about the same time as *en* (Mohler and Vani, 1992; P. Beachy, pers. commun.), presumably in the same cells and by the same mechanisms. *gsb* is initiated in cells that express *prd*. (See text for references)

5.1 Initiation of the Segment Pattern

The first signs of segmentation appear as cell membrane enclose the nuclei of the syncytial blastoderm (stage 5; see Table 1). Overlapping domains of pair-rule gene expression control the initiation of segment-periodicity stripes of *en*, *hh*, *wg* and *gsb* transcription (Fig. 8). *en* expression is initiated in those nuclei that express both *paired* and *fushi tarazu* or *even-skipped*. It also requires *odd-paired* (DiNardo and O'Farrell 1987; Ingham et al. 1988; Martinez-Arias et al. 1988). *wg* expression is initiated in nuclei which express neither *ftz* nor *eve*, and it requires *paired* and *odd-paired* (Ingham et al. 1988). *hh* expression is initiated in the same cells that express *en* (Mohler and Vani, 1992). *gsb* expression is initiated in stripes that overlap both the *en* and *wg* stripes, although the exact registration is uncertain (Bopp et al. 1986; Baumgartner et al. 1987; Côté et al. 1987). This striped expression of *en*, *hh*, *wg*, and *gsb* flanks the incipient parasegment border. It serves as a starting point from which the segment polarity genes interact amongst themselves to organize the proliferating cells of the segment.

5.2 Regulation of the Domain of *en/hh* Expression

Germ-band elongation (stage 8) rearranges *en*-expressing cells into jagged discontinuous stripes (Bejsovec and Martinez-Arias, 1991). Within the next 40 min (stage 9) the pattern of *en* expression resolves into coherent stripes. Some non-*en* expressing cells initiate transcription of *en* de novo, and other cells lose *en* expression entirely (Vincent and O'Farrell, 1992; A. Martinez-Arias and G. Technau, pers. comm.). Inhibition of *en* transcription by *nkd* and *sgg* and maintenance by the Wg signal ensures that any cells misplaced during the movements of germ-band elongation will be entrained to the correct fate at the incipient parasegment border. The arguments for the current view of regulation of *en* are presented below. The working model (summarized in schematic form in Figs. 9A, 10) is, at best, incomplete since many genes remain to be intergrated into the model.

5.2.1 *nkd* and *sgg* Repress *en* Expression Away from the Parasegment Border

In *nkd* or *sgg* mutant embryos, *en* transcription expands during stage 9 to fill the anterior half of the parasegment (DiNardo et al. 1988; Martinez-Arias et al. 1988; E. Siegfried and N. Perrimon, pers. comm.; Fig. 7). Therefore *nkd* and *sgg* activities must normally repress *en* in the anterior half of the parasegment. How *nkd* and *sgg* repress *en* is unknown. *sgg* encodes a putative protein kinase (Bourouis et al. 1990; Siegfried et al. 1990) which could be involved in signal transduction. The *nkd* gene has not yet been cloned. We do not understand why *en* derepression is limited to the anterior half of the parasegment in *nkd* and *sgg* mutants. Additional constraints in the posterior half of the parasegment, or additional permissive conditions in the anterior half of the parasegment, must limit *en* expression in *nkd* mutants.

5.2.2 A Wg Signal Promotes *en* and *hh* Expression Near the Parasegment Border

wg is expressed in cells immediately anterior to the incipient parasegment border, but *wg* function is required for *en* transcription to persist in cells posterior to the parasegment border (Figs. 9A). Although initiated normally, *en* expression begins to fade during stage 9 in *wg* mutant embryos (DiNardo et al. 1988; Martinez-Arias et al. 1988; Bejsovec and Martinez-Arias 1991). The Wingless protein (WG) probably carries the signal required to maintain *en* expression in neighboring cells. It is secreted from the cells in which it is made, and then precipitates on the extracellular matrix or is taken up by neighboring cells (van den Heuvel et al. 1989). Immunohistochemical studies show that WG moves two or three cells from its source (Gonzalez et al. 1991), traversing the width of the *en* stripe. The correlation between

28 Joan E. Hooper and Matthew P. Scott

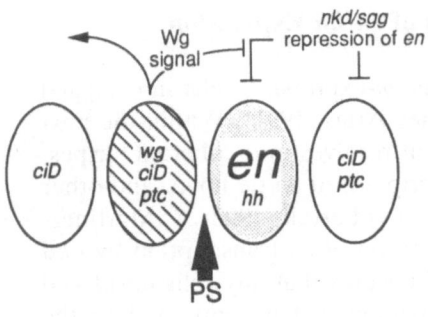

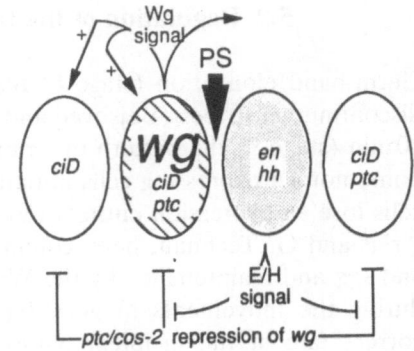

A) Control of *en* expression

B) Control of *wg* expression

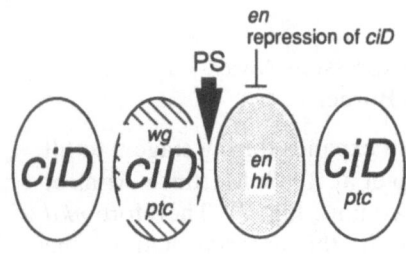

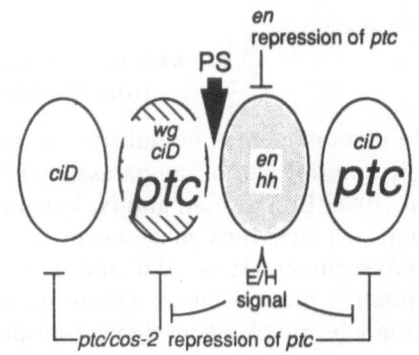

C) Control of *ciD* expression

D) Control of *ptc* expression

Fig. 9. Regulation of *en*, *wg*, *ciD*, and *ptc* following their initiation. Currently known gene interactions are summarized for the ventral ectoderm. Dorsal-ventral differences are ignored. The *four ovals* represent different cell states of the segmental repeat, identified by the combination of segment polarity genes which they express. *Hatched cells* had *wg* expression initiated at the blastoderm stage; *shaded cells* had *en* and *hh* initiated at the blastoderm stage (Fig. 8). The parasegment border (*heavy arrow*; **PS**) is the interface between *en/hh* and *wg* cells. *Small arrows* represent activation. *T-bars* represent inhibition. *Italics* indicate the gene or its transcription; *uppercase* indicates the protein product. **A** Control of *en* expression. *nkd* and *sgg* repress transcription of *en*. *wg*-expressing cells secrete a short-range Wg signal that overcomes *nkd* and *sgg* repression, but only near *wg*-expressing cells. Thus, *en* is expressed in near neighbors of *wg*-expressing cells. **B** Control of *wg* expression. *ptc* and *cos-2* repress transcription of *wg*. *en/hh*-expressing cells produce a short-range E/H signal that is dependent on *en*, *smo*, and *hh* function. The E/H signal overcomes *ptc* and *cos-2* repression of *wg* transcription. The secreted Wg signal is an autocrine activator of *wg* transcription. Thus, *wg* is expressed in neighbors of *en/hh* cells that also receive the Wg signal. **C** Control of *ciD* expression. *ciD* is expressed in all cells that do not express *en*. **D** Control of *ptc* expression. Like *ciD*, *ptc* transcription is repressed in cells that express *en*. Like *wg*, *ptc* transcription is also repressed by *ptc* and *cos-2*, a repression that is relieved next to *en/hh* cells by the E/H signal. The net effect is *ptc* expression only in cells flanking *en/hh* cells. (See text for references)

WG distribution and *en* expression suggests that WG itself carries the signal that determines whether or not a cell will express *en* during stage 9.

arm, *dsh*, and *porc* are all required to transmit and/or receive the Wg signal (Fig. 10). Embryos mutant for *arm*, *dsh* and *porc* are indistinguishable from *wg* mutant embryos, both morphologically and with respect to *en* expression (Perrimon and Mahowald 1987; Peifer et al. 1991). *porc* is involved in presenting the Wg signal: WG is confined to the cells in which it is made in embryos mutant for the single extant allele of *porc* (J. Klingensmith and N. Perrimon, pers. comm.). *arm* and *dsh* may act at the receiving end of the Wg signalling pathway since cells mutant for *arm* or *dsh* are not rescued by neighboring wild-type cells in mosaic flies (Gergen and Wieschaus 1986; J. Klingensmith and N. Perrimon, pers. comm.). *arm* is essential for cell-viability since null mutations are cell-lethal. Hypomorphic (partial loss of function) *arm* alleles give the segment polarity phenotype (Wieschaus and Riggleman 1987; Klingensmith et al. 1989). *arm* encodes a protein related to plakoglobin (Riggleman et al. 1989), an abundant cytoplasmic protein which accumulates at desmosomes. Armadillo protein (ARM) is ubiquitous, but accumulates at the plasma membrane in the posterior third of the parasegment, in and near cells that make WG. The accumulation of ARM at the plasma membrane requires *wg*, *dsh*, and *porc* activity (Riggleman et al. 1990). Therefore, ARM changes its subcellular localization in response to the Wg signal. Whether ARM localization is important for ARM's role in receiving or transducing the Wg signal is unclear. How *dsh* participates in the reception of the Wg signal remains to be seen.

Two observations suggest that *wg* maintains *en* transcription by relieving the inhibition mediated by *sgg* and *nkd* (Figs. 9A, 10). First, WG produced in all cells under control of a heat-inducible promoter mimics loss of *nkd* function and permits *en* expression throughout the anterior half of the parasegment (J. Nordermeer, P. Johnston, F. Rijsewijk, R. Nusse and P. Lawrence, submitted). Second, *dsh* is only required for *en* expression if *sgg* is also functional; in embryos doubly mutant for *dsh* and *sgg*, *en* expression is derepressed just as in *sgg* single mutants (E. Siegfried and N. Perrimon, pers. comm.). Whether other players in the Wg-signaling pathway become dispensable for *en* expression in *nkd* or *sgg* mutant backgrounds is not yet clear. If so, then transduction of the Wg signal in *en/hh*-expressing cells relieves inhibition of *en* and *hh* transcription by *nkd* and *sgg*.

5.2.3 Transcriptional Regulation by the EN Protein

The Engrailed protein (EN) represses transcription of *ptc* and *ciD* (Figs. 9C,D, 10). During gastrulation and germ-band elongation (stages 6–8), *ptc* and *ciD* mRNA patterns resolve from homogeneous expression into broad stripes, exactly complementary to the *en* stripes (Hooper and Scott 1989; Nakano et al. 1989; Eaton and Kornberg 1990; Orenic et al. 1990). By stage 9, cells express EN or express *ptc* and *ciD*. In *en* mutants, *ptc* and *ciD* remain homogeneously expressed (Eaton and Kornberg 1990; Hidalgo and Ingham 1990; Orenic et al. 1990). In *nkd* mutants, *en* expression expands posteriorly to cover half the segment (Fig. 7), and *ptc* expression retracts to maintain the complementary relationship (Hidalgo and Ingham 1990). EN is nuclear localized, can bind specific DNA sequences, and can repress transcription of target genes in transfected cultured cells (Desplan et al. 1985; Fjose et al. 1985; Kornberg et al. 1985; Jaynes and O'Farrell 1988). It is therefore likely that the interstripes (nonexpressing cells) in the *ptc* and *ciD* patterns are the direct result of transcriptional repression by EN.

EN can turn off transcription of *wg*. *wg* and EN are not normally expressed in the same cells, but EN produced in all cells under control of a heat-inducible promoter abolishes *wg* expression (Heemskerk et al. 1991). EN might repress *wg* transcription directly. Alternatively, EN might affect *wg* through repressing *ciD*, a gene required for *wg* expression (Sect. 5.3.3). Either mechanism assures that *wg* and *en* expression will be mutually exclusive, so there will be no ambiguity about cell identity at the parasegment border.

hh and *en* are expressed in the same cells, and both require the Wg signal (Mohler and Vani, 1992). Both are initiated in single-cell wide stripes at the end of cellular blastoderm (Fig. 8). From the time parasegment grooves appear (stage 10; 90 min after gastrulation), *en* and *hh* expression patterns are in register. They are probably expressed in the same ectodermal cells throughout embryogenesis. *hh* expression requires *wg* and *arm* function, presumably because *hh* expression requires the same Wg signal that is required for *en* expression (Fig. 9A). *hh* expression is **not** controlled by *en*; *hh* expression is normal through stage 12 in embryos lacking *en* function (Mohler and Vani, 1992). This falsifies a generally held tacit assumption that *en* expression is necessary and sufficient to specify all properties of the most anterior cells of the parasegment, of the *en* expression domain.

5.3 Regulation of the Domain of *wg* Expression

Between 80 and 140 min after gastrulation (stage 10, which includes the second postblastoderm mitosis), the *wg* pattern changes. Ventrally and dorsally, *wg* stripes narrow to only one to two cells wide, just anterior to the parasegment groove. Laterally, *wg* disappears entirely (Martinez-Arias et al. 1988; van den Heuvel et al. 1989; Bejsovec and Martinez-Arias 1991). Two conditions must be met during stage 10 for a cell to express *wg*; it must be adjacent to an *en/hh*-expressing cell, and the Wg signalling pathway must be intact (Figs. 9B, 10). *en/hh* cells generate a signal (the E/H signal) that activates *wg* transcription. *ptc* and *cos-2* repress *wg* expression in the absence of that signal. The Wg-signalling pathway is also required to maintain *wg* expression. The requirement for the Wg signal gives cells a "memory" and the requirement for the E/H signal ensures that only cells at the parasegment border generate the Wg signal during the ensuing stages of development. We will now consider these interactions in detail.

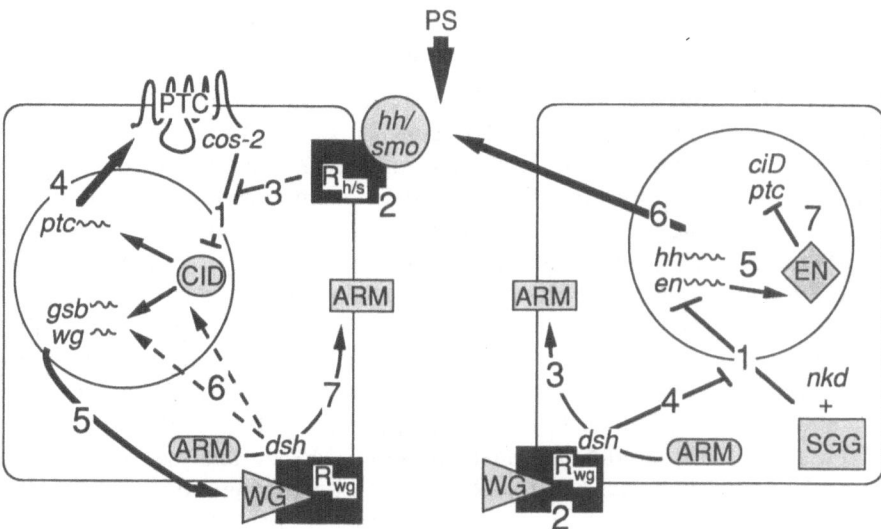

Fig. 10. Schematic summary of molecular interactions that regulate segment polarity gene expression. Details of segment polarity gene interactions within the *wg* cell (*left*) and the *en/hh* cell (*right*). The parasegment border (*heavy arrow at top*; **PS**) will form between these two cells. Protein products of known subcellular location and/or biochemical function are indicated by *shaded shapes* with **uppercase names**. Genes whose products and locations are unknown are indicated in *italics*. Transcription is indicated by a *squiggle* following the italicized gene name.

The *wg* cell: The activities of PTC protein at the plasma membrane and *cos-2* at an unknown location block transcription within the nucleus of *ptc*, *gsb*, and *wg* (**1**), perhaps through inactivation of CID protein. An extracellular signal directed by *hh* and *smo* (**2**) binds to a hypothetical cell surface receptor ($R_{h/s}$). Activation of this receptor blocks PTC + *cos-2* repression (**3**), establishing a permissive state for *ptc*, *gsb*, and *wg* transcription. This derepression is sufficient for *ptc* transcription, and production of PTC protein (**4**). *wg* and *gsb* transcription require additional activation from the Wg signal (**5**). Wingless protein activates its hypothetical receptor (R_{wg}), creating conditions permissive for *wg* and *gsb* transcription (**6**). The target of the Wg signal response pathway in the nucleus is not understood. Both *dsh* and *arm* activity are required in the Wg signal response pathway. Activation of the Wingless receptor induces ARM protein to relocate from cytoplasm to plasma membrane (**7**).

The *en* cell (*right*): *nkd* + SGG activities repress transcription of *en* (**1**) in the nucleus. The nature and location of the *nkd* gene product are unknown; SGG is a putative protein kinase. This repression is blocked by the Wg signal as follows. Extracellular WG protein binds to a hypothetical Wingless receptor (R_{wg}). Activation of the receptor (**2**) induces ARM protein to relocate from cytoplasm to plasma membrane (**3**). It also blocks *nkd* + SGG-mediated repression of *en* transcription (**4**). *dsh* is required for both effects of the Wg signal. Derepression of *en* transcription results in EN protein synthesis (**5**). Active EN protein represses transcription of *ciD* and *ptc* (**6**). The arguments and references for this model are presented in Section 5.2

5.3.1 *ptc* and *cos-2* Repress *wg* Expression Away
from the Parasegment Border

In *ptc* or *cos-2* mutants, *wg* expression expands anteriorly to fill the posterior half of the parasegment (Martinez-Arias et al. 1988; Z. Forbes and P. Ingham, pers. comm.). Therefore, *ptc* and *cos-2* activities must normally repress *wg* transcription (Fig. 9B). Transient overexpression of *ptc*, directed by a heat-inducible promoter, temporarily shuts off *wg* expression (Sampedro and Guerrero 1991), and prolonged over-expression of *ptc* can cause a *wg*-phenotype (K. Schuske, J. Hooper and M. Scott, unpub. data). *ptc* encodes a putative integral membrane protein with multiple hydrophobic domains (Hooper and Scott 1989; Nakano et al. 1989). It probably spans the membrane bilayer between 7 and 14 times. By immunohistochemistry, The Patched protein (PTC) is found at the periphery of cells (Ingham et al. 1991), presumably at the plasma membrane. Mosaic analysis of imaginal discs suggests that *ptc* is involved both in receiving and transmitting signals (Phillips et al. 1990). The PTC protein is probably a receptor for some signal which modifies transmission of yet another signal. The mechanism by which PTC protein communicates with the nucleus is unknown. The *cos-2* gene has not yet been cloned.

5.3.2 An E/H Signal Promotes *wg* Expression Near
the Parasegment Border

en and *hh* are expressed in cells immediately posterior to the incipient parasegment border, but *en* and *hh* functions are required for *wg* transcription to persist in cells **anterior** to the parasegment border (Fig. 9B). Although initiated normally, *wg* expression disappears completely during stage 9 in *hh* mutant embryos and during stage 10 in *en* mutant embryos (Martinez-Arias et al. 1988; Hidalgo and Ingham 1990; Bejsovec and Martinez-Arias 1991; J. Hooper, unpubl. observ.). We call this signal, which is dependent on both *en* and *hh*, the E/H signal. The E/H signal alleviates the *ptc*- and *cos-2*-mediated inhibition of *wg* transcription. The E/H signal has no effect in a *ptc* mutant background; embryos doubly mutant for *ptc* and for *en* or *hh* have the *ptc* phenotype (E. Wieschaus, pers. comm.; Ingham et al. 1991; J. Hooper, unpubl. observ.). Without *ptc*, *wg* cannot be repressed. In the double mutants *hh* and *en* become irrelevant because they have no *ptc* to act upon. *smo* probably functions in the same pathway. Like *hh*, *smo* is required during stage 9/10 to maintain *wg* expression, and like *hh*, *smo* becomes irrelevant in a *ptc* mutant background (J. Hooper, unpubl. observ.). EN is a nuclear-localized transcription factor (e.g. DiNardo et al. 1985), and so must affect neighboring cells indirectly. *hh* is nonautonomous in mosaics (Mohler 1988), suggesting that *hh* is involved in sending a signal. Whether *hh* encodes the signal, or whether it is required to produce the signal, awaits the characterization and immunolocalization of the Hedgehog protein (HH). The HH sequence contains a possible signal sequence and transmembrane domain (Mohler and Vani, 1992). The *smo* gene has not yet been cloned.

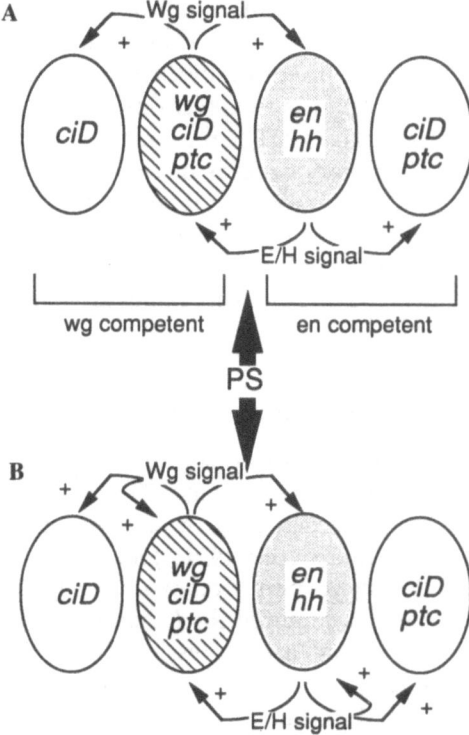

Fig. 11. Models for the polarized response to bidirectional signals. WG protein moves both anteriorly and posteriorly from *wg*-expressing cells, but only posterior neighbors respond by making *en*. *en/hh* cells signal in both directions to influence *ptc* expression (Fig. 9D), but only anterior neighbors respond by making *wg* and *ptc*. Model **A**: Cells in the anterior half of the parasegment are "competent" to express *en*. Cells in the posterior half of the parasegment are "competent" to express *wg*. If cells are in the anterior half of the parasegment and receive the Wg signal, then *en* is expressed; if cells are in the posterior half of the parasegment and receive the E/H signal, then *wg* is expressed. "Competence" is bestowed by pair-rule genes at the blastoderm stage. Pair-rule proteins are undetectable when *wg* or *en* expression is being modulated, so the mechanism for competence is unclear (Ingham et al. 1991). Model **B**: The polarized response to bidirectional signals follows from the interactions of the segment polarity genes and their products, an inevitable consequence of expression initiated at the blastoderm stage. *wg* expression is limited by a dual requirement for autoactivation and induction (the Wg signal and the E/H signals). The E/H signal does not induce *wg* expression posterior to the *en* domain, because this is beyond the range of the Wg signal. The Wg signal does not induce *wg* expression anterior to the *wg* domain because this is beyond the range of the E/H signal. *en* expression is limited by similar dual requirements for autoactivation and induction. Cells immediately adjacent to the parasegment border receive both signals but respond differently because of differential activity of *hh*, *smo*, *ciD* and/or *gsb*

Ingham et al. (1991) proposed that PTC is a receptor whose constitutive activity in the unoccupied state represses transcription of *wg* (Figs. 9, 10). He further proposed that HH produced by *en*-expressing cells binds to and inactivates PTC, thereby permitting expression of *wg* in cells immediately adjacent to *en*-expressing cells. Is PTC a receptor that is constitutively active in the unoccupied state? Primary cultures of dissociated *wg*-expressing cells from early embryos retain *wg* expression even though they cannot be receiving the E/H signal (M. Krasnow, pers. comm.). These preliminary results suggest that *wg* repression by *ptc* requires a signal different from the Wg signal and from the E/H signal. No candidate genes for such a signal have been identified.

5.3.3 The Wg Signal Is Required as an Autocrine Factor for *wg* Expression

The Wg-signalling pathway, employing *wg*, *arm*, and *dsh*, is required to maintain *wg* expression during stages 9 and 10 (Fig. 9B). *wg* mRNA and protein disappear from ectodermal cells in embryos mutant for *wg*, *arm*, or *dsh* (Bejsovec and Martinez-Arias 1991; Hidalgo and Ingham 1991; J. Klingensmith and N. Perrimon, pers. comm.). The requirement could be indirect since the Wg signal is required for *hh/en* expression (Fig. 9A), and the E/H signal is required for *wg* expression (Fig. 9B). The following observations argue against this. First, *wg* and *hh* fade at about the same time in *wg* mutants (Bejsovec and Martinez-Arias 1991; Mohler and Vani, 1992), rather than serially as would be predicted by a relay mechanism. Double-labeling experiments for *hh* and *wg* in *wg* mutants will be required to prove the temporal relationship. Second, if the requirement of *wg* for its own activity were through the E/H signal then, like *hh* mutants, *wg* mutants should be sensitive to the presence or absence of *ptc* function. However, the requirement for *wg* is independent of *ptc*; embryos doubly mutant for *wg* and *ptc* lose *wg* expression despite the absence of repression via *ptc* (J. Hooper, unpubl. observ.). Third, contrary to the experimental observation (M. Krasnow, pers. comm.), the relay mechanism predicts that dissociated *wg*-expressing cells should require an exogenous signal to maintain their *wg* expression. Apparently, *wg*-expressing cells autonomously receive and interpret the Wg signal as an autocrine activator of *wg* expression.

ciD and *fu* are also required for *wg* expression during stages 9 and 10 (Fig. 10). By stage 10, *wg* expression disappears from ectodermal cells in embryos missing *ciD* or *fu* function (Limbourg-Bouchon et al. 1991; J. Hooper, unpubl. observ.). The requirement for *ciD* is independent of *ptc*; embryos doubly mutant for *ptc* and *Ce* (a near-null allele of *ciD*) lose *wg* expression despite the absence of repression via *ptc* (J. Hooper, unpubl. observ.). *ciD* encodes a putative transcription factor (CID), a zinc-finger protein similar to the human oncogene, GLI (Orenic et al. 1990). *ciD* is expressed in all cells that do not express *en* (Eaton and Kornberg 1990; Orenic et al. 1990). Perhaps CID is the transcription factor that is activated by the Wg-signaling pathway, and whose activation results in transcription of *wg*. *fu* encodes a serine-threonine kinase (Préat et al. 1990) with an essential role in the autocrine and paracrine pathways required for *wg* expression. The mechanism of *fu* action is unknown.

5.4 The Gooseberry Paradox

gsb is a member of the *wg* phenotypic group (Fig. 6). Like all members of the *wg* group, embryos homozygous for *gsb* deficiencies develop denticles at the expense of naked cuticle. Unlike other members of the *wg* group, *gsb* is not required to maintain the parasegment border. Apparently *gsb* is only required during and after stage 11, after the parasegment border has stabilized and *en* expression becomes independent of the *wg* signal (Bejsovec and Martinez-Arias 1991; Heemskerk et al. 1991). During stage 11, *wg* expression fades away in embryos homozygous for *gsb* deficiencies; *en* is not affected in *gsb* mutants (Hidalgo and Ingham 1990), presumably because it no longer requires the Wg signal. *gsb* expression is initiated in broad stripes (Fig. 8) that are progressively refined during stages 8–10 so that *gsb* and *wg* expression become largely coincident in the ventral ectoderm (J. Hooper, unpubl. observ.). *gsb* encodes a putative transcription factor with both a homeobox and a paired box (Bopp et al. 1986). Apparently, it positively regulates *wg* expression, but only after stage 10. The paradox: ubiquitous overexpression of *gsb* during stages 8 or 9, under the control of a heat-inducible promoter, causes ectopic expression of both *wg* and *en* in a *nkd*-like pattern (A. Ungar and R. Holmgren, pers. comm.). Therefore, *gsb* can drive ectopic *en* and *wg* expression during stages 8 and 9. Why should *gsb* be expressed in complex and ever-changing patterns during stages 6–10 if it has no function then? Why should the embryo risk expressing *gsb* during stages 6–10 if this potent activator does not play some vital role?

5.5 A Missing Link; Polarized Responses to Bidirectional Signals

WG protein moves both anteriorly and posteriorly from *wg*-expressing cells, but only posterior neghbors respond by making *en* (Figs. 9A, B). *en/hh* cells signal in both directions to influence *ptc* expression (Fig. 9D), but only anterior neighbors respond by making *wg* (Fig. 9B). Why are responses to these signals limited to neighbors on only one side? Ingham et al. (1991) proposed a latent capacity to express *en* in the anterior half of the para-segment, and a latent capacity to express *wg* in the posterior half of the parasegment (Fig. 11A). If cells in the anterior half of the parasegment receive the Wg signal then *en* is expressed; if cells in the posterior half of the parasegment receive the E/H signal then *wg* is expressed. The latent capacity is bestowed by pair-rule genes at the blastoderm stage. Pair-rule proteins are undetectable when *wg* or *en* expression is being modulated, so either a low level of pair-rule product persists, or some "memory" would have to exist.

As an alternative (Fig. 11B), the polarized response to bidirectional signals might follow from the interactions of the segment polarity genes and their products (Figs. 9, 10). The "memory" might be the cascading effects of segment polarity gene interactions, the inevitable consequence of expression initiated at the blastoderm stage. A dual requirement for autoactivation and induction (the Wg signal and the E/H signals) could account for the precisely limited *wg* domain. The E/H signal does not induce *wg* expression posterior to the *en* domain, because this is out of range of the Wg signal, as genetically defined by cells where *nkd* represses *en* transcription (Sect. 5.2.2). The Wg signal does not induce *wg* expression anterior to the *wg* domain because this is beyond the range of the E/H signal. A similar dual requirement of *en* expression for autoactivation and induction could account for the precise limits of the *en* domain. Autoactivation of *en* is clear when the Wg signal is no longer required (Heemskerk et al. 1991), and might also act earlier.

In this model, cells on both sides of the parasegment border receive both signals. Then why do cells anterior to the parasegment border respond by expressing *wg* while posterior cells respond by expressing *en*? Localized activity of *hh*, *smo*, *ciD*, and *gsb* might predispose the *wg* and *en* domains to respond differently to the same signals. For instance: *en* cells turn off *ciD* (Sect. 5.2.3, Fig. 9C), which is essential for *wg* expression (Sect. 5.3.3, Fig. 10). Therefore, *en* cells are incapable of *wg* transcription, even though they receive both the E/H and Wg signals. Given the lack of detailed information about mechanisms, many models could be proposed by which segment polarity genes, acting amongst themselves, generate the final patterns.

6 Segment Polarity Genes and Morphogenesis

Stabilizing the parasegment border is the earliest function for most segment polarity genes, but there is more to pattern formation within the segment than establishing borders. After early stage 11, many segment polarity genes devote their efforts to other morphogenetic processes. *en* expression becomes independent of the Wg signal (Bejsovec and Martinez-Arias 1991; Heemskerk et al. 1991), *wg* expression becomes dependent on *gsb* (Hidalgo and Ingham 1990), and the broad stripes of *ptc* expression split into two single-cell wide stripes marking the edges of the preceding broad stripe (Hooper and Scott 1989; Nakano et al. 1989; Hidalgo and Ingham 1990). Morphogenesis (described in Campos-Ortega and Hartenstein 1985) begins during stage 9 as unique neuroblasts are specified at unique positions in every parasegment. During stage 10, shallow grooves form at the para- segment border (Martinez-Arias and Lawrence 1985) and tracheal placodes form laterally in the middle of each parasegment. During stage 12, sensory precursor cells begin to be specified, and by stage 13 imaginal discs begin to segregate (Bate and Martinez-Arias 1991). How do segment polarity genes specify unique positions for each of these events?

6.1 A Late Function of *wg* Specifies Naked Cuticle

After 6h (mid-stage 11), the Wg signal is no longer required for *en* and *hh* expression and for stability of the parasegment border (Bejsovec and Martinez-Arias 1991; Heemskerk et al. 1991). Instead it directs the choice between naked cuticle and denticles in the ventral epidermis. This was demonstrated using the temperature-sensitive allele of *wg*, shifting embryos to or from the nonpermissive temperature and monitoring the resulting *en* expression and cuticle pattern (Bejsovec and Martinez-Arias 1991: Heemskerk et al. 1991). Lateral cells acquire stable *en* expression by 1h after gastrulation, dorsal cells after 1.5h, and ventral cells 3h after gastrulation. After mid-stage 11 (3h after gastrulation), shifts to a nonpermissive temperature cause denticles to develop in place of the naked cuticle that should develop around the parasegment border (Baker 1988; Bejsovec and Martinez-Arias 1991). The extent of naked cuticle in wild-type correlates approximately with the range over which WG protein is detected by immunohistochemistry (Gonzalez et al. 1991). The Wg signal may therefore directly influence one aspect of differentiation, the choice between naked cuticle and denticles.

6.2 Specification of Neuroblasts and Sensory Structures

Genes of the *achaete-scute* complex specify different neural fates in different regions of the segment (Moscoso del Prado and Garcia-Bellido 1984; Cabrera et al. 1987; Romani et al. 1989; Simpson 1990). Small groups of epidermal cells express specific combinations of *achaete-scute* genes, presumably under direct control of transcription factors encoded by segment polarity genes and dorsal-ventral genes. Within each group of cells, lateral inhibition by cells becoming neuroblasts returns most cells to the epidermal pathway of development (Moscoso del Prado and Garcia-Bellido 1984; Doe and Goodman 1985). The lateral inhibition is mediated by neurogenic genes including *Notch* (reviewed in Artavanis-Tsakonas 1988). A single unique neuroblast within each group of cells is specified by this process, and a crude pattern is refined into a precise one.

6.3 Specification of Imaginal Discs

Imaginal discs originate at the parasegment border in a lateral position in embryos 9–10h old (stage 13; Wieschaus and Gehring 1976; Bate and Martinez-Arias 1991; Cohen et al. 1991). *wg* is required either for the specification or survival of imaginal disc primordia, since imaginal discs cannot be recovered and cultured from embryos mutant for *wg* (Simcox et al. 1989). The dorsal-ventral gene, *decapentaplegic*, is also required (Spencer et al. 1982). The first sign of imaginal disc primordia is localized expression of *Distal-less* (also called *Brista*), which encodes homeodomain protein required for leg development (Cohen and Jürgens 1989; Cohen 1990). *Distal-less* expression appears at intersections of longitudinal stripes of the dorsal-ventral gene *decapentaplegic* (St. Johnston and Gelbart 1987) and transverse stripes of *wg* during stage 11 (Cohen 1990; S. Cohen, pers. comm.). Apparently, the position of imaginal disc precursors is combinatorially specified by *wg* and *decapentaplegic*, representing two orthogonal positional information systems in the embryo.

7 Segment Polarity Genes, Positional Information, and Future Prospects

Segment polarity genes establish and communicate positional information. The expression patterns of many segment polarity genes change rapidly during development and are dependent on intracellular and intercellular interactions between segment polarity genes and their products. Other segment polarity genes (e.g. the Fused and Shaggy kinases) are ubiquitously expressed, but are probably locally activated by other segment polarity genes. The localized activities of segment polarity genes and their products determine patterns of differentiation and hence the ultimate morphology. Therefore, the interactions among segment polarity genes that modulate their activity patterns also organize positional information within segments.

Cell communication events mediated by segment polarity gene products provide cellular and molecular mechanisms for establishing and maintaining positional information. But how are the high-resolution patterns within larval segments and imaginal discs generated so reproducibly? Some theories have posited that positional information is a continuously graded quantity (Fig. 4A). This mechanism requires graded distributions of something, either a morphogen such as a diffusible small molecule or a graded distribution of a physiological property such as the level of phosphorylation of some protein. In either case, all pattern evolves from a single system, so genes should exist whose product is everywhere, whose mutants abolish all pattern, and whose mutant effects are nonautonomous in mosaics. None of the known genes fit these criteria. Most segment polarity mutants only interfere with part of the pattern, suggesting that different groups of cells are specified by the activities of different genes. Pattern elements are then specified by lateral interactions between cells of the same or different states as exemplified by parasegment grooves (Sect. 4.4) or neuroblasts (Sect. 6.1).

The study of segment polarity genes is just beginning. Immediate goals include cloning and characterizing the products and expression patterns of known segment polarity genes, identifying new maternal/zygotic segment polarity genes, demonstrating biochemically the inferred signaling pathways, and identifying the "downstream" functions by which positional information is translated into morphogenesis. It is already clear that what we learn studying positional information in insect segments applies widely in metazoan development. Vertebrate homologues for *en*, *wg*, *sgg*, and *ciD* are intimately involved in growth control (Kinzler et al. 1988; Bourouis et al. 1990; Siegfried et al. 1990) and development (Joyner et al. 1985; Rijsewijk et al. 1987; McMahon and Bradley 1990; Thomas and Capecchi 1990; Christian et al. 1991; Joyner et al. 1991; Sokol et al. 1991). Functional relationships between these molecules have probably been conserved through evolution. Therefore, studies of cell communication directed by segment polarity genes should have significance far beyond insect systems.

Acknowledgments

We dedicate this chapter to the memory of Carlos Cabrera, a valued colleague in the study of segment polarity and a spirited scientist and friend.

We are very grateful to Drs. Phil Beachy, Stephen Cohen, Susan Cumberledge, Stephen DiNardo, Zandy Forbes, Robert Holmgren, Phil Ingham, John Klingen-smith, Peter Lawrence, Jym Mohler, Jasprin Nordermeer, Roel Nusse, Norbert Perrimon, Esther Siegfried, and Marcel van den Heuvel for communication of unpublished results and to Drs. Phil Ingham, Mark Krasnow, Alfonso Martinez Arias, Roel Nusse, Kim Schuske, Eric Wieschaus, and Wenlin Zeng for helpful discussions. We thank Iva Greenwald and Robert Holmgren for suggestions for improvements in the manuscript. We thank Yvette Gonzalez for help checking the proofs. Research in the authors' laboratories was supported by grants from the N.I.H. (#24584 to M.P.S. and #R29 GM45396 to J.E.H.) and the American Cancer Society (#NP-784).

References

Abbott LC, Karpen GH, Schubiger G (1981) Compartmental restrictions and blastema formation during pattern regulation in *Drosophila* imaginal leg discs. Dev Biol 87:64–75

Akam M (1987) The molecular basis for metameric pattern in the *Drosophila* embryo. Development 101:1–22

Anderson K (1987) Dorsal-ventral embryonic pattern genes of *Drosophila*. Trends Genet 3:91–97

Artavanis-Tsakonas S (1988) The molecular biology of the *Notch* locus and the fine tuning of differentiation in *Drosophila*. Trends Genet 4:95–100

Baker NE (1987) Molecular cloning of sequences from *wingless*, a segment polarity gene in *Drosophila*: the spatial distribution of a transcript in embryos. EMBO J 6:1765–1773

Baker NE (1988a) Embryonic and imaginal requirements for *wingless*, a segment-polarity gene in *Drosophila*. Dev Biol 125:96–108

Baker NE (1988b) Localization of transcripts from the *wingless* gene in whole *Drosophila* embryos. Development 103:289–98

Baker NE (1988c) Transcription of the segment-polarity gene *wingless* in the imaginal discs of *Drosophila*, and the phenotype of a pupal-lethal *wingless* mutation. Development 102:489–97

Bate M, Martinez Arias A (1991) The embryonic origin of imaginal discs in *Drosophila*. Development 112:755–761

Baumgartner S, Bopp D, Burri M, Noll M (1987) Structure of two genes at the *gooseberry* locus related to the *paired* gene and their spatial expression during *Drosophila* embryogenesis. Genes Dev 1:1247–1267

Baumgartner S, Noll M (1991) Network of interactions among pair-rule genes regulating *paired* expression during primordial segmentation of *Drosophila*. Mech Dev 33:1–18

Bejsovec A, Martinez-Arias A (1991) Roles of *wingless* in patterning the larval epidermis of *Drosophila*. Development 113:471–485

Bopp D, Burri M, Baumgartner S, Frigerio G, Noll M (1986) Conservation of a large protein domain in the segmentation gene *paired* and in functionally related genes of *Drosophila*. Cell 47:1033–1040

Bourouis M, Heitzler P, El Messal M, Simpson P (1989) Mutant *Drosophila* embryos in which all cells adopt a neural fate. Nature (Lond) 341:442–444

Bourouis M, Moore P, Ruel L, Grau Y, Heitzler P, Simpson P (1990) An early embryonic product of the gene *shaggy* encodes a serine/threonine protein kinase related to the CDC28/cdc2+ subfamily. EMBO J 9:2877–2884

Bryant PJ (1975) Pattern formation in the imaginal wing disc of *Drosophila melanogaster*: fate map, regeneration and duplication. J Exp Zool 193:49–77

Bryant PJ (1987) Experimental and genetic analysis of growth and cell proliferation in *Drosophila* imaginal discs. In: Genetic regulation of development. Alan R Liss, New York

Bryant PJ, Fraser SE (1988) Wound healing, cell communication, and DNA synthesis during imaginal disc regeneration in *Drosophila*. Dev Biol 127: 197–208

Bryant SV, French V, Bryant PJ (1981) Distal regeneration and symmetry. Science 212:993–1002

Busson D, Limbourg-Bouchon B, Mariol M-C, Préat T, Lamour-Isnard C (1988) Genetic analysis of viable and lethal *fused* mutants of *Drosophila melanogaster*. Wilhelm Roux's Arch Dev Biol 197:221–230

Cabrera C, Martinez-Arias A, Bate M (1987) The expression of three members of the *achaete-scute* gene complex correlates with neuroblast segregation in *Drosophila*. Cell 50:425–433

Campbell GL, Caveney S (1989) *engrailed* gene expression in the abdominal segment of *Oncopeltus*: gradients and cell states in the insect segment. Dev Biol 106: 727–737

Campbell GL, Shelton PMJ (1987) Cell behavior during postembryonic pattern regulation in the insect abdomen (*Oncopeltus fasciatus*) I. Regeneration of the segment borders. Development 101:221–235

Campos-Ortega JA, Hartenstein V (1985) The embryonic development of *Drosophila melanogaster*. Springer, Berlin Heidelberg New York

Carroll SB (1990) Zebra patterns in fly embryos: activation of stripes or repression of interstripes? Cell 60:9–16

Carroll SB, Scott MP (1986) Zygotically-active genes that affect the spatial expression of the *fushi tarazu* segmentation gene during early *Drosophila* embryogenesis. Cell 45:113–126

Carroll SB, Vavra SH (1989) The zygotic control of *Drosophila* pair-rule gene expression. II. Spatial repression by gap and pair-rule gene products. Dev 107:673–683

Chasan R, Anderson KV (1989) The role of *easter*, an apparent serine protease, in organizing the dorsal-ventral pattern of the *Drosophila* embryo. Cell 56:391–400

Cohen B, Wimmer EA, Cohen SM (1991) Early development of the leg and wing primordia in the *Drosophila* embryo. Mech Dev 33:229–240

Cohen SM, Bronner G, Kuttner F, Jürgens G, Jäckle H (1989) *Distal-less* encodes a homeodomain protein required for limb development in *Drosophila*. Nature (Lond) 338:432–434

Cohen SM (1990) Specification of limb development in the *Drosophila* embryo by positional cues from segmentation genes. Nature (Lond) 343:173–177

Côté S, Preiss A, Haller J, Schuh R, Kienlin A, Seifert A, Jäckle H (1987) The *gooseberry-zipper* region of *Drosophila*: five genes encode spatially restricted transcripts in the embryo. EMBO J 6:2793–2801

Crick FH, Lawrence PA (1975) Compartments and polyclones in insect development. Science 189:340–347

Dale L, Bownes M (1980) Is regeneration in *Drosophila* the result of epimorphic regulation? Wilhelm Roux's Arch Dev Biol 189:91–96

DeLotto R, Spierer P (1986) A gene required for the specification of dorsal-ventral pattern of *Drosophila* appears to encode a serine protease. Nature (Lond) 323:688–692

Denell RE, Hummels KR, Wakimoto B, Kaufman TC (1981) A developmental genetic analysis of the lethal syndrome associated with the *Antennapedia* locus of *Drosophila melanogaster*. Dev Biol 81:43–50

Desplan C, Theis J, O'Farrell PH (1985) The *Drosophila* developmental gene, *engrailed*, encodes sequence specific DNA binding activity. Nature (Lond) 318:630–635

DiNardo S, O'Farrell PH (1987) Establishment and refinement of segmental pattern in the *Drosophila* embryo: spatial control of *engrailed* expression by pair-rule genes. Genes Dev 1:1212–1225

DiNardo S, Kuner JM, Theis J, O'Farrell PH (1985) Development of embryonic pattern in *Drosophila melanogaster* as revealed by accumulation of the nuclear *engrailed* protein. Cell 43:59–69

DiNardo S, Sher E, Heemskerk-Jorgens J, Kassis J, O'Farrell P (1988) Two-tiered regulation of spatially patterned *engrailed* gene expression during *Drosophila* embryogenesis. Nature (Lond) 332:604–609

Doe CQ, Goodman CS (1985) Early events in insect neurogenesis. II. The role of cell interactions and cell lineage in the determination of neuronal precursor cells. Dev Biol 111:206–219

Driever W, Nüsslein-Volhard C (1989) The *bicoid* protein is a positive regulator of *hunchback* transcription in the early *Drosophila* embryo. Nature (Lond) 337:138–143

Eaton S, Kornberg TB (1990) Repression of *ci-D* in posterior compartments of *Drosophila* by *engrailed*. Genes Dev 4:1068–1077

Eldon ED, Pirotta V (1991) Interactions of the *Drosophila* gap gene *giant* with maternal and zygotic pattern-forming genes. Development 111:367–378

Fausto-Sterling A (1971) On the timing and place of action during embryogenesis of the female-sterile mutants *fused* and *rudimentary Drosophila melanogaster*. Dev Biol 26:452–463

Fausto-Sterling A (1978) Pattern formation in the wing veins of the *fused* mutant (*Drosophila melanogaster*). Dev Biol 63:358–369

Fjose A, McGinnis WJ, Gehring WJ (1985) Isolation of a homeobox-containing gene from the *engrailed* region of *Drosophila* and the spatial distribution of its transcript. Nature (Lond) 313:284–289

Foe VE (1989) Mitotic domains reveal early commitment of cells in *Drosophila* embryos. Development 107:1–22

Foe VE, Alberts BM (1983) Studies of nuclear and cytoplasmic behaviour during the five mitotic cycles that precede gastrulation in *Drosophila* embryogenesis. Cell Sci J 61:31–70

French V, Bryant PJ, Bryant SV (1976) Pattern regulation in epimorphic fields. Science 193:969–981

Frasch M, Levine M (1987) Complementary patterns of *even-skipped* and *fushi tarazu* expression involve their differential regulation by a common set of segmentation genes in *Drosophila*. Genes Dev 1:981–995

Frigerio G, Burri M, Bopp D, Baumgartner M, Noll M (1986) Structure of the segmentation gene paired and the *Drosophila* PRD gene set as part of a gene network. Cell 47:735–746

Frischer LE, Hagen FS, Garber RL (1986) An inversion that disrupts the *Antennapedia* gene causes abnormal structure and localization of RNAs. Cell 47:1017–1023

Garcia-Bellido A (1975) Genetic control of wing disc development in *Drosophila*. Ciba Found Symp 29:161–182

Garcia-Bellido A (1977) Homeotic and atavic mutations in insects. Am Zool 17: 613–629

Garcia-Bellido A, Lawrence PA, Morata G (1979) Compartments in animal development. Sci Am 241:102–110

Garcia-Bellido A, Ripoll P, Morata G (1973) Developmental compartmentalization of the wing disk of Drosophila. Nature (Lond) New Biol 245:251–253

Gergen JP, Wieschaus EF (1986) Localized requirements for gene activity in segmentation of Drosophila embryos: analysis of armadillo, fused, giant and unpaired mutations in mosaic embryos. Wilhelm Roux's Arch Dev Biol 195: 49–62

Gergen JP, Coulter D, Wieschaus E (1986) Segmental pattern and blastoderm cell identities. In: Subtelny S (ed) Gametogenesis and the early embryo. Liss Symp Soc Dev Biol, New York, pp 195–220

Gierer A, Meinhardt H (1972) A theory of biological pattern formation. Kybernetik 12:30–39

González F, Swales L, Bejsovec A, Skaer H, Martinez-Arias A (1991) Secretion and movenment of wingless protein in the epidermis of the Drosophila embryo. Mech Dev 35:43–54

Grau Y, Simpson P (1987) The segment polarity gene costal-2 in Drosophila. I. The organization of both primary and secondary embryonic fields may be affected. Dev Biol 122:186–200

Harding K, Levine M (1988) Gap genes define the limits of Antennapedia and bithorax gene expression during early development in Drosophila EMBO J 7:205–214

Harding K, Hoey T, Warrior R, Levine M (1989) Autoregulatory and gap gene response elements of the even-skipped promoter of Drosophila. EMBO J 8:1205–1212

Harding K, Rushlow C, Doyle HJ, Hoey T, Levine L (1986) Cross-regulatory interactions among pair-rule genes in Drosophila. Science 233:953–959

Hashimoto C, Hudson KL, Anderson KV (1988) The Toll gene of Drosophila, required for dorsal-ventral embryonic polarity, appears to encode a transmembrane protein. Cell 52:269–279

Hazelrigg T, Kaufman TC (1983) Revertants of dominant mutations associated with the Antennapedia gene complex of Drosophila melanogaster. Genetics 105: 581–600

Heemskerk J, DiNardo S, Kostriken R, O'Farrell PH (1991) Multiple modes of engrailed regulation in the progression towards cell fate determination. Nature (Lond) 352:404–410

Hidalgo A (1991) Interactions between segment polarity genes and the generation of the segmental pattern in Drosophila. Mech Dev 35:77–87

Hidalgo A, Ingham P (1990) Cell patterning in the Drosophila segment: spatial regulation of the segment polarity gene patched. Development 110:291–301

Hiromi Y, Gehring WJ (1987) Regulation and function of the Drosophila segmentation gene fushi tarazu. Cell 50:963–974

Hooper JE, Scott MP (1989) The Drosophila patched gene encodes a putative membrane protein required for segmental patterning. Cell 59:751–765

Hooper KL, Parkhurst SM, Ish-Horowicz D (1989) Spatial control of hairy protein expression during embryogenesis. Development 107:489–504

Hoch M, Seifert E, Jäckle H (1991) Gene expression mediated by cis-acting sequences of the Krüppel gene in response to the Drosophila morphogens bicoid and hunchback. EMBO J 10:2267–2278

Howard K, Ingham P (1986) Regulatory interactions between the segmentation genes fushi tarazu, hairy, and engrailed in the Drosophila blastoderm. Cell 44:949–957

Howard K, Ingham P, Rushlow C (1988) Region-specific alleles of the Drosophila segmentation gene hairy. Genes Dev 2:1037–1046

Hülskamp M, Pfeifle C, Tautz D (1990) A morphogenetic gradient of *hunchback* protein organizes the expression of the gap genes *Krüppel* and *knirps* in the early *Drosophila* embryo. Nature (Lond) 346:646–648

ImmergLück K, Lawrence PA, Bienz M (1990) Induction across germ layers in *Drosophila* mediated by a genetic cascade. Cell 62:261–268

Ingham PW (1988) The molecular genetics of embryonic pattern formation in *Drosophila*. Nature (Lond) 335:25–34

Ingham P, Martinez-Arias A, Lawrence PA, Howard K (1985) Expression of *engrailed* in the parasegment of *Drosophila*. Nature (Lond) 317:634–636

Ingham PW, Martinez-Arias A (1986) The correct activation of *Antennapedia* and bithorax complex genes requires the *fushi tarazu* gene. Nature (Lond) 324: 592–597

Ingham PW, Baker NE, Martinez-Arias A (1988) Regulation of segment polarity genes in the *Drosophila* blastoderm by *fushi tarazu* and *even skipped*. Nature (Lond) 331:73–75

Ingham PW, Taylor AM, Nakano Y (1991) Role of the *Drosophila patched* gene in positional signalling. Nature (Lond) 353:184–187

Irish VF, Martinez-Arias A, Akam M (1989) Spatial regulation of the *Antennapedia* and *Ultrabithorax* genes during *Drosophila* early development. EMBO J 8:1527–1537

Jaynes JB, O'Farrell PH (1988) Activation and repression of transcription by homeodomain-containing proteins that bind a common site. Nature (Lond) 336:744–749

Jiang J, Hoey T, Levine M (1991) Autoregulation of a segmentation gene in *Drosophila*: Combinatorial interactions of the *even-skipped* homeo box protein with a distal enhancer element. Genes Dev 5:265–277

St Johnston DR, Gelbart WM (1987) Decapentaplegic transcripts are localized along the dorsal-ventral axis of the *Drosophila* embryo. EMBO J 6:2785–2791

Joyner AL, Herrup BA, Auerbach CA, Rossant DJ (1991) Subtle cerebellar phenotype in mice homozygous for a targeted deletion of the *En-2* homeobox. Science 251:1239–1243

Joyner AL, Kornberg T, Coleman KG, Cox RD, Martin GR (1985) Expression during embryogenesis of a mouse gene with sequence homology to the *Drosophila engrailed* gene. Cell 43:29–37

Jürgens G, Wieschaus E, Nüsslein-Volhard C, Kluding H (1984) Mutations affecting the pattern of the larval cuticle in *Drosophila melanogaster*. II. Zygotic loci on the third chromosome. Wilhelm Roux's Arch Dev Biol 196:141–157

Kaufman TC, Lewis R, Wakimoto B (1980) Cytogenetic analysis of chromosome 3 in *Drosophila melanogaster*: the homeotic gene complex in polytene chromosomal interval 84A, B. Genetics 94:115–133

Klingensmith J, Noll E, Perrimon N (1989) The segment polarity phenotype of *Drosophila* involves differential tendencies toward transformation and cell death. Dev Biol 134:130–145

Kornberg T (1981) *engrailed*: a gene controlling compartment and segment formation in *Drosophila*. Proc Natl Acad Sci USA 78:1095–1099

Kornberg T, Siden I, O'Farrell P, Simon M (1985) The *engrailed* locus of *Drosophila*: in situ localization of transcripts reveals compartment-specific expression. Cell 40:45–53

Krasnow MA, Saffman EE, Kornfeld K, Hogness DS (1989) Transcriptional activation and repression by *Ultrabithorax* proteins in cultured *Drosophila* cells. Cell 57:1031–1043

Kraut R, Levine M (1991) Spatial regulation of the gap gene *giant* during *Drosophila* development. Development 111:601–609

Kuner JM, Nakanishi M, Ali Z, Drees B, Gustavson E, Theis J, Kauvar L, Kornberg T, O'Farrell PH (1985) Molecular cloning of *engrailed*: a gene involved in the development of pattern in *Drosophila melanogaster*. Cell 42:309–316

Laughon A, Scott MP (1984) Sequence of a *Drosophila* segmentation gene: protein structure homology with DNA-binding proteins. Nature (Lond) 310:25–31

Lawrence PA (1966) Gradients in the insect segment: the orientation of hairs in the milkweed bug *Oncopeltus fasciatus*. J Exp Biol 44:607–620

Lawrence PA (1981) The cellular basis of segmentation in insects. Cell 26:3–10

Lawrence PA (1982) Cell lineage of the thoracic muscles of *Drosophila*. Cell 29: 493–503

Lawrence PA (1989) Cell lineage and cell states in the *Drosophila* embryo. In: Evered D, Marsh J (eds) Cellular basis of morphogenesis, 144. John Wiley, New York, pp 131–149

Lawrence PA, Johnston P (1984) On the role of the *engrailed+* gene in the internal organs of *Drosophila*. EMBO J 3:2839–2844

Lawrence PA, Morata G (1976) Compartments in the wing of *Drosophila*: a study of the *engrailed* gene. Dev Biol 50:321–337

Lawrence PA, Morata G (1983) The elements of the bithorax complex. Cell 35: 595–601

Lawrence PA, Johnston P, Macdonald P, Struhl G (1987) Borders of parasegments in *Drosophila* embryos are delimited by the *fushi tarazu* and *even-skipped* genes. Nature (Lond) 328:440–442

Lewis EB (1978) A gene complex controlling segmentation in *Drosophila*. Nature (Lond) 276:565–570

Limbourg-Bouchon B, Busson D, Lamour-Isnard C (1991) Interactions between *fused*, a segment polarity gene in *Drosophila*, and other segmentation genes. Development 112:417–429

Lindsley DL, Grell HE (1968) Genetic variations of *Drosophila melanogaster*. Carnegie Inst Washington

Locke M (1959) The cuticular pattern in an insect, *Rhodnius prolixus*, stal. J Exp Biol 36:459–477

Locke, M (1966) The cuticular pattern in an insect: the behavior of grafts in segmented appendages. J Insect Physiol 12:397–402

Locke M, Huie P (1981) Epidermal feet in insect morphogenesis. Nature (Lond) 293:733–735

Lohs-Schardin M, Sander K, Cremer C, Cremer T, Zorn C (1979) Localized ultraviolet laser microbeam irradiation of early *Drosophila* embryos: fate maps based on location and frequency of adult defects. Dev Biol 68:533–545

Macdonald PM, Ingham P, Struhl G (1986) Isolation, structure and expression of *even-skipped*: a second pair-rule gene of *Drosophila* containing a homeobox. Cell 47:721–734

Mahaffey JW, Kaufman TC (1988) The homeotic genes of the *Antennapedia* complex and the bithorax complex of *Drosophila*. In: Malacinski GM (ed) Developmental genetics of higher organisms: a primer in developmental biology. Macmillan, New York, pp 329–360

Marcus W (1962) Untersuchiungen über die Polarität der Rumpfhaut von Schmetterlingen. Wilhelm Roux's Arch Entwicklungsmech Org 154:56–102

Mariol M-C, Préat T, Limbourg-Bouchon B (1987) Molecular cloning of *fused*, a gene required for normal segmentation in the *Drosophila melanogaster* embryo. Mol Cell Biol 7:3244–3251

Martinez-Arias A (1985) The development of *fused*-embryos of *Drosophila melanogaster*. J Embryol Exp Morphol 87:99–114

Martinez-Arias A (1989) A cellular basis for pattern formation in the insect epidermis. TIG 5:262–267

Martinez-Arias A, Ingham PW (1985) The origin of pattern duplications in segment polarity mutants of *Drosophila melanogaster*. J Embryol Exp Morphol 87: 129–135

Martinez-Arias A, Lawrence PA (1985) Parasegments and compartments in the *Drosophila* embryo. Nature (Lond) 313:639–642

Martinez-Arias A, White RAH (1988) *Ultrabithorax* and *engrailed* expression in *Drosophila* embryos mutant for segmentation genes of the pair-rule class. Development 102:325–338

Martinez-Arias A, Baker NE, Ingham PW (1988) Role of segment polarity genes in the definition and maintenance of cell states in the *Drosophila* embryo. Development 103:157–170

Meinhardt H (1983) Cell determination boundaries as organizing regions for secondary embryonic fields. Dev Biol 96:375–385

Meinhardt H (1986) Hierarchical inductions of cell states: a model for segmentation in *Drosophila*. J Cell Sci (Suppl) 4:357–381

Meinhardt H, Gierer A (1980) Generation and regeneration of sequences of structures during morphogenesis. J Theor Biol 85:429–450

Mittenthal JE (1981) The rule of normal neighbors: a hypothesis for morphogenetic pattern regulation. Dev Biol 88:15–26

Mohler J (1988) Requirements for *hedgehog*, a segmental polarity gene, in patterning larval and adult cuticle of *Drosophila*. Genetics 120:1061–1072

Mohler J, Eldon ED, Pirotta V (1989) A novel spatial transcription pattern associated with the segmentation gene *giant* of *Drosophila*. EMBO J 8: 1539–1548

Mohler J, Vani K (1992) Molecular organization and embryonic expression of the *hedgehog* gene in cell-cell communication in segmental patterning of *Drosophila*. Development, in press

Morata G, Lawrence PA (1975) Control of compartment development by the *engrailed* gene in *Drosophila*. Nature (Lond) 255:614–617

Morata G, Lawrence PA (1977a) The development of *wingless*, a homeotic mutation of *Drosophila*. Dev Biol 56:227–240

Morata G, Lawrence PA (1977b) Homoeotic genes, compartments and cell determination in *Drosophila*. Nature (Lond) 265:211–216

Moscoso del Prado J, Garcia-Bellido A (1984) Genetic regulation of the *Achaete-scute* complex of *Drosophila melanogaster*. Wilhelm Roux's Arch Dev Biol 193:242–245

Nakano Y, Guerrero I, Hidalgo A, Taylor A, Whittle JRS, Ingham PW (1989) The *Drosophila* segment polarity gene *patched* encodes a protein with multiple potential membrane spanning domains. Nature (Lond) 341:508–513

Nardi JB, Kafatos FC (1976) Polarity and gradients in lepidopteran wing epidermis. J Embryol Exp Morphol 36:469–487

Nardi JB, Magee-Adams SM (1986) Formation of scale spacing patterns in a moth wing. I. Epithelial feet may mediate cell rearrangement. Dev Biol 116:278–290

Nauber U, Pankratz MJ, Kienlin A, Seifert E, Klemm U, Jackle H (1988) Abdominal segmentation of the *Drosophila* embryo requires a hormone receptor-like protein encoded by the gap gene *knirps*. Nature (Lond) 336: 489–492

Nübler-Jung K (1974) Cell migration during pattern reconstitution by intercalary regeneration and cell sorting in *Dysdercus intermedius* (Dist.). Nature (Lond) 248:610–611

Nübler-Jung K (1977) Pattern stability in the insect segment. I. Pattern reconstitution by intercalary regeneration and cell sorting in *Dysdercus interemedius* Dist. Wilhelm Roux's Arch Dev Biol 183:17–40

Nüsslein-Volhard C, Roth S (1989) Axis determination in insect embryos. In: Evered D, Marsh J (eds) Cellular basis of morphogenesis, 144. John Wiley, New York, pp 37–64

Nüsslein-Volhard C, Wieschaus E (1980) Mutations affecting segment number and polarity in *Drosophila*. Nature (Lond) 287:795–801

Nüsslein-Volhard C, Lohs-Schardin M, Cremer C (1980) A dorso-ventral shift of embryonic primordia in a new maternal-effect mutant of *Drosophila*. Nature (Lond) 238:474–476

Nüsslein-Volhard C, Wieschaus E, Kluding H (1984) Mutations affecting the pattern of the larval cuticle in *Drosophila melanogaster*. I. Zygotic loci on the second chromosome. Wilhelm Roux's Arch Dev Biol 193:267–282

Orenic T, Chidsey J, Holmgren R (1987) *Cell* and *cubitus interruptus Dominant*; two segment polarity genes on the fourth chromosome in *Drosophila*. Dev Biol 124:50–56

Orenic TV, Slusarski C, Kroll KL, Holmgren RA (1990) Cloning and characterization of the segment polarity gene *cubitus interruptus Dominant* of *Drosophila*. Genes Dev 4:1053-1067

Pankratz MJ, Seifert E, Gerwin N, Billi B, Nauber U, Jäckle H (1990) Gradients of *Krüppel* and *knirps* gene products direct pair-rule gene stripe patterning in the posterior region of the *Drosophila* embryo. Cell 61:309-317

Patel NH, Kornberg T, Goodman CS (1989a) Expression of *engrailed* during segmentation in grasshopper and crayfish. Development 107:201-212

Patel NH, Schafer B, Goodman CS, Holmgren R (1989b) The role of segment polarity genes during *Drosophila* neurogenesis. Genes Dev 3:890-904

Peifer M, Wieschaus E (1990) The segment polarity gene *armadillo* encodes a functionally modular protein that is the *Drosophila* homologue of human plakoglobin. Cell 63:1167-1178

Peifer M, Rauskolb C, Williams M, Riggleman B, Wieschaus E (1991) The segment polarity gene *armadillo* interacts with the *wingless* signalling pathway in both embryonic and adult pattern formation. Development 111:1029-1043

Perrimon N, Mahowald AP (1987) Multiple functions of segment polarity genes in *Drosophila*. Dev Biol 119:587-600

Perrimon N, Smouse D (1989) Multiple functions of a *Drosophila* homeotic gene, *zeste-white 3*, during segmentation and neurogenesis. Dev Biol 135:287-305

Perrimon N, Engstrom L, Mahowald AP (1989) Zygotic lethals with specific maternal effect phenotypes in *Drosophila melanogaster*. I. Loci on the X chromosome. Genetics 121:333-352

Phillips RG, Roberts IJH, Ingham PW, Whittle JRS (1990) The *Drosophila* segment polarity gene *patched* is involved in a position-signalling mechanism in imaginal discs. Development 110:105-114

Piepho H (1955) Über die Ausrichtung der Schuppenbälge und Schuppen am Schmetterlingsrumpf. 42:22

Piepho, H. and C. Hintze-Podufal (1971) Zur Polarität des Insektensegments. I. Induktion des Segmenthinterrandes bei *Galleria mellonella* L. Biol Zentralbl 90:419-431

Pignoni F, Baldarelli RM, Steingrimsson E, Diaz RJ, Patapoutian A, Merriam JR, Lengyel JL (1990) The *Drosophila* gene *tailless* is expressed at the embryonic termini and is a member of the steroid receptor superfamily. Cell 62:151-163

Poole SJ, Kauvar LM, Drees B, Kornberg T (1985) The *engrailed* locus of *Drosophila*: structural analysis of an embryonic transcript. Cell 40:37-43

Postlethwait JH, Schneiderman HA (1969) A clonal analysis of determination in *Antennapedia*, a homeotic mutant of *Drosophila melanogaster*. Proc Natl Acad Sci USA 64:176-183

Préat T, Thérond P, Lamour-Isnard C, Limbourg-Bouchon B, Tricoire H, Erk I, Mariol M-C, Busson D (1990) A putative serine/threonine protein kinase encoded by the segment polarity *fused* gene of *Drosophila*. Nature (Lond) 347:87-89

Price JV, Clifford RJ, Schüpbach T (1989) The maternal ventralizing locus *torpedo* is allelic to *faint little ball*, an embryonic lethal, and encodes the *Drosophila* EGF receptor homologue. Cell 56:1085-1092

Reinitz J, Levine M (1990) Control of the initiation of homeotic gene expression by the gap genes *giant* and *tailless* in *Drosophila*. Dev Biol 140:57-72

Reuter R, Panganiban GEF, Hoffman FM, Scott MP (1990) Homeotic genes regulate the spatial expression of putative growth factors in the visceral mesoderm of *Drosophila* embryos. Development 110:1031-1040

Riggleman B, Schedl P, Wieschaus E (1990) Spatial expression of the *Drosophila* segment polarity gene *armadillo* is posttranscriptionally regulated by *Wingless*. Cell 63:549-560

Riggleman B, Wieschaus E, Schedl P (1989) Molecular analysis of the *armadillo* locus: uniformly distributed transcripts and a protein with novel internal repeats are associated with a *Drosophila* segment polarity gene. Genes Dev 3:96-113

Rijsewijk F, Schuermann M, Wagenaar E, Parren P, Weigel D, Nusse R (1987) The *Drosophila* homolog of the mouse mammary oncogene *int-1* is identical to the segment polarity gene *Wingless*. Cell 50:649–657

Riley PD, Carroll SB, Scott MP (1987) The expression and regulation of *Sex combs reduced* protein in *Drosophila* embryos. Genes Dev 1:716–730

Rosenberg UB, Schroder C, Preiss A, Kienlin A, Côte S, Riede I, Jäckle H (1986) Structural homology of the product of the *Drosophila Krüppel* gene with *Xenopus* transcription factor IIIA. Nature (Lond) 319:336–339

Rushlow CA, Hogan A, Pinchin SM, Howe KM, Lardelli M, Ish-Horowicz D (1989) The *Drosophila hairy* protein acts in both segmentation and bristle patterning and shows homology to N-*myc*. EMBO J 8:3095–3103

Russell MA (1974) Pattern formation in the imaginal discs of a temperature-sensitive cell-lethal mutant of *Drosophila melanogaster*. Dev Biol 40:24–39

Sampedro J, Guerrero I (1991) Unrestricted expression of the *Drosophila* gene *patched* allows normal segment polarity. Nature (Lond) 353:187–190

Samson M-L, Jackson-Grusby L, Brent R (1989) Gene activation and DNA binding by *Drosophila Ultrabithorax* and *abdominal A* proteins. Cell 57:1045–1052

Schneuwly S, Kuroiwa A, Gehring WJ (1987) Molecular analysis of the dominant homeotic *Antennapedia* phenotype. EMBO J 6:201–206

Scott MP, Carroll SB (1987) The segmentation and homeotic gene network in early *Drosophila* development. Cell 51:689–698

Scott MP, O'Farrell PH (1986) Spatial programming of gene expression in early *Drosophila* embryogenesis. Annu Rev Cell Biol 2:49–80

Sharma RP (1973) *wingless*, a new mutant in *Drosophila melanogaster*. Drosophila Inf Serv 50:134

Siegfried E, Perkins LA, Capaci TM, Perrimon N (1990) Putative protein kinase product of the *Drosophila* segment-polarity gene *zeste-white3*. Nature (Lond) 345:825–829

Simcox AA, Sang JH (1983) When does determination occur in *Drosophila* embryos? Dev Biol 97:212–221

Simcox AA, Roberts IJH, Hersperger E, Gribbin MC, Shearn A, Whittle JRS (1989) Imaginal discs can be recovered from cultured embryos mutant for the segment polarity genes *engrailed*, *naked* and *patched* but not from *wingless*. Development 107:715–722

Simpson P (1990) Lateral inhibition and the development of the sensory bristles of the adult peripheral nervous system. Development 109:509–519

Simpson P, Crau Y (1987) The segment polarity gene *costal-2* in *Drosophila* II. The origin of imaginal pattern duplications. Dev Biol 122:201–209

Simpson P, El Messal M, Moscoso Del Prado J, Ripoll P (1988) Stripes of positional homologies across the wing blade of *Drosophila melanogaster*. Dev Biol 103:391–401

Small S, Kraut R, Hoey T, Warrior R, Levine M (1991) Transcriptional regulation of a pair-rule stripe in *Drosophila*. Genes Dev 5:827–839

Sokol S, Christian JL, Moon RT, Melton DA (1991) Injected Wnt RNA induces a complete body axis in *Xenopus* embryos. Cell 67:741–752

Spencer FA, Hoffman M, Gelbart WM (1982) Decapentaplegic: a gene complex affecting morphogenesis in *Drosophila melanogaster*. Cell 28:451–461

Stanojevic D, Hoey T, Levine M (1989) Sequence-specific DNA-binding activities of the gap proteins *hunchback*, and *Krüppel* in *Drosophila*. Nature (Lond) 341:331–335

Stanojevic D, Small D, Levine M (1991) Regulation of a segmentation stripe by overlapping activators and repressors in the *Drosophila* embryo. Science 254:1385–1387

Steward R (1987) *Dorsal*, an embryonic polarity gene in *Drosophila*, is homologous to the vertebrate proto-oncogene, *c-rel*. Science 238:692–694

Struhl G (1981) A homeotic mutation transforming leg to antenna in *Drosophila*. Nature (Lond) 292:635–638

Struhl G, Struhl K, Macdonald PM (1989) The gradient morphogen bicoid is a concentration dependent transcriptional activator. Cell 57:1259–1273

Stumpf H (1968) Further studies on gradient-dependent diversification in the pupal cuticle of *Galleria mellonella*. J Exp Biol 49:49–59

Tautz D, Lehmann R, Schnurch H, Schuh R, Seifert E, Kienlin K, Jäckle H (1987) Finger protein of novel structure encoded by *hunchback*, a second member of the gap class of *Drosophila* segmentation genes. Nature (Lond) 327:383–389

Thali M, Müller M, DeLorenzi M, Matthias P, Bienz M (1988) *Drosophila* homeotic genes encode transcriptional activators similar to mammalian OTF-2. Nature (Lond) 336:598–601

Thomas KR, Capecchi MR (1990) Targeted disruption of the murine *int-1* proto-oncogene resulting in severe abnormalities in midbrain and cerebellar development. Nature (Lond) 346:847–850

Tiong SYK, Nash D (1990) Genetic analysis of the *adenosine3* (*Gart*) region of the second chromosome of *Drosophila melanogaster*. Genetics 124:889–897

Treisman J, Desplan C (1989) The products of the *Drosophila* gap genes *hunchback* and *Krüppel* bind to the *hunchback* promoter. Nature (Lond) 341:335–337

van den Heuvel M, Nusse R, Johnston P, Lawrence PA (1989) Distribution of the *wingless* gene product in *Drosophila* embryos; a protein involved in cell-cell communication. Cell 59:739–749

Vincent J-P, O'Farrell PH (1992) The state of *engrailed* expression is not clonally transmitted during early *Drosophila* development. Cell 68:923–931

Warrior R, Levine M (1990) Dose-dependent regulation of pair-rule stripes by gap proteins and the initiation of segment polarity. Development 110:759–767

Weiss PA (1939) Principles in development. Henry Holt

White RA, Lehmann R (1986) A gap gene, *hunchback*, regulates the spatial expression of *Ultrabithorax*. Cell 47:311–321

Whittle JRS (1976) Clonal analysis of a genetically caused duplication of the anterior wing in *Drosophila melanogaster*. Dev Biol 51:257–268

Wieschaus E, Noell E (1986) Specificity of embryonic lethal mutations in *Drosophila* analyzed in germ line clones. Wilhelm Roux's Arch Dev Biol 195:63–73

Wieschaus E, Gehring W (1976) Clonal analysis of primordial disc cells in the early embryo of *Drosophila melanogaster*. Dev Biol 50:249–263

Wieschaus E, Riggleman R (1987) Autonomous requirements for the segment polarity gene *armadillo* during *Drosophila* embryogenesis. Cell 49:177–184

Wieschaus E, Nüsslein-Volhard C, Jürgens G (1984) Mutations affecting the pattern of the larval cuticle in *Drosophila melanogaster*. I. Zygotic loci on the X chromosome and the fourth chromosome. Wilhelm Roux's Arch Dev Biol 193:267–282

Wigglesworth VB (1940) Local and general factors in the development of "pattern" in *Rhodnius prolixus* (Hemiptera). J Exp Biol 36:180–200

Winslow GM, Hayashi S, Krasnow M, Hogness DS, Scott MP (1989) Transcriptional activation by the *Antennapedia* and *fushi tarazu* proteins in cultured *Drosophila* cells. Cell 57:1017–1030

Wolpert L (1969) Positional information and the spatial pattern of cellular differentiation. J Theor Biol 25:1–47

Wright D, Lawrence PA (1981) Regeneration of the segment boundary in *Oncopeltus*. Dev Biol 85:317–327

Zuo P, Stanojevic D, Colgan J, Han K, Levine M, Manley JL (1991) Activation and repression of transcription by the gap proteins *hunchback* and *Krüppel* in cultured *Drosophila* cells. Genes Dev 5:254–264

2 Embryonic Development in *Caenorhabditis elegans*

1 Introduction

Nematodes were first used to study of embryogenesis more than 100 years ago, and this in part led to the concepts of cell-autonomous differentiation and localized cytoplasmic determinants. More recently, the techniques of genetics, experimental and descriptive embryology, and molecular biology have been combined to study the development of the small, free-living nematode *Caenorhabditis elegans* (Brenner 1974, 1988). This chapter focuses on embryonic development and is intended as a general overview of *C. elegans* embryogenesis, illustrating the experimental techniques available for this organism and the conclusions that can be drawn. Excellent reviews on postembryonic development (i.e. after hatching) in *C. elegans* and most other aspects of the worm's development, genetics and biology can be found in Wood (1988a). This book includes extensive appendices detailing techniques and anatomy and includes phenotypic descriptions of all mutants known at the time of publication. Other reviews of *C. elegans* embryogenesis can be found in Kemphues (1989), Wood (1988b), Schierenberg (1989) and Strome (1989).

2 The System

There are many aspects of *C. elegans* biology that make it ideal for studies of embryogenesis. The short life cycle facilitates the isolation and analysis of embryonic mutants. Because the worm is transparent at all stages of its life cycle, Nomarski differential interference contrast microscopy can be used to observe the anatomy of living animals and embryos in exquisite detail. In addition, recent advances in molecular biology have greatly simplified cloning of developmentally interesting genes.

Department of Medical Biochemistry, University of Calgary 3330 Hospital Drive N.W. Calgary, Alberta T2N 4N1, Canada

Results and Problems in Cell Differentiation 18
W. Hennig (Ed.)
Early Embryonic Development of Animals
© Springer-Verlag Berlin Heidelberg 1992

2.1 Life Cycle and Genetics

C. elegans is a small, free-living soil nematode that can be cultured on petri dishes using a lawn of *Escherichia coli* as a food source. The adult is only 1.2 mm long and has a generation time of just 3.5 days (all developmental times are for 20 °C). This allows large populations to be screened for rare events since thousands of individuals can be grown on a single petri dish. Embryogenesis, the time from fertilization to hatching, lasts only 15 h and is followed by four larval stages, designated L1 through L4, separated by molts of the cuticle. The final molt results in the sexually mature adult. Starved animals can form dauer larvae, an alternative L3 stage resistant to desiccation and starvation. Upon refeeding, the animals emerge as L4's.

    C. elegans has two sexes, hermaphrodites and males (Fig. 1). The L4 hermaphrodite makes sperm, which are stored in the spermatheca. After the worm reaches adulthood, the hermaphrodite's germ line switches from spermatogenesis to oogenesis, and the stored sperm are used for internal self-fertilization. The diploid hermaphrodite has five pairs of autosomes and a pair of X chromosomes, and rare nondisjunctional events (approximately 1/700 gametes) lead to X0 male progeny (Hodgkin et al. 1979). Males only make sperm and have a tail specialized for mating with hermaphrodites (Fig. 1). Conveniently, mated hermaphrodites preferentially use male sperm to fertilize their oocytes (Ward and Carrel 1979). Self-fertilizing hermaphrodites are limited by the number of stored sperm to about 300 progeny, but mated hermaphrodites can produce many more offspring. This sexual system is ideal for genetic analysis; *C. elegans* can be propagated either by self-fertilization or as a dioecious sexual species using males to transfer genetic markers between hermaphrodites.

    The extensive genetic map is constantly updated by the curators of the *Caenorhabditis* Genetics Center using data contributed by *C. elegans* researchers. Presently, the map includes 831 mapped genes and lists 225 chromosomal deficiencies and 62 duplications (Edgley and Riddle 1990).

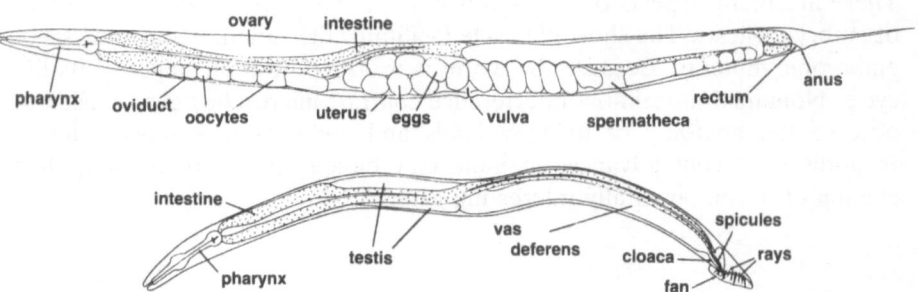

Fig. 1. Drawings of the lateral view of an adult hermaphrodite (*above*) and an adult male (*below*). Dorsal is to the top. Note that in the anterior of the hermaphrodite, the intestine lies to the left of the ovary but the situation is reversed in the posterior (After Sulston and Horvitz 1977)

The *Caenorhabditis* Genetics Center currently maintains over 1400 strains frozen in liquid nitrogen. This collection includes representative mutations of 690 genes as well as 143 chromosomal rearrangements.

2.2 Molecular Techniques

C. elegans is also amenable to molecular analysis. It has a relatively small genome (10^5 kb per haploid genome) with little repetitive DNA (Sulston and Brenner 1974; Emmons 1988). Genes can be cloned by transposon tagging (Greenwald 1985; Moerman et al. 1986; Collins et al. 1987; Mori et al. 1988) or by using the physical map.

The nearly completed *C. elegans* physical map is serving as a model for the assembly of large physical maps. Sequencing of the entire *C. elegans* genome is being done as a preliminary step in the human genome project (Coulson et al. 1986, 1988; Roberts 1990). The physical map simplifies the cloning of genes represented by mutations: once a gene has been genetically mapped between markers located on the physical map, genomic clones that span the area of interest, in either cosmids or yearst artificial chromosomes, can be obtained by mail. This "clone by phone" strategy eliminates the need for tedious chromosomal walks. The DNA that actually contains the gene can then be identified by transformation rescue of the mutant phenotype (Stinchcomb et al. 1985; Fire 1986). Clones are microinjected into the hermaphrodite gonad, transgenic strains are established, and the presence of the wild-type allele in the exogenous DNA is determined by the rescue of the mutant phenotype.

2.3 A Description of Early Embryogenesis

The hermaphrodite contains a double-armed gonad whose proximal regions open into a common vulva (Fig. 1). Germ cells in the distal end form a mitotically dividing syncytium, and developing oocytes separate into individual cells and begin meiosis as they reach the region where the gonad reflexes upon itself (Fig. 1; Hirsh et al. 1976; Strome 1986a). As an oocyte matures, its cytoplasm increases dramatically in volume, due in part to vitellogenins transported from the intestine (Kimble and Sharrock 1983). The nuclear envelope of the oocyte breaks down prior to fertilization, and the oocyte arrests at diakinesis of the first meiotic division. The oocyte is then squeezed into the spermatheca with the end that will become the posterior of the embryo leading. Sperm entry is usually at the posterior, on the side opposite the maternal chromosomes. The maternal chromosomes then complete meiosis, two polar bodies are extruded (these serve as convenient anterior markers later during embryonic development), and the eggshell begins to form (Nigon et al. 1960; Hirsh et al. 1976; Ward and Carrel 1979; Albertson 1984). Once the eggshell is well formed (about 20 min after fertilization) embryos can be dissected from gravid hermaphrodites for better observation. The embryos are laid when there are about 28 cells, approximately 2 h after fertilization.

Approximately 20 min after fertilization, the maternal and paternal pronuclei form and begin migrating toward one another (Fig. 2A–B; Nigon

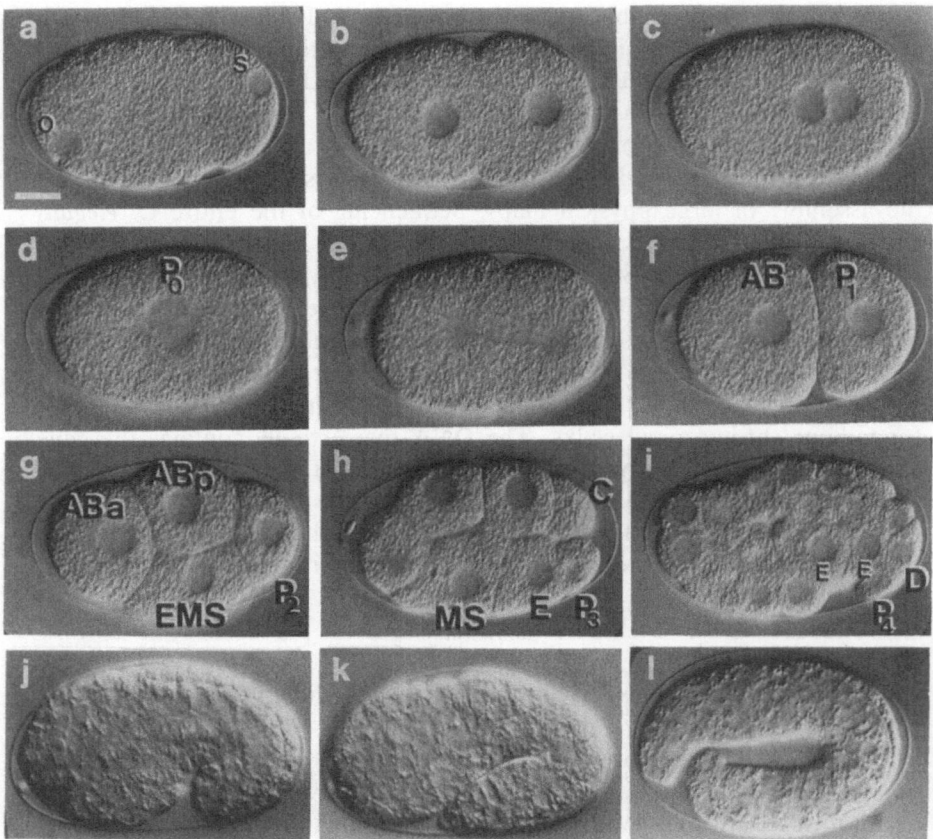

Fig. 2. Nomarski photomicrographs of living *C. elegans* embryos. In all cases, anterior is to the *left* and dorsal to the *top*. **a** Formation of the pronuclei from the oocyte (*o*) and sperm (*s*). **b** Pseudocleavage and pronuclear migration. **c** Pronuclear meeting in the posterior of the embryo. **d** The pronuclei have migrated back toward the center of the embryo and are undergoing a 90° rotation immediately prior to the division of P_0. **e** Beginning of the first cleavage. The spindle is in the region of cytoplasm devoid of granules. **f** Two-cell embryo with the larger (anterior) founder cell *AB* and the smaller stem cell P_1. **g** AB has divided equally into *ABa* and *ABp* while P_1 has divided unequally into the founder cell *EMS* and stem cell P_2. **h** P_2 has divided unequally into *C* and P_3. EMS has divided into two more founder cells, *MS* and *E*. Note that several nuclei are now out of the focal plane. **i** Beginning of gastrulation. P_3 has undergone the final stem-cell division into P_4 and *D*. The *E* cell has divided and its daughters have begun their migration into the interior of the embryo. **j** Beginning of morphogenesis. This is the "comma" stage. **k** "Tadpole". **l** "Pretzel". *Bar* = 10 μm. (Schierenberg 1986)

et al. 1960; Hirsh et al. 1976; Strome and Wood 1983; Albertson 1984; Kirby et al. 1990). During this microtubule-mediated movement, the embryo undergoes a pseudocleavage as a membrane furrow starts to form and then retracts (Fig. 2B–C). The maternal pronucleus migrates furthest and meets the paternal pronucleus in the posterior of the egg (Fig. 2C). The two pronuclei then move toward the center of the embryo, rotate 90° and fuse (Fig. 2D). The nuclear membranes immediately break down, and the first mitotic spindle forms along the anterior-posterior axis, slightly posterior to the center of the embryo (Fig. 2E).

The fertilized embryo, P_0, undergoes a series of stem-cell-like asymmetric divisions; the daughters differ in both their size and their future division patterns (Fig. 3). The larger daughter of each stem-cell division is referred to as a somatic or founder cell (AB, EMS, C and D), and their future divisions produce equally sized daughters that will divide synchronously with cell cycle times characteristic of each lineage. The smaller daughters of the stem-cell divisions (P_1 through P_4) always divide later than their somatic sisters and behave like stem cells, dividing asymmetrically into another founder cell and another stem cell. [The daughters of each founder cell retain their parent's name with the added suffixes a (anterior), p (posterior), l (left) or r (right) depending on the direction of the division that gave rise to them. For example ABalp is the posterior daughter of the left daughter of the anterior daughter of AB (Fig. 3)].

P_0 cleaves into the larger anterior AB cell and the smaller posterior P_1 cell (Figs. 2F and 3; Deppe et al. 1978; Sulston et al. 1983). AB then divides symmetrically. Its spindle is initially parallel to the dorsal-ventral axis of the embryo (Fig. 4A); but as the spindle grows, it begins to exceed the dimensions of the eggshell. As a result, by telophase one or the other end of the dividing AB cell tilts forward producing one daughter at the anterior of the embryo and the other on the dorsal side (Fig. 2G). When the AB division is half completed, the P_1 cell begins its stem-cell division. Its spindle is initially situated along the anterior-posterior axis, but the movement of the dividing AB cell forces the anterior region to skew ventrally (Figs. 2G and 4A). The larger anterior daughter of P_1, on the ventral side of the embryo, is the EMS cell (EMSt in the older literature) while the posterior daughter is designated P_2. EMS is an exceptional founder cell because it divides unequally to give rise to two more founder cells, MS and E. P_2 next undergoes an asymmetric division, yielding the founder cell C and the stem cell P_3, and P_3 gives rise to D and P_4. The P_4 cell is the ultimate stem cell, giving rise to the entire germ line.

After the last stem-cell division, the embryo enters gastrulation (Sulston et al. 1983). This is relatively early compared to other organisms, commencing about 2 h after fertilization when there are 28 cells. The two E cells, which give rise to the gut, are on the ventral surface and sink into the embryo. P_4 and the MS progeny follow. Later, the D cell descendants and the progeny of C that will form muscle and those of AB that give rise to the pharynx enter the embryo through the ventral gastrulation

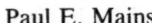

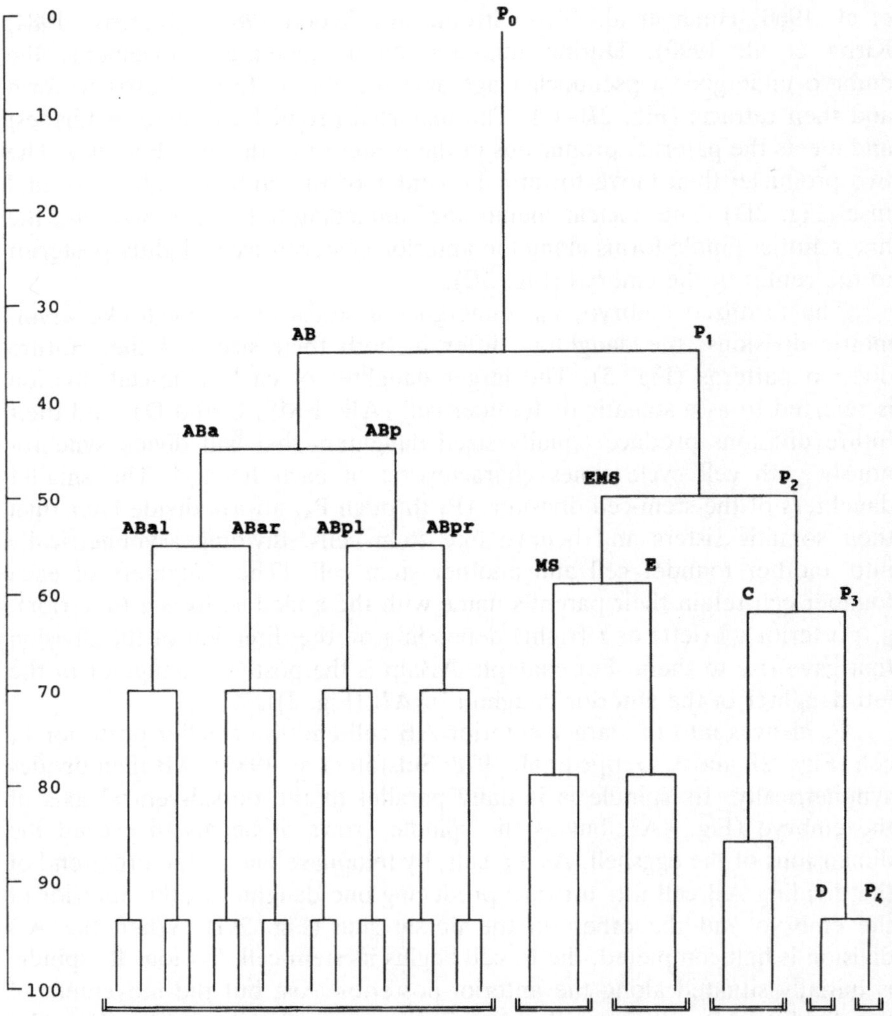

Hypodermal	72			13		
Neuronal	254	13		2		
Muscle	23	48		32	20	
Gut			20			
Germ line						2
Others	40	19				
Deaths	98	14		1		

Fig. 3. Early embryonic lineages. The time scale (in min) is for development at 20°C; *0* is the time of fertilization. *Horizontal lines* indicate cell divisions with the daughter cells named as indicated. *Below* is a listing of the number of differentiated cell types present at hatching derived from each founder lineage

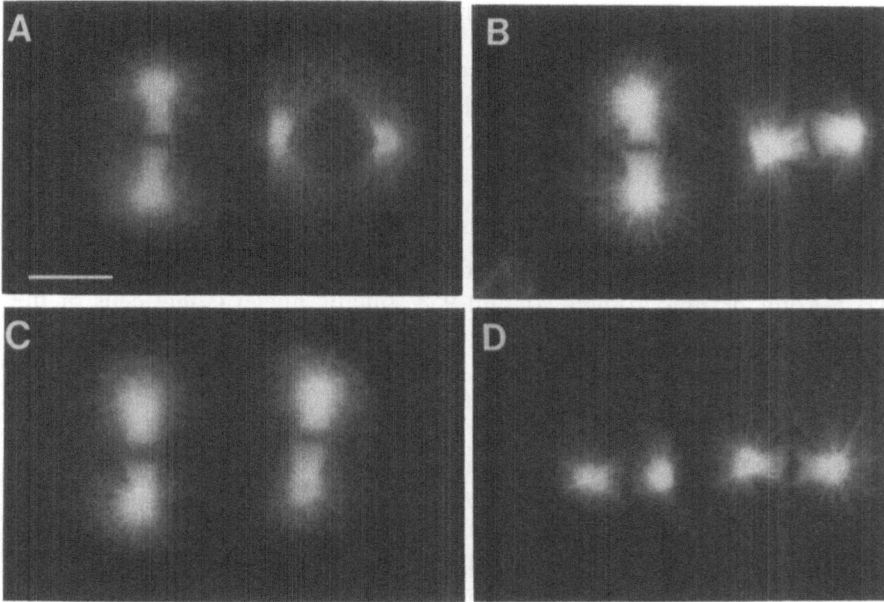

Fig. 4. Indirect immunofluorescence photomicrographs of wild-type and *par* mutant two-cell embryos fixed and then stained with anti-tubulin antibody to visualize the spindle and asters. Anterior is to the *left* and dorsal is to the *top*. **A** Wild type. The AB spindle is dorsal-ventral, but it will later skew toward the anterior-posterior axis as it grows beyond the dorsal-ventral eggshell dimensions, producing the AB daughters as situated in Fig. 2g. In this photograph, the spindle in AB is more developed than in P_1 since the AB division occurs earlier. Due to the later movement of AB, the P_1 spindle will skew from its current anterior-posterior orientation toward the dorsal-ventral axis (note the positions of EMS and P_2 in Fig. 2g). **B** *par-1* embryo. The spindles of this embryo are in the proper orientation; but as indicated by the equal development of their spindles, both cells will divide simultaneously. **C** *par-2* embryo. The cells are dividing synchronously with both spindles oriented parallel to the dorsal-ventral axis. **D** *par-3* embryo. The cells are dividing synchronously with both spindles oriented along the dorsal-ventral axis. *Bar* = 10 µm. (Kemphues et al. 1988b)

furrow. Gastrulation is completed 5.5 h after fertilization. The surface of the embryo becomes covered with those descendents of AB and C that form the hypodermis, which later secretes the collagenous cuticle. At this point, there are about 650 cells whose rate of division decreases dramatically. A large number of programmed cell deaths take place as development proceeds, decreasing the total number of cells to 558 at hatching.

Morphogenesis begins after the end of gastrulation and is completed during the second half of embryogenesis (Fig. 2J–L). The embryo is squeezed circumferentially by microfilaments, transforming it from a ball of cells into a worm curled within the eggshell (Priess and Hirsh 1986). Differentiated cell types become apparent; muscles become active and the developing worm twitches and rolls within its shell.

At hatching, all organs except the gonad are functional. During larval growth, the worm increases fivefold in length and the number of somatic cells doubles (Sulston and Horvitz 1977). Most somatic postembryonic cell divisions expand the hypodermis and the somatic portion of the gonad; this primarily creates structures for secondary sexual characteristics, including the hermaphrodite vulva and the specialized male tail (Fig. 1). (The postembryonic blast cells are given one or two letter designations which have no relation to those of the founder cells. For instance, the blast cells P1 through P12 have no relationship to P_{0-4} but rather are descendants of AB.) Adult hermaphrodites contain 958 somatic nuclei (some cells of the hypodermis and gut are multinucleate) and adult males have 1057. The greatest postembryonic increase in cell number occurs in the germ line, which expands from two cells at hatching to 2500 in adult hermaphrodites (Hirsh et al. 1976).

The zygotic genome is active during early embryogenesis. Hecht et al. (1981) determined that nuclear poly(A)$^+$ RNA (measured by in situ hybridization) began to accumulate at about the 100-cell stage. Using a perhaps more sensitive run-on transcription assay, Schauer and Wood (1990) found that pregastrulation embryos were, on a per nucleus basis, as transcriptionally active as later-stage embryos. Normalized to the cytoplasmic volume, transcription levels are initially quite low but increase rapidly with the near exponential increase in cell number. These authors estimated that the majority of genes in C. elegans are active by late embryogenesis, and in fact most genes appear to be active (albeit at very low levels) prior to gastrulation. The parasitic nematode Ascaris also appears to be transcriptionally active at an early time (Cleavinger et al. 1989).

3 Mechanisms of Cell Determination

The cell lineage of C. elegans is essentially invariant; the timing and orientation of cell divisions do not differ significantly between individuals. This reproducibility has allowed the complete lineage from fertilization to adulthood to be determined (Sulston and Horvitz 1977; Deppe et al. 1978; Kimble and Hirsh 1979; Sulston et al. 1983). Each cell at every stage of the life cycle can be identified, and its past history and future fate are known. The types of differentiated cells derived from each founder cell lineage at hatching are enumerated in Fig. 3. Because of this detailed knowledge of normal development, subtle deviations from the wild-type pattern caused by physical, chemical or genetic manipulations can be attributed to the experimentalist rather than random variation between individual embryos.

We will now describe the mechanisms that are likely used to make developmental decisions in C. elegans and will outline how these mechanisms have been determined experimentally. The general conclusion is that differentiation depends on both cell autonomous and nonautonomous mechanisms. Localized cytoplasmic determinants are inherited in a lineal manner causing cell autonomous differentiation: a cell's fate depends upon its ancestry rather than its neighbors (Sects. 3.1 and 3.2). This pattern is overlaid with cases of cell-cell interactions guiding development: the fates of some cells are determined nonautonomously, depending on neighboring cells rather than ancestry (Sects. 3.3 and 3.4).

3.1 Cell Autonomous Differentiation

An invariant lineage is usually (but not necessarily) indicative of a cell-autonomous mode of development. The fate of an individual cell depends upon internal pre-programming rather than on its external environment; each lineage may inherit spcialized regions of the egg's cytoplasm containing specific "determinants" that will guide its future differentiation (reviewed in Davidson 1986, 1990). This is termed "mosaic" development because the animal is essentially a patchwork of independent developmental programs. Killing a particular cell affects only its descendants, while other cells differentiate the same in its presence or absence. In contrast, differentiation in higher organisms usually depends on interactions between cells. This often allows greater flxibility since missing cells can be replaced by recruiting others to take their place, often after additional cell proliferation. This is termed "regulative" development.

Nematodes are considered a classic example of mosaic development, but an invariant lineage can also accommodate nonautonomous modes of differentiation. One outcome of an invariant lineage is a reproducible pattern of cell-cell contacts that could, in theory, result in a reproducible pattern of cell-cell signaling. Experimental manipulations are therefore required to determine if a particular developmental decision is cell autonomous or nonautonomous. Indeed, development in early sea urchin, gastropod mollusc and ascidian embryos is invariant but results from a reproducible pattern of cell-cell interactions (reviewed in Davidson 1990).

A simple way to show cell autonomous development is to ablate a particular cell and then to note the absence (or presence) of particular structures later in development; the missing structures should be only those normally derived from descendants of the ablated cell. Cell ablation is easily done using a laser microbeam during late-embryonic and post-embryonic *C. elegans* development, and the results are almost always consistent with cell autonomous differentiation (reviewed in Sulston 1988). Unfortunately, laser microsurgery cannot be used on early embryos because of the large size of the blastomeres, but as described below, cell autonomy has been demonstrated by a number of other methods that accomplish the same goals.

Several aspects of cell differentiation can be assayed in *C. elegans* embryos, including the execution of the proper lineage patterns and the appearance of tissue-specific differentiation markers. Gut differentiation can be detected by histological staining of an intestinal-specific esterase (Edgar and McGhee 1986) or by noting the autofluorescence or birefringence in polarized light of "gut granules" that result from tryptophan breakdown products (Babu 1974). Other products of differentiation have been scored with antisera directed against components of muscle or cuticle. However, since the gut is the only somatic tissue derived from a single founder cell (Fig. 3), most of the work on cell-autonomous differentiation has been done on the E cell and its descendants.

3.1.1 Autonomous Expression of Tissue-Specific Differentiation Markers

Laufer et al. (1980) showed that gut differentiation could occur in isolated blastomeres. Pressure was applied to a cover slip causing blastomeres of early embryos to burst at random. Embryos with only one intact cell were followed and were found to continue to divide normally. Gut granules were later seen at the appropriate time (4–6 h postfertilization), but only when an E-cell precursor, P_1 or P_2, remained intact. Similar experiments were performed by Edgar and McGhee (1986) who found that in embryos burst at the two-cell stage, P_1 but not AB could go on to express the gut-specific esterase activity. Laufer et al. (1980) and Edgar and McGhee (1986) also exposed embryos to cytochalasin, a treatment that prevented future interactions between cells by blocking the division of cells (but not nuclei). Expression of gut granules or esterase was seen only within those multinucleated cells that included cytoplasm that, under normal circumstances, would have been included in the E cell. Thus, the factors required for the expression of gut-specific markers are intrinsic to the E-cell lineage. These determinants are probably present at fertilization and are either localized to the region of the embryo where the E cell will later form, or they may initially have a wider distribution but are then actively segregated to the E-cell precursor at each division.

It is possible that the cytoplasm required for expression of the gut-specific markers does not induce transcription of the appropriate genes in the zygote. Rather, it could contain maternally encoded precursors of the marker, either as mRNA or protein, that are then activated at the appropriate time, independent of any zygotic transcription. This can be ruled out since expression of both gut granules and esterase requires de novo transcription. Edgar and McGhee (1986) showed that expression of these markers is sensitive to the RNA polymerase inhibitor α-amanitin. The same conclusion can be drawn by showing that the special cytoplasm responsible for expression of gut granules requires a nucleus (Wood et al. 1984; Schierenberg 1985). A laser microbeam was used to make a hole in the eggshell, and the P_1 nucleus was extruded through the hole by applying pressure to the cover slip. The P_1 cytoplast remaining within the eggshell did not later express gut granules. However, the nucleus of an AB descendant could be reprogrammed to express gut granules if it was fused to the P_1 cytoplast using a laser microbeam.

Determination of the gut might be unusual since it is the only somatic tissue that is exclusively clonally derived (Fig. 3). However, Cowan and McIntosh (1985) demonstrated cell autonomous expression of markers outside the E lineage. In cleavage-blocked embryos, a hypodermis-specific marker showed the expected pattern of expression in P_0, AB and P_2 (cf. Fig. 3). However, some patterns of marker expression were different from what would have been predicted by the lineage. Although P_1 could express markers for the hypodermis, body wall muscle and gut, only one of these markers was expressed in any given embryo. This "exclusivity" may indicate that only one differentiation program can be carried out in a particular cell; perhaps the programs compete for some limiting factor(s) or some type of binary switch governs expression of the different programs. Another deviation from the pattern predicted by the lineage was that P_0 did not express a marker specific for body wall muscle or gut granules even though P_1 did. Edgar and McGhee (1986) found similar results for gut granules and the gut-specific esterase. This could indicate that inhibitors of gut and body wall muscle development are present in P_0 that are segregated to AB at the first cleavage, allowing expression of the markers in P_1. Alternatively, activation of gut or muscle determinants might require a round of cell division, which would not occur in cleavage-blocked embryos.

Aamodt and McGhee (1991) have shown that in addition to positive factors that induce certain developmental programs, there are negative factors that prevent expression of gut-specific markers in a lineage-specific manner. These authors conducted a deletion analysis of the 5' region of the gut-specific esterase gene and, as expected, found regions required for expression in the E lineage. However, they also found that deletion of other regions often resulted in ectopic expression of esterase in developing embryos; furthermore, this abnormal expression was specific to the MS or C lineages (the "sister" and "cousin" lineages of E, respectively). It appeared that in addition to regions that interact with E lineage-specific activators, the 5' esterase sequence contains areas that respond to transcriptional repressors specific to the MS and C lineages.

3.1.2 Autonomous Execution of the Lineage Pattern

The pattern of the cell divisions is a unique characteristic of each lineage and can be examined for dependence on external or internal cues. During postembryonic development, this is often considered an expression of cell fate since the division pattern and the final differentiation of the daughter cells usually correlate strongly (Sternberg and Horvitz 1984). Several aspects of the cell division pattern of the early blastomeres can be scored, including the time of division, the orientation of the cleavage spindle and furrow and the stem-cell (asymmetric) nature of the division.

Schierenberg (1987, 1988) has studied the lineage patterns of isolated blastomeres and cell fragments by extruding regions of the embryo through laser-induced holes in the eggshell. He found that cells inside or outside the eggshell continued to proliferate and underwent the symmetric or asymmetric cleavage patterns appropriate to each blastomere (Fig. 5). Laufer et al. (1980) observed similar results with isolated blastomeres in burst embryos. Thus, the patterns of division are consistent with cell autonomy, and this supports the notion that the early embryo contains determinants that become localized to specific lineages where they direct the future pattern of cell division and differentiation.

Schierenberg (1988) noted that there seems to be specialized cytoplasm in the posterior of the embryo required for the execution of the asymmetric cleavages of the stem cells. When this cytoplasm was extruded from P_0, the remainder of the cell within the eggshell cleaved equally rather than unequally. Such equal cleavage did not happen when anterior cytoplasm was instead eliminated. Similar conclusions were drawn when the P_0 nucleus

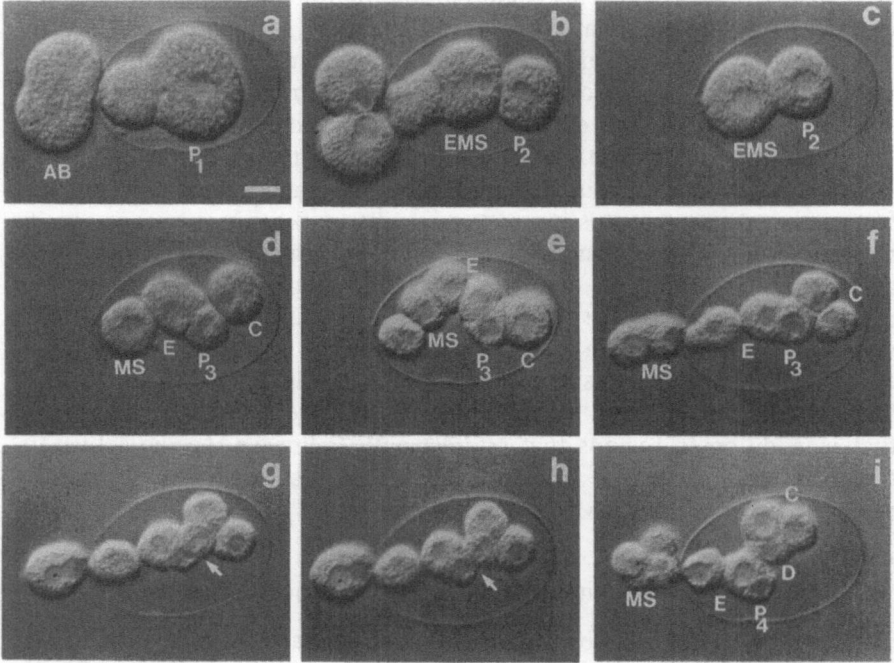

Fig. 5. Division pattern of the P lineage after extrusion of the AB cell through a laser-induced hole in the anterior of the eggshell. Anterior is to the *left* and dorsal to the *top*. **a** Most of the *AB* cell has been extruded leaving P_1 in the eggshell. **b** P_1 has divided unequally along the anterior-posterior axis into the anterior cell *EMS* and posterior cell P_2. The extruded AB fragment has divided equally. **c** The remainder of AB has been extruded from the shell. **d** P_2 has divided asymmetrically, but in contrast to the earlier divisions within the P lineage, the stem cell daughter, P_3, is anterior while the founder cell daughter, *C*, is posterior (compare to P_2 and *EMS* in **b**). EMS has divided into *MS* and *E*. **e** *MS* and *E* have both divided. **f** The *MS* cells have been extruded from the shell and *C* has divided. **g,h** Unequal division (*arrows*) of P_3 into **i** the anterior P_4 cell and posterior *D*. *Bar* = 10 µm. (Schierenberg 1987)

was included in the external fragments: if the nucleus was extruded from the eggshell with anterior cytoplasm, the external cell fragment cleaved symmetrically, but if the nucleus was extruded with posterior cytoplasm, it underwent stem-cell-like divisions. The requirement for localized determinants for stem-cell divisions could be tested further by noting that, while the posterior regions of P_0 and P_1 give rise to the stem cell (Fig. 5A–B), it is the anterior regions of P_2 and P_3 that do so (Fig. 5D, I). When Schierenberg (1987) extruded the P_0 or P_1 nuclei with the posterior region of those cells, the isolated fragments divided asymmetrically. If the P_2 nucleus was extruded with its posterior cytoplasm early in the cell cycle, it also divided asymmetrically. However, if this was done later in the cell cycle, the isolated P_2 fragment instead divided symmetrically. This indicates a "polarity reversal" during the P_2 cell cycle; the determinants required for the stem-cell fate migrate from the posterior to the anterior of the cell, the region where the P_3 cell will form.

Schierenberg and Wood (1985) have shown that the timing of cell division seems to depend on localized determinants that are partitioned at different concentrations to each founder cell lineage. In contrast to the determinants that have been previously discussed, these acted in a quantitative rather than a qualitative fashion: when cells with different cycling times were fused, the hybrid divided at an intermediate time. Furthermore, if a cell's volume was decreased by extruding a portion of cytoplasm, the nucleus still divided on schedule, indicating that it is the concentration of some factor, rather than the absolute amount or the ratio of DNA to cytoplasm, that determines the rate of cell division.

3.1.3 Nuclear Localization of Determinants

It is usually assumed that lineage-specific determinants are cytoplasmic, but at some stage they must regulate transcription in the nucleus. As mentioned earlier (Sect. 3.1.1), cleavage-blocked embryos expressed gut-specific markers only in the mutlinucleated cells that included the regions that, under normal circumstances, would have formed the E cells. When Edgar and McGhee (1988) also blocked DNA synthesis with aphidicolin, they still observed expression of gut markers, but only if the E cell had begun DNA synthesis for its first round of division (80 min postfertilization; Fig. 3). This round of DNA synthesis could reflect the binding of determinants (transcription factors for gut-specific functions?) to the chromosomes, an event that might require a prior round of DNA synthesis. The results were not due to an indirect inhibition of transcription by aphidicolin since experiments with α-amanitin showed that the mRNAs for the gut markers were not synthesized till much later. The expression of a marker for muscle and one for hypodermis were also sensitive to inhibition of DNA synthesis prior to their sensitivity to inhibition to RNA synthesis, indicating that the observations might be more general.

Certain cytosine residues are methylated differently in the male and female germ lines during mammalian gametogenesis, resulting in differential expression of some paternal and maternal genes in certain tissues (Monk 1990). Although there is no methylated cytosine in *C. elegans* (Simpson et al. 1986), it is possible that the paternal and maternal chromosomes are differentially preprogrammed by other covalent or noncovalent modifications. If this occurs, some of the paternal or maternal DNA strands could be segregated to specific blastomeres during the early cleavages. To test this, Ito and McGhee (1987) differentially labeled the paternal or maternal chromosomes by growing males or hermaphrodites, respectively, in the presence of bromodeoxyuridine. These animals were crossed to unlabeled individuals, and the locations of the labeled DNA strands contributed by the sperm or oocyte were examined in the outcross embryos. The distribution of labeled chromosomes was random, indicating parental DNA strands were not segregated in a lineage-specific manner. Therefore, determinants are apparently not irreversibly bound to the chromosomes at fertilization, but it is still possible that they might bind and be nonrandomly segregated at later times.

3.2 The P Granule, a Possible Cytoplasmic Determinant

The biochemical nature of a cytoplasmic determinant would be very interesting; the only physically defined candidate in *C. elegans* is the P granule. These serologically defined structures are so named because they are always localized to the P cell at each division (Strome and Wood 1982, 1983; Yamaguchi et al. 1983). P granules are found throughout the cytoplasm at fertilization, but they segregate to the posterior of the embryo where the P_1 cell will form (Fig. 6). This pattern of segregation is repeated for each stem-cell division, and ultimately P granules are found only in the P_4 cell (and germ cells at later stages). Thus, P granules are an obvious candidate for the determinant causing the stem-cell cleavage pattern or for inducing the descendants of P_4 to become germ cells (these two processes may or may not be caused by the same determinant).

In many organisms, the determination of the germ line may depend on localized determinants. The case is strongest for *Drosophila*. Transplantation experiments demonstrate that pole plasm, the cytoplasm in the region where the germ cells form, can induce ectopic differentiation of functional germ cells when transplanted to other regions of the embryo (Ilmensee and Mahowald 1974). In *Drosophila*, other insects, and amphibians (the latter representing highly regulative organisms), the cytoplasm destined to become the germ line is visibly different, containing large inclusions referred to as "nuage" (reviewed in Eddy 1975). This is true in nematodes as well. The cytoplasm segregated to the P lineage in the parasitic nematode *Parascaris* is distinct, and Boveri (1910) proposed that P cytoplasm contains factors that

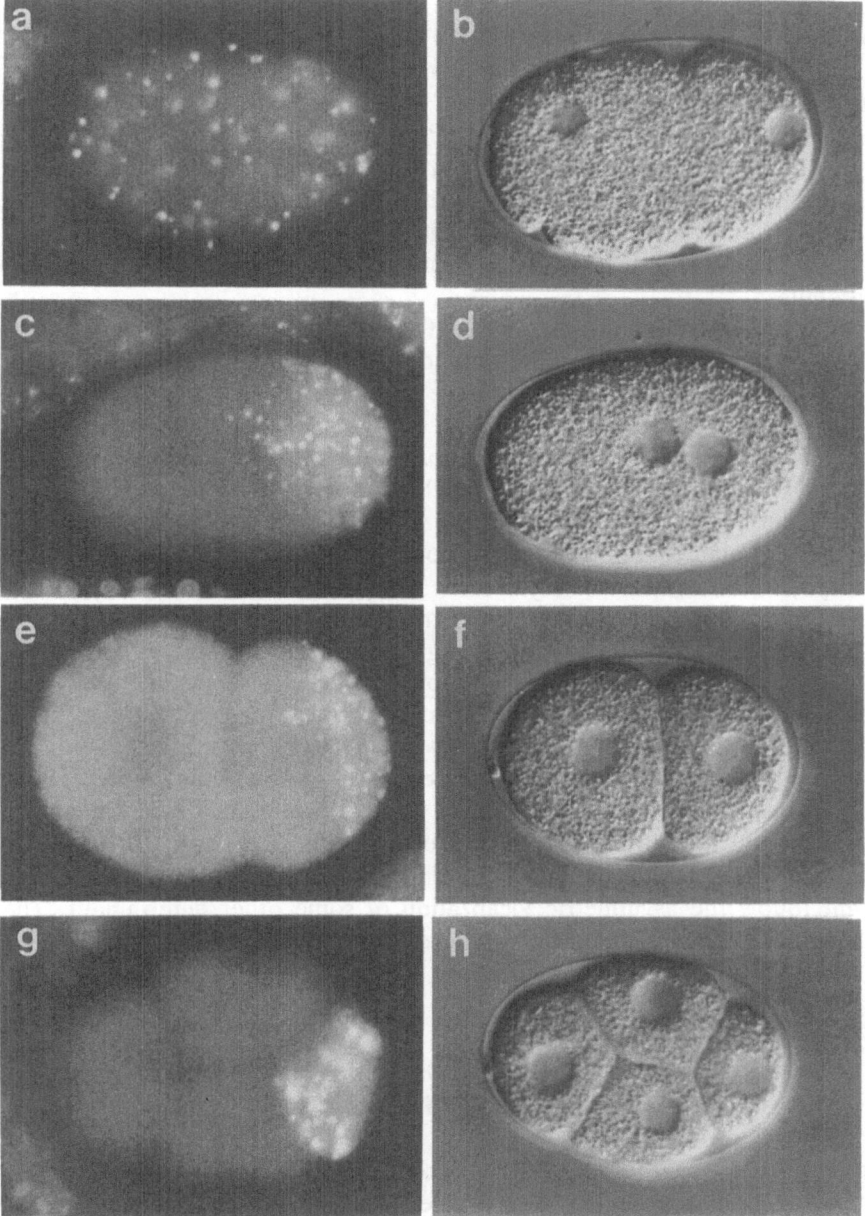

Fig. 6. Localization of P granules, visualized by indirect immunofluorescence. Anterior is to the *left* and dorsal to the *top*. *Left-hand panels* are fixed embryos stained with anti-P granule antibody and *right-hand panels* are Nomarski images of embryos at the same developmental stage. **a,b** P granules are found throughout the cytoplasm shortly after formation of the pronuclei. **c,d** By the time of pronuclear meeting, the P granules have segregated to the posterior of P_0. **e,f** Two-cell embryo with P granules in P_1. **g,h** Four-cell embryo, the P granules have segregated to P_2. (Kemphues 1989)

prevent chromosome diminution in the germ line. [In many nematodes, but *not* in *C. elegans* (Emmons et al. 1979; Albertson and Thomson 1982), chromosomes in the somatic blastomeres undergo diminution, fragmenting into hundreds of pieces, discarding mainly interstitial repetitive DNA sequences (Bennett and Ward 1986; Pimpinelli and Goday 1989)]. Boveri (1910) found that centrifugation of *Parascaris* embryos caused redistribution of germ plasm to additional cells and that diminution did not occur in these cells. Wolf et al. (1983) identified nuage with the electron microscope in *C. elegans*; these structures are likely to contain the antigenic determinants for P granules.

While P granules are an excellent candidate for a cytoplasmic determinant for germ-cell differentiation and/or the stem-cell division pattern of the P lineage, cause and effect have not been demonstrated for either of these fates. In fact, Schierenberg (1988) showed that P granules are not sufficient for the stem-cell division pattern. Posterior cytoplasm was extruded with the paternal pronucleus before P-granule segregation (equivalent to the stage shown in Fig. 6A, B). Even though this isolated fragment contained only a small proportion of all the P granules in the embryo, it still underwent asymmetric divisions. A similar conclusion was drawn when Hill and Strome (1990) treated embryos with cytochalasin, a procedure that prevents P-granule segregation (Strome and Wood 1983). This treatment often caused the anterior "AB" cell to cleave unequally, but this unequal cleavage did not happen every time that P granules were mislocalized to its cytoplasm. However, all cells that cleaved unequally included mislocalized P granules. Therefore, P granules may be required at some specific concentration to cause a stem-cell division, or they might be necessary but not sufficient for the process. Alternatively, P granules may not be required for the asymmetric stem-cell divisions but their movements might follow the asymmetric localization of some other determinant. The role of P granules in the other property of P cells, the formation of germ cells per se, remains an open question.

3.3 Localized Determinants Alone Cannot Explain
C. elegans Differentiation

Localized determinants clearly play a role in *C. elegans* development. The simplest model for development guided by localized determinants in *C. elegans* would result in a strictly clonal pattern of determination; each founder cell would inherit a specific set of determinants that would limit its descendants to a set of related fates. Different subsets of these determinants could then be segregated to different cells during subsequent divisions, further narrowing their fates. This simple mechanism cannot be the rule in *C. elegans*. Only two tissues show exclusively clonal derivation: the gut derives from E and the germ line derives from P_4 (Fig. 3). Lineal and functional boundaries do not coincide for other tissues. For example, while muscle is the only product of D, muscle is also derived from other lineages. The pharynx is made from both AB and MS descendants, but the origin of each of the cell types found within the pharynx (e.g. muscle, neuron, gland) is not restricted to one or the other lineage. Most of the founder cells are not even restricted to a single germ layer: AB, MS and C all give rise to both ectoderm and mesoderm, and there are cases where a terminal cell division can give rise to a muscle cell and a neuron (Sulston et al. 1983). It is difficult to imagine how localized determinants and cell autonomous differentiation alone could result in these patterns of development, and so it is likely that extrinsic signals are also required to determine some fates.

The primary function of the stereotyped lineage may be to place cells in specific positions rather than strictly to guide differentiation through the segregation of localized determinants. The relationship between cell fates and lineage in *C. elegans* is rather baroque and may represent cell autonomous determination caused by segregation of localized determinants overlaid by cell nonautonomous inductions. Each founder cell does give rise to mainly (in some cases exclusively) one type of tissue: AB, hypodermis; MS, C and D, muscle; E, intestine and P_4, germ line (Fig. 3). These fates could result from the action of cytoplasmic determinants, while other fates within each founder cell lineage could result from inductive events overlaid upon this more basic (primitive?) program (Strome 1989). A cell's localized determinants might dictate one fate, but its position in the embryo might serve a better purpose; for example, a cell that would normally become a neuron might be positioned where a muscle cell is needed. An inductive event from a nearby cell(s) could override the default muscle program. In some ways, this combination of mechanisms could be considered more efficient since it would eliminate the need for migration of a muscle cell from elsewhere in the embryo; and, in fact, there is little cell migration during *C. elegans* development compared to more regulative organisms (Sulston et al. 1983). We will now describe cases of nonautonomous differentiation resulting from extracellular cues in *C. elegans*.

3.4 Nonautonomy and Cell-Cell Signaling

There are a number of examples of nonautonomous patterns of cell differentiation in *C. elegans*; perhaps the most intensively studied being the postembryonic development of the vulva (reviewed in Sulston 1988; Horvitz 1988). Initially, two cells interact to determine which will become the anchor cell (AC) and which will become the ventral uterine cell (VU). The choice is random in any particular individual, but the final decision depends upon signals sent between the cells. Signal reception is mediated by the product of the gene *lin-12* (*lin*eage defective). Later in the development of the vulva, the AC sends a graded signal to the six vulval precursor cells (VPCs) that lie along the ventral hypodermis. These cells then divide and differentiate according to the strength of the perceived signal; the three possible patterns of division and differentiation are termed the 1°, 2° and 3° fates. The closer a VPC is to the AC, the stronger the perceived signal and the higher the fate (1° being the highest). Normally, VPCs adopt fates in the anterior-posterior order 3° 3° 2° 1° 2° 3° with the 1° cell nearest the AC. In the normal situation, each VPC takes on its specific fate due to a consistent and reproducible pattern of cellular interactions, but under the proper circumstances, any of the VPCs can adopt any of the three fates. The six cells represent an "equivalence group". If the AC is laser-ablated, all VPCs adopt the 3° fate, which represents the ground state. If the VPC nearest the AC is laser-ablated, an adjacent VPC moves into the ablated cell's position and switches from the 2° to the 1° fate; likewise a 3° cell can adopt the 2° fate if the latter cell is absent. An equivalence group is similar to the morphogenetic field seen in higher organisms, except that in an equivalence group, only one part of the pattern is reformed, at the expense of another part.

In addition to the graded signal sent by the AC, the refinement of the pattern of VPC differentiation depends upon lateral interactions between the cells, and this again requires the activity of the *lin-12* gene product (Sternberg and Horvitz 1989).

An example of cell signaling during early embryogenesis was elegantly demonstrated by Priess and coworkers (Priess and Thomson 1987; Priess et al. 1987). They showed that the EMS cell (or its descendants) induces the ABa cell (or its descendants) to differentiate into parts of the pharynx. The descendants of ABa and ABp always differentiate along different paths, but experiments to be described below demonstrate that the AB cell division generates two cells that have equal potentials; ABa and ABp form an equivalence group. The difference in their fates depends not on the asymmetric segregation of determinants, but on subsequent interactions with other cells.

In the two cell embryo, the AB cleavage spindle is initially perpendicular to the anterior-posterior axis, but as the spindle elongates it is forced to slip diagonally due to eggshell constraints (Figs. 2 and 4). Is the direction of the tilt random or nonrandom? If prior segregation of cytoplasmic determinants is responsible for the differing ABa and ABp fates, then the direction of the tilt must be predetermined and so the region destined to become ABa must take the forward position. However, the direction could be random with the later differences between ABa and ABp resulting from external cues. Priess and Thomson (1987) showed that the latter is true. After the skewing of the AB spindle began, they used a micromanipulator to push the elongating cell in the opposite direction, causing the cell that would have become ABa to take the posterior position. Priess and Thomson found that these embryos developed normally; the cell that in the absence of their manipulations would have become ABa takes on the normal ABp fate and vice versa. This clearly indicates that differing ABa and ABp fates must result from external cues rather than relying exclusively on prelocalized determinants.

Priess and Thomson (1987) went on to show that EMS or its descendants signal the ABa lineage to differentiate into the anterior region of the pharynx. The pharynx is derived from descendants of both ABa and EMS. When the precursor of ABa (AB) was extruded from the two-cell embryo, the portion of the pharynx derived from EMS developed at least to a limited extent, indicating some degree of autonomy. However, the converse was not true: if EMS or its precursor P_1 was extruded, the descendants of ABa did not differentiate into pharyngeal tissue. This indicates that the ABa lineage requires cues from the EMS lineage. At the 28-cell stage, the precursors of the pharynx consist of three descendants of ABa and two of MS, and these cells are grouped together. If any of these cells were ablated at this time, the remainder differentiated normally, indicating that the inductive process had been completed.

A gene involved in the induction of the ABa lineage has been identified by Priess et al. (1987). Maternal-effect mutations of *glp-1* cause defects similar to removing EMS from the embryo; the descendants of ABa fail to differentiate into pharyngeal tissue and instead resemble hypodermal cells (this might represent the default fate, in part resulting from the action of cytoplasmic determinants). Consistent with the previous work, the temperature-sensitive period of temperature-sensitive *glp-1* alleles was between the 2- and 28-cell stages. The molecular cloning and sequencing of the gene predicted a membrane-bound receptor similar to other genes that include epidermal growth factor-like repeats that are also involved in signal transduction (Austin and Kimble 1989; Yochem and Greenwald 1989). Included in this family is the *Drosophila* gene *Notch*, which is involved in the choice of epidermal and neuronal fates in the early embryo. Another member of this family is the *C. elegans* gene *lin-12*, which also is involved in cell-cell signaling, as discussed earlier in this section (Greenwald 1985). Furthermore, *lin-12* and *glp-1* are genetically tightly linked and probably arose as a tandem duplication followed by divergence into genes with similar, but nonidentical, functions.

Another parallel between the genes *lin-12* and *glp-1* is that they both function multiple times during development. As described above, *lin-12* functions in several steps of vulval development and, in fact, also participates in at least eight other sets of lineage decisions (Horvitz 1988). *glp-1* also has an additional function, regulating the choice between meiosis and mitosis in the germ line. Its name stands for *g*erm *l*ine *p*roliferation (Austin and Kimble 1987). The distal tip cells are somatic cells situated at the distal end of each gonadal arm (one cell on each arm) that sends a signal to the germ cells preventing them from entering meiosis. Germ cells that receive a strong signal from a distal tip cell proliferate mitotically, but as they move away, the signal presumably weakens, and the germ cells enter meiosis. Eliminating the signal by laser-ablating the distal tip cell (Kimble and White 1981) or by removing the putative receptor with a *glp-1* mutation causes the germ cells to prematurely cease mitotic proliferation and enter meiosis (Austin and Kimble 1987). As might be expected of membrane-bound receptors, it has been shown that both the *lin-12* (Seydoux and Greenwald 1989) and *glp-1* products (Austin and Kimble 1987) probably function in receiving rather than sending signals between cells. Hence, two structurally similar genes, *lin-12* and *glp-1*, both participate in the reception of extracelluar cues and both function multiple times during development.

Wood (1991) has shown that the daughters of ABa and ABp also form equivalence groups. The second division round of the AB lineage occurs along the left-right axis of the embryo, but the daughters on the left are slightly anterior to their sisters on the right, and the left-right sisters have different fates. Using a micromanipulator to reverse the left-right asymmetry of the embryo after the divisions of ABa and ABp, Wood found that the embryos developed normally. The ABal and ABar cells switched fates, as did ABpl and ABpr. Therefore, the different left-right fates result not from cytoplasmic determinants but must instead require external cues. These experiments also slowed that early embryonic left-right asymmetry is responsible for adult left-right asymmetries. For example, the anterior arm of the hermaphrodite gonad is normally to the right of the intestine while in the posterior the gonad is to the left (Fig. 1). The animals that grew from Wood's manipulated embryos showed the mirror-image orientations.

Another embryonic induction, between EMS and P_2, has been described by Schierenberg (1987). EMS is the only founder cell that undergoes an unequal cleavage, giving rise to two additional founder cells, MS and E. The P_2 cell (or one of its descendants) is apparently required to cause MS and E to differentiate properly. After P_2 was extruded from the embryo, the EMS cell underwent its normal asymmetric division, but subsequently cell cycle times within the E lineage were not retarded with respect to MS as they normally are. In addition, the E lineage went through extra divisions, and neither gut granules were produced from E nor did muscle twitching develop from MS. Schierenberg pointed out that this induction may require the reversal of polarity in the P lineage that was described earlier (Sect. 3.1.2); otherwise P_3 and P_4 would not lie adjacent to the E cells. Interestingly, this reversal of polarity does not occur in the nematode *Bradynema rigidum*, but P_4 compensates by migrating to a position next to the E cells (Schierenberg 1987).

The induction of EMS by the P lineage (Schierenberg 1987) indicates nonautonomous expression of gut-specific markers, but the work of Laufer et al. (1980), Cowan and McIntosh (1985) and Edgar and McGhee (1986) with cleavage-blocked P_1 blastomeres demonstrated autonomous development prior to the EMS-P_2 division (Sect. 3.1.1). The differing results lead to the conclusion that either the signal can also occur intracellularly within the cleavage-blocked P_1 cell (the cell could express both the signal and receptor) or that development of the intestine is at first autonomous and later becomes nonautonomous. A trivial explanation is that extrusion of the P_2 cell damaged EMS enough to block differentiation; indeed, Priess and Thomson (1987) did observe gut granules after P_2 extrusion. However, it might be difficult to imagine how damage could prevent differentiation without also inhibiting cell division. Another possibility is that positive interactions between P_2 and EMS are needed to overcome negative interactions between other cells and EMS. Since neither would occur in cleavage-blocked P_1 cells, gut-specific markers can be expressed. In support of this, combinations of positive and negative cell-cell interactions have been observed during postembryonic development in *C. elegans* (Herman and Hedgecock 1990; Waring ahd Kenyon 1990). A perhaps related observation was made by Cowan and McIntosh (1985) who found that cleavage-blocked P_1 cells could produce the body wall muscle marker, but its descendant, EMS, could not.

3.5 Long-Range Patterning in the *C. elegans* Embryo

Cell fate in *C. elegans* clearly appears to result from the combination of both localized determinants mediating autonomous differentiation and cell-cell signaling guiding nonautonomous development. How do these mechanisms contribute to global (long-range) positional information? In *Drosophila*, global patterning of the early embryo relies in part on positional information imparted by the concentration gradient of the *bicoid* gene product. This system contributes to the determination of a number of different fates, depending upon the specific concentration of the *bicoid* protein in a given region (Driever and Nüsslein-Volhard 1988a,b). In contrast, the known cases of determinants in *C. elegans* appear to be localized in a discrete manner and seem to each induce only one type of fate. In fact, a *bicoid*-like gradient, resulting from simple diffusion from a source, could probably not form in *C. elegans* because of the small size of the embryo (approximately $50\,\mu m$ long, about one-tenth the length of the *Drosophila* egg). Making the assumption that the physical parameters describing diffusion in the two embryos is similar, McGhee (1992) calculated that the molecular weight of a worm equivalent would need to be several orders of magnitude larger than that of the *bicoid* protein to produce a similarly shaped gradient.

Global positional information might be created in a piecemeal fashion from a series of short-range cell-cell communications rather than by long-range *bicoid*-like gradients (e.g. see Davidson 1990; Hooper and Scott, this Vol.). It is possible that a discretely localized determinant autonomously specifies the fate of a specific cell, and this cell in turn signals its neighbors to nonautonomously specify their fate. These cells in turn could signal their neighbors to create a longer-ranged pattern, or combinations of signals emanating from multiple, autonomously determined centers could build a larger pattern. These types of mechanisms might opperate in the *C. elegans* embryo. As we have seen, genes such as *lin-12* and *glp-1* do contribute to localized patterning in *C. elegans* by cell-cell communication. Another well-studied example of pattern formation in *C. elegans*, also mediated by cell-cell communication, contributes to the structure of the adult tail (Kenyon 1986; Waring and Kenyon 1990).

4 Control of the Orientation of the Early Cleavages

Both authonomous and nonautonomous developmental programs require a
precise execution of the early cleavages to ensure the proper positioning
of cells for both the segregation of cytoplasmic determinants and the
reproducible set of cell-cell interactions. Many factors are required for the
proper temporal and spatial execution of the early cleavages. Asymmetry
must be established and maintained along the anterior-posterior axis (Sect.
4.1), and this must then influence the orientation of the cell divisions and
the segregation of determinants (Sect. 4.2). We will now consider what is
known about these processes.

4.1 Anterior-Posterior Asymmetries

The first asymmetry visible in the developing oocyte is the movement of the
maternal nucleus to the future anterior of the embryo prior to meiosis
and fertilization (Ward and Carrel 1979; Albertson 1984). There is little
evidence for any other asymmetries at this time (Strome 1986b). However,
after fertilization, many anterior-posterior asymmetries are seen: ruffling of
the anterior membrane occurs during pseudocleavage, pronuclear meeting is
in the posterior, P granules segregate to the posterior and the cleavage
spindle and cytokinesis are slightly posterior to the center of the embryo.
Hill and Strome (1988) have shown that all of these events are sensitive to
the microfilament inhibitor cytochalasin during the same specific period of
the first cell cycle. They proposed that the actin microfilament network may
serve as a scaffold for the localization of factors that either cause or respond
to the global anterior-posterior asymmetry of the embryo.

Filamentous actin forms a cortical meshwork throughout the oocyte, but
during pronuclear migration, it transiently concentrates in the anterior of
the embryo until the beginning of the first cleavage (Strome 1986a). Brief
cytochalasin treatment rapidly disrupts the actin cytoskeleton. If a pulse of
this drug was administered during a 5–10 min window during pronuclear
migration and pseudocleavage, all asymmetry was lost: pseudocleavage,
the anterior ruffling of the membrane and P-granule segregation did not
occur. Furthermore, pronuclear meeting and the first cleavage spindle and
furrow occurred in the center of the embryo (Hill and Strome 1988). If the
cytochalasin pulse was either before or after this window, none of these
asymmetric events were affected. Most of these events normally occur
during the time of drug sensitivity, so their disruption is not surprising.
However, the asymmetric placement of the first cleavage spindle and furrow
happen at a later time but nevertheless showed the same temporal sensitivity
as the other events. Therefore, a critical cytochalasin-sensitive step appears
necessary for the establishment, but not for the maintenance, of the anterior-
posterior asymmetry.

In a second study, Hill and Strome (1990) found that cytochalasin treatment during pseudocleavage and pronuclear migration resulted in mis-localization of the factor(s) responsible for the stem-cell cleavage pattern. In some cases, embryonic polarity seemed to be reversed with the anterior cell cleaving like P_1 and the posterior cell like AB. In other cases, both cells behaved like P_1 (a posterior duplication) or both cells cleaved like AB (an anterior duplication).

4.2 Orientation of Cell Division

The orientation of cell divisions must respond to the same asymmetries described above so that the cellular anatomy can be coordinated with the segregation of determinants. The relative sizes of the daughter cells and their subsequent cleavage rhythms are characteristic of either the P or AB cell lineages, and we have seen that these differences depend upon localized cytoplasmic determinants (Sect. 3.1.2). The cytochalasin experiments of Hill and Strome (1990) showed that the orientation of the cleavage spindles was coordinated with equal or unequal cleavages: in the reversed polarity and the anterior or posterior duplication embryos mentioned above (Sect. 4.1), the P_1-like cells showed *both* an anterior-posterior spindle and an asymmetrically placed cleavage furrow; likewise, the AB-like cells exhibited *both* a dorsal-ventral spindle and a symmetric cleavage.

The polarity reversal and the posterior duplications observed by Hill and Strome (1990) represent neither a complete nor an accurate reorganization of the embryo along a different coordinate system. Even though reversed polarity and posterior duplication embryos had P_1-like blastomeres, markers for gut or muscle differentiation were rarely expressed by their descendants. This may indicate that the determinants necessary for the stem-cell cleavage pattern and those required for differentiation of other markers are sensitive to cytochalasin in different ways.

Spindle orientation is determined by the location of the spindle poles, which in turn is determined by the positions of the centrosomes. The migrations of the centrosomes in *C. elegans* embryos foreshadow the position of the cleavage spindles. The centrosome is contributed by the sperm and is initially found behind the sperm pronucleus (Fig. 7A; Albertson 1984; Strome 1986b). During pronuclear migration, the daughter centrosomes move to positions parallel to the dorsal-ventral axis between the oocyte and sperm pronucleus (Fig. 7B). When the pronuclei meet, the centrosomes and the pronuclei rotate 90° in either a clockwise or a counterclockwise direction (Fig. 7C). The centrosomes now lie on the anterior-posterior axis to orient the P_0 spindle.

The succeeding cleavage spindles of the AB lineage occur along the orthogonal axes (Sulston et al. 1983). The AB spindle is dorsal-ventral, those for its daughters (ABa and ABp) are left-right, and the spindles of the next division round are anterior-posterior. In the P lineage, succeeding

72   Paul E. Mains

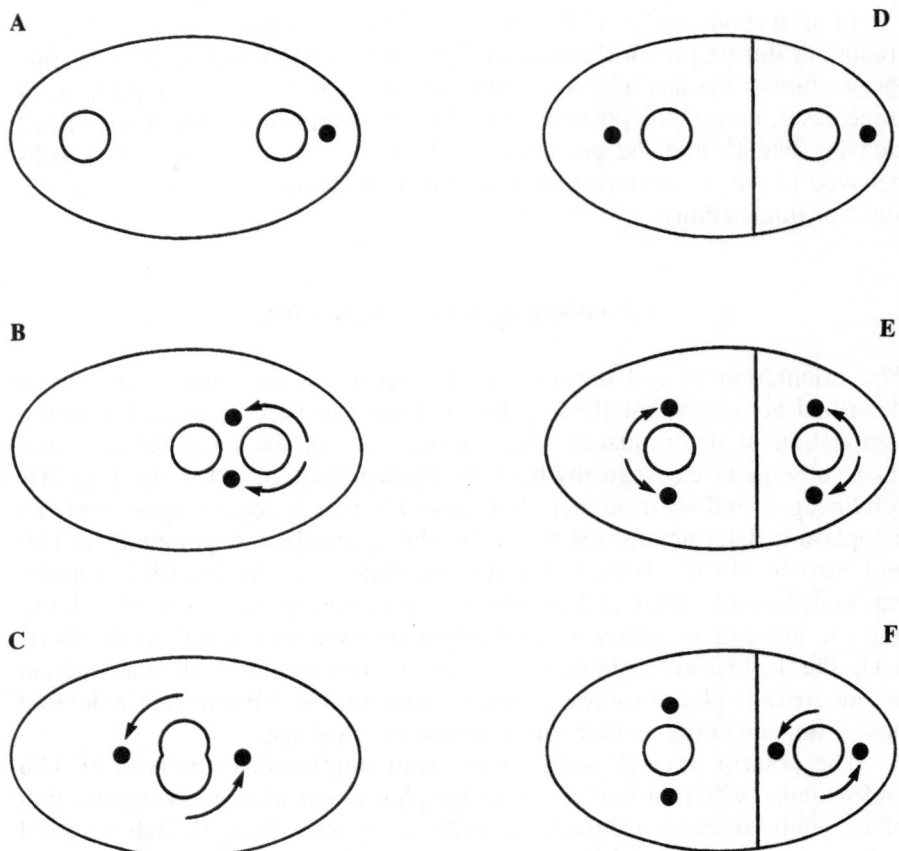

Fig. 7. Diagram of centrosome movements in the one- and two-cell embryos. Centrosomes are indicated by the *black dots* and anterior is to the *left*. **A** The centrosome is contributed by the sperm and is initially situated behind the paternal pronucleus. **B** During pronuclear migration, the daughter centrosomes migrate to a position between the pronuclei. **C** The centrosomes and the joined pronuclei have rotated 90° during their migration back toward the center of the embryo. The centrosomes now lie along the anterior-posterior axis for the first cleavage. **D** Two-cell embryo. **E** In both AB and P_1, the daughter centrosomes have migrated in opposite directions to be aligned parallel to the dorsal-ventral axis. **F** The centrosomes in the P_1 cell have undergone an additional 90° rotation, this time in the same direction, to lie along the anterior-posterior axis

spindles tend to form along the anterior-posterior axis. These patterns are most clearly seen when the constraints of the eggshell or other cells are removed in partial embryos (e.g. Fig. 5; Laufer et al. 1980; Schierenberg 1987).

Hyman and White (1987) and Hyman (1989) noted that the pattern of centrosomal movements differs between the AB and P lineages in accordance with their differing patterns of spindle orientations. The AB centrosome is located on the anterior side of the nucleus; after duplication, the centrosome daughters migrate 90° in opposite directions so that they lie parallel to the dorsal-ventral axis for the division (Fig. 7D, E). This type of migration is repeated during subsequent divisions, resulting in the orthogonal AB lineage cleavage pattern. In the P_1 cell, centrosomal migration is initially similar. The centrosome lies between the nucleus and the posterior of the embryo, and as seen for the AB cell, its daughters migrate 90° to positions parallel to the dorsal-ventral axis (Fig. 7D, E). However, they undergo a second 90° rotation, but this time they both move in the same direction (either clockwise or counterclockwise) to lie along the anterior-posterior axis (Fig. 7F). This two-part migration pattern is repeated in the subsequent stem-cell cleavages. The centrosomal movements appear to require microtubuler connections with the egg cortex to generate force for the migrations. The rotations (Fig. 7F), but not the preceding migration (Fig. 7E) of the centrosomes in the P_1 and P_2 cells may require or respond to a system involving microfilaments since these movements are sensitive to cytochalasin. Curiously, this is not the case for the P_0 centrosomes.

5 Genetic Analysis

Up to this point, we have seen how embryogenesis has been studied using physical and chemical manipulations. However, these techniques involve rather gross insults to the embryo and are therefore potentially difficult to interpret since they could affect many unrelated processes. A different approach uses mutations to remove gene products one at a time, perhaps a more specific "ablation". The genetic analysis complements and extends the other approaches. Few assumptions about underlying mechanisms are necessary when looking for mutations affecting a process, and the cloning of the gene can rapidly lead to a mechanistic understanding. However, the interpretation of the effects of a mutation is not without complications. A gene can participate in multiple processes and may be active multiple times during development. In addition, not all mutations completely eliminate the activity of a gene product: some alleles may decrease but not abolish the activity of a gene, while others may increase or alter its action (see Herman 1988 for a discussion).

We will first describe the general principles and conclusions that have been drawn from the genetic analysis of *C. elegans* embryogenesis (Sect. 5.1) and then we will describe specific mutations (Sect. 5.2). This latter section is relatively short since suprisingly few of the known embryonic mutations are yet informative. The possible reasons for this observation will be discussed in the final section (Sect. 5.3).

5.1 Maternal Control of Early Embryogenesis

The maternal genome controls most of early embryogenesis in *C. elegans*; in addition to raw materials (energy stores, amino acids, etc.), mRNA and protein are packed into the oocyte prior to fertilization for later use. In addition, the orientation and timing of the pregastrulation lineages are primarily under maternal control.

The reliance of early development on the maternal genome is indicated in part by the fact that most embryonic-lethal mutations (those resulting in death prior to hatching) show maternal effects. Only the genotype of the hermaphrodite matters; the zygote's genotype is irrelevant since the critical gene function must be expressed prior to fertilization. Maternal or zygotic (the latter is sometimes referred to as "nonmaternal") activity of a particular locus can be demonstrated by a number of genetic tests. Consider an arbitrary recessive mutation *m* (Table 1; Wood 1988b). If the product must be provided by the zygote's genome, then one-fourth of the embryos that result from self-fertilization of an *m/+* hemaphrodite will die because they are *m/m* (Table 1, column 2). However, if the product can be provided maternally, then these *m/m* embryos will be viable, but all of their progeny will die. This will also be true of any outcross *m/+* progeny if the *m/m* hermaphrodite is mated to wild-type males; the genotypes of the outcross embryos are irrelevant because the + allele has to be expressed in the oocyte prior to fertilization (Table 1, column 3). There are also genes whose products can be provided *either* maternally *or* zygotically. Like true maternal-effect mutations, all progeny of an *m/+* hermaphrodite survive, but progeny of an *m/m* hermaphrodite can be rescued by mating to wild-type males (Table 1, column 4). In contrast to strict maternal-effect alleles that show no male rescue, these mutations are said to show partial maternal effects. Loci whose contributions are required specifically in the oocyte to direct development are called pure maternal genes (Kemphues et al. 1988a); but as we shall later see, strict maternal-effect mutations do not always represent pure maternal genes.

Many of embryonic-lethal mutations in *C. elegans* were isolated in screens for temperature-sensitive (*ts*) alleles. Animals can be grown efficiently between 15 and 25 °C, giving a fairly wide difference between permissive and nonpermissive temperatures. *ts* alleles have an advantage over nonconditional mutations since they can be maintained as homozygotes at the permissive temperature; otherwise, the allele must be maintained in *m/+* stocks. The tests described earlier can be performed on *m/+* hermaphrodites at the restrictive temperature, but additional tests are also possible with animals homozygous for *ts* alleles if they are raised at the permissive temperature and then upshifted after reaching adulthood.

Since all progeny of an adult *m/m* hermaphrodite shifted to the nonpermissive temperature will die, one cannot tell if this death results from the lack of maternal or zygotic expression. However, if the *m/m* hermaphrodite is mated to wild-type males at the restrictive temperature, mutations of zygotically required genes will be rescued while genes required maternally will not (Table 1). *ts* alleles can also detect a class of genes that could not be distinguished with nonconditional mutations, namely those requiring *both* maternal and zygotic expression. Like strict zygotic mutations, the one-fourth *m/m* self-progeny of *m/+* hermaphrodites die at the restrictive temperature. These *m/m* animals can be reared at the permissive temperature; but in contrast to strict zygotic mutations, none of their progeny will be rescued by mating to wild-type males at the restrictive temperature. Only the zygotic activity of this type of gene would have been detected with a non-*ts* allele: the *m/m* hermaphrodite would have died before the allele could have been tested for maternal effects.

　　ts embryonic lethal mutations have been isolated in a number of screens (Hirsh and Vanderslice 1976; Vanderslice and Hirsh 1976; Miwa et al. 1980; Schierenberg et al. 1980; Wood et al. 1980; Cassada et al. 1981; Isnenghi et al. 1983; Denich et al. 1984; Hirsh et al. 1985, compiled in Wood 1988b; only screens that reported all of the embryonic lethal mutations, rather than only those with specific phenotypes, are included in the following discussion). These screens were designed to identify mutations with either maternal or zygotic effects. Hermaphrodites were mutagenized and putative *m/+* F_1 progeny were cultured at the permissive temperature. F_2 homozygotes were obtained by self-fertilization (consider how complicated this step would be in an organism that was not self-fertile), and these *m/m* adults were allowed to lay eggs at both temperatures. Those that showed hatching at 15 °C but not 25 °C were saved.

　　These collections of mutations showed that there is indeed a large maternal contribution to embryogenesis; 44 of the 47 genes analyzed had mutations showing at least partial maternal effects (last line of Table 1; Wood 1988b). Of these, over half (28/44) showed strict maternal effects. The temperature-sensitive periods (TSPs) for embryonic viability of the strict maternal-effect mutations often occurred partly or exclusively after fertilization, demonstrating that maternally contributed products are indeed active in the zygote. Of the remaining genes, the wild-type products of 11 genes could be supplied either maternally or zygotically, and products of

Table 1. Percent of embryos hatching for mutations in genes with different modes of expression

Hermaphrodite	Zygotic	Maternal	Zygotic or maternal	Zygotic and maternal
m/+ selfed	75%	100%	100%	75%
m/m selfed	0[a]	0	0	0[a]
m/m × +/+ male	100[a]	0	100	0[a]
Number of genes[b]	3	28	11	5

[a] These tests are only possible for *ts* alleles.
[b] Compiled by Wood (1988b).

five genes were required both maternally and zygotically (interestingly, two genes fell into both of these categories, indicating that different mutations of the same gene can have quite different properties, a concept that we will return to later). Mutations in only three genes had no maternal affects, being required strictly zygotically.

The relative unimportance of the embryo's own genome has been shown in a number of other ways. Mutations in most zygotically active genes result in death after hatching, indicating that the oocyte contains enough maternal product to complete embryonic development. A large number of non-conditional (that is, non-*ts*) zygotic-lethal mutations have been isolated in *C. elegans* without regard to time of arrest; a mutation is scored as lethal if an individual dies any time prior to adulthood (Herman 1978; Meneely and Herman 1979; Rogalski et al. 1982; Sigurdson et al. 1984; Rosenbluth et al. 1988; Howell and Rose 1990). Only 5% (9/179) of the genes with zygotic-lethal alleles showed embryonic lethality; the rest arrested some time after hatching. Most of these nonembryonic lethal mutations are likely to represent "housekeeping genes" active throughout the life cycle. These genes are primarily involved in basic metabolic processes and are apparently provided maternally in sufficient quantities to last at least through hatching. In *Drosophila*, 24% of zygotic-lethal mutations result in embryonic lethality (Wieschaus et al. 1984), perhaps suggesting a greater reliance on the zygotic genome.

It must be kept in mind that the 5% of zygotic genes that can cause embryonic lethality represent several hundred of the estimated 2000–4000 genes of *C. elegans* (Brenner 1974; Herman 1988; Howell and Rose 1990). However, relatively few of these are needed exclusively for early development. As mentioned earlier (Sect. 2.3), Schauer and Wood (1990) found that most genes are active in embryos, but they estimated that only 20 to 30 are preferentially transcribed early.

It has been shown that the patterning of the early cleavages is primarily, if not exclusively, under maternal control. Storfer (1990) examined the timing and orientation of the early cleavages of embryos homozygous for deficiencies covering different regions of the genome. The earliest embryonic defects should be characteristic of the first gene whose zygotic activity is required; this assumption has held true for *Drosophila* (Merrill et al. 1988; Wieschaus and Sweeton 1988; Vavra and Carroll 1989). However, no visible defects were observed in *C. elegans* embryos up to the beginning of gastrulation (2 h post-fertilization when there are 28 cells) using deficiencies that removed approximately 50% of the genome in total. Storfer (1990) concluded, within 95% confidence limits, that there were four or fewer genes in the rest of the genome required for the early cleavage pattern.

Therefore, even though the majority of *C. elegans* genes are transcribed prior to hatching, sufficient quantities of most can be provided maternally. Zygotic transcription of few genes, if any, is required to execute the visible pattern of the early cleavages. This is not to say that the zygotic genome plays no role during early embryogenesis; cells of the early embryo could very well be expressing genes required to make crucial developmental decisions whose effects are not visible until much later.

5.2 Specific Mutations

The screens for *ts* embryonic-lethal cited earlier included a survey of general effects caused by the mutations. The mutations were tested for maternal versus zygotic activities, TSPs and early cleavage defects. We will now consider the mutations that have been studied in greater detail and see how they can be interpreted in view of the nongenetic analyses of *C. elegans* development. One of the most notable mutations, *glp-1*, has already been described (Sect. 3.4).

Kemphues and colleagues have described a set of maternal-effect mutations that disrupt cytoplasmic organization of fertilized embryos and prevent proper segregation of cytoplasmic determinants; the genes are named *par* for *par*titioning defective (Kemphues et al. 1988b; Kirby et al. 1990). As can be seen in Table 2, these mutations affect several aspects of embryogenesis that we have seen may require or respond to localized determinants or to

Table 2. Phenotypes of *par* mutations[a]

	par-1	*par-2*	*par-3*	*par-4*
Pseudocleavage[b]	Anterior	Posterior	Variable	Variable
Pronuclear meeting[c]	Normal	Medial	Medial	Medial
Microfilament cap	Abnormal	Abnormal or absent	Absent	Abnormal
Furrow of first cleavage[d]	Medial	Medial	Medial	Normal
Spindles of second cleavage[e]	$\updownarrow \updownarrow$ (25%)[f]	$\updownarrow \updownarrow$	$\updownarrow \leftrightarrow$ $\leftrightarrow\leftrightarrow$ $\leftrightarrow \updownarrow$	$\updownarrow \updownarrow$ (20%)[f]
Timing of second cleavage[g]	Synchronous	Synchronous	Synchronous	Synchronous
P-granule segregation	None	Weak	Weak	Weak
Grandchildless survivors	No survivors	Yes	Yes	Yes
Percent expressing:[h]				
Gut granules	0	20	79	0
Pharyngeal myosin	94	92	95	43
Body wall myosin	94	100	100	13

[a] Compiled from Kemphues et al. (1988) and Kirby et al. (1990).
[b] Medial in wild type.
[c] Posterior in wild type.
[d] Posterior in wild type.
[e] $\updownarrow \leftrightarrow$ In wild type (left arrow indicates spindle orientation in AB, right arrow in P_1).
[f] Remainder are wild type.
[g] Asynchronous in wild type.
[h] Percent of embryos expressing marker.

anterior-posterior asymmetries present in the early embryo. Affected are
the asymmetries present during the first cell cycle, the stem-cell division
pattern, the differing cell cycle times in different lineages, the segregation of
P granules and the differentiation of the gut. Germ-line development is also
disrupted as evidenced by a high frequency of agametic adults among the
survivors of "leaky" alleles. This is considered a "grandchildless" phenotype
(it may be significant that the gonads of these sterile animals do contain
P granules). Different *par* genes display different subsets of phenotypes,
perhaps indicating that the different markers respond to different but overlapping sets of cues. The phenotypes, especially those of *par-3*, are reminis-
cent of cytochalasin treatment of wild-type embryos (Sect. 4.1). This could
imply that some of the *par* gene products are part of the actin cytoskeleton
or interact with it (Kemphues et al. 1988b; Hill and Strome 1990), however,
none of the *pars* map near the *C. elegans* actin genes.

Schnabel and Schnabel (1990a) described maternal-effect mutations in
the gene *cib-1* (changed *i*dentity of *b*lastomeres), which interfere with the
stem-cell division pattern. Strong alleles of *cib-1* cause the P_1 cell to skip a
cell cycle, and when the cell resumes division, it behaves like a founder cell,
dividing into equally sized daughters. Its subsequent pattern of division
and differentiation resembles EMS, the somatic daughter of P_1. In weaker
alleles, the P_1 pauses for less than a full cell cycle and divides unequally as it
should. Later, either P_2 or P_3 divide equally, behaving like their somatic
daughters C and D, respectively. The *cib-1* gene product may therefore be
necessary for the execution of the stem-cell divisions of the P lineage, and in
its absence, the underlying somatic cell fate is expressed. It is interesting
that this change in fate is coupled to skipping or delaying a cell division. P_0
is unaffected by these mutations, so *cib-1* may act after the *par* gene
products.

A number of *ts* maternal-effect mutations affect the orientation and
functioning of the mitotic apparatus of the early embryo. Mutations of *zyg-9*
(*zyg*ote defective), *mei-1* (*mei*osis defective) and *mel-26* (*m*aternal-*e*ffect
*l*ethal) result in a short dorsal-ventral spindle in the posterior of the embryo
(Wood et al. 1980; Albertson 1984; Kemphues et al. 1986; Mains et al.
1990a). Since mutations in these genes result in similar phenotypes, their
products might be required for the same process. This idea was supported
by the genetic enhancement between the mutations; double mutants resulted
in a much more severe phenotype than expected, indicating that the products might interact (Mains et al. 1990b). Other alleles of *mei-1* and mutations of *mei-2* block the formation of the meiotic rather than the mitotic
spindle and are able to suppress (correct) the defects resulting from the
mitotic-defective *mei-1* and *mel-26* mutations (Mains et al. 1990b). The
products of *zyg-9*, *mei-1*, *mei-2* and *mel-26* may be rquired for some of the
features that distinguish the meiotic from the mitotic spindle, or they may be
involved in the transition from one mode of division to the other. The
mitotic defects caused by these mutations can be phenocopied in wild-type
embryos by low doses of the microtubule-inhibitor nocodazole, suggesting

that microtubule-based structures might be affected (Strome and Wood 1983; Kemphues et al. 1986).

Meiosis is delayed and the first cleavage furrow occurs at variable positions along the anterior-posterior axis in maternal-effect mutations of *zyg-11* (Hirsh and Vanderslice 1976; Wood et al. 1980; Kemphues et al. 1986). This gene has been cloned, but shows no similarity to other known genes (Carter et al. 1990). Curiously, even though the genetic analysis shows that *zyg-11* is a pure maternal gene and the TSP for *ts* alleles is restricted to the first cleavage, expression of the *zyg-11* transcript is not restricted to oogenesis. Perhaps the *zyg-11* product is nonessential or redundant at other developmental stages or in other tissues.

emb-29 (*emb*ryogenesis defective) represents one of the few embryonic-lethal mutations where zygotic gene function is both necessary and sufficient (Cassada et al. 1981; Isnenghi et al. 1983; Denich et al. 1984). The gene product appears necessary for the transition from the G2 phase of the cell cycle into mitosis, and mutant embryos arrest with approximately 140 cells (Hecht et al. 1987). This stage also marks the beginning of the TSP, so prior to this time the embryos must be using a maternally supplied gene product. Apparently, this maternal product is not sufficient for completion of embryogenesis. As might be expected for a general cell cycle (i.e. housekeeping) function, there is a requirement for the product outside of embryogenesis: individuals upshifted at other points in the life cycle arrest as larva or sterile adults with abnormal gonads.

As described earlier, *glp-1* is required maternally to induce ABa to form parts of the pharynx, while the rest of this organ develops in an at least partially cell-autonomous manner from EMS (Sect. 3.4; Priess and Thomson 1987; Priess et al. 1987). The zygotic activity of the gene *pha-1* (*pha*rynx defective) appears to be required for the terminal differentiation of the pharynx (Schnabel and Schnabel 1990b). This is true of all pharyngeal cell types, regardless of their germ layer or lineal origin. Development of *pha-1* mutants is initially normal; at the beginning of morphogenesis, they contain the correct number of pharyngeal cells enclosed in a basement membrane. However, they subsequently fail to undergo the morphogenesis or express several markers characteristic of the pharynx. It is not yet known whether *pha-1* functions autonomously in all lineages that give rise to the pharynx or whether a specific cell(s) signals the terminal differentiation of the cells which will form the organ. *pha-1* represents an excellent example of a zygotically active gene required early in embryogenesis but whose effects are not seen until a much later stage: the TSP begins early, at the initiation of gastrulation, and extends to the start of pharyngeal morphogenesis, but the embryonic defects are only apparent late.

Mutations in *spe-11* (*spe*rm defective) are unique because they in show paternal rather than maternal effects (L'Hernault et al. 1988; Hill et al. 1989) When a wild-type oocyte is fertilized with a sperm from a *spe-11* male, activation of the developmental program is aberrant: meiosis often fails, eggshell formation is incomplete and a dorsal-ventral spindle forms. Activation is not entirely deficient since embryogenesis is initiated, polyspermy does not result and P granules segregate normally.

5.3 Where Are the Genes Affecting Early Development?

Considering the sophistication of *C. elegans* genetics, there seems to be a relative paucity of informative mutations affecting early embryogensis. Most of this review has focused on descriptive embryology and physical or chemical perturbations of wild-type development, rather than on analysis of mutant phenotypes. In contrast, similar reviews of *Drosophila* development (for example, the accompanying chapter by Hooper and Scott, this Vol.) are based primarily on the analysis of and interactions between a much larger collection of embryonic-specific mutations.

In part, this lack of informative embryonic-lethal mutations could be a problem in deciding which *C. elegans* mutations are likely to represent developmentally interesting genes and which are not. The phenotype scored in a search for embryonic-lethal mutations in *C. elegans* is simply dead eggs, but obviously a large number (probably the majority) of the organism's genes are required either maternally or zygotically during embryogenesis. Most of these will involve housekeeping genes, which are not of much interest to a developmental biologist. Some criteria are therefore needed to exclude these genes. One method is to employ some sort of rapid secondary screen to identify those mutations that specifically disrupt development. Choosing the criteria for the secondary screen is often difficult and intro-duces biases, but it is a necessary step when sorting through a large number of mutations, many of which will ultimately turn out to be uninteresting. As we shall see, this may be a difficult task in *C. elegans*.

In *Drosophila*, the secondary screen used to decide which embryonic lethal warranted further attention was straightforward. In the saturation screen, performed by Nüsslein-Volhard, Wieschaus and colleagues for zygotic lethal mutations affecting embryonic pattern formation, 586 of 4358 mutations showed visible defects (Wieschaus et al. 1984). Morphogenetic landmarks form fairly early in *Drosophila*, thus the terminal phenotypes of embryos that had early defects could be compared to the well-defined (and quickly scored) anatomy of the hatching larva (Nüsslein-Volhard and Wieschaus 1980). Interesting defects such as missing or altered segments were easily detected. Since morphogenesis does not occur until fairly late in *C. elegans*, the terminal phenotype of most embryonic-lethal mutations is an amorphous mass of cells. Differentiated cell types such as muscle, pharynx, and gut can be scored, which is rather limited compared to the landmarks of the *Drosophila* embryo.

An alternative strategy for *C. elegans* is the examination of the early embryonic cleavages in living embryos. This potentially includes a great deal of information, but it lacks the convenience of being a single time point. Tracing the lineage of an embryo just to the beginning of gastrulation requires up to 2 h and scoring the lineages at later times becomes exceedingly tedious. Since few, if any, zygotic-lethal mutations will show visible defects prior to gastrulation; they may be difficult to analyze by this method. Most maternal-effect mutations in *C. elegans* do result in defects visible prior to the onset of gastrulation, but unfortunately, the number of interesting phenotypes was limited (keeping in mind that "interesting" is a subjective term). The most common phenotypes were general defects in cytokinesis or slowing of cell cycle times; these are likely to represent housekeeping genes. Since mutations with these phenotypes may greatly outnumber mutations in genes required specifically for embryogensis, secondary screens based on examination of early lineages are slow and might be impractical when considering hundreds (or thousands) of individual mutations.

Some criteria are needed to eliminate probable mutations in housekeeping genes. One strategy considers genes required *only* during early embryogenesis, that is pure maternal genes (Kemphues et al. 1988a). A *ts* mutation in a housekeeping gene will likely cause maternal effects on embryonic viability, but this type of mutation can be eliminated from further consideration since its function is required outside of embryogenesis. Zygotic developmental abnormalities will result from upshift of a *ts* mutation at any point in the life cycle; maternal effects are seen only if the animal is upshifted after reaching adulthood. In contrast, strict maternal-effect mutations develop normally as long as the upshift occurred after the embryonic TSP. Table 1 indicates that mutations in 28 genes showed strict maternal effects on embryonic viability, but 16 of these also resulted in either arrested larvae or adults with abnormal gonads if the eggs were allowed to hatch into L1's and then shifted to the restrictive temperature.

By considering mutations that affect only embryogenesis, most mutations in housekeeping genes can be eliminated. However, this does not completely solve the problem since housekeeping genes can mutate to rare maternal-effect lethal with no effects outside embryogenesis (Perimon et al. 1986; Kemphues et al. 1988a). As described above, loss of a gene product required outside of embryogenesis should result in zygotic arrest. This will be true of mutations that cause the complete absence of a gene's activity (null alleles). However, if a particular mutation results only in a decrease rather than the complete absence of a gene's function (a hypomorphic allele), the homozygote might survive to adulthood and then show maternal effects. This results because embryogenesis is the time of the most rapid cell proliferation, so it is the point in the life cycle that would be most sensitive to a decrease in the level of a housekeeping product (gonadogenesis, a common arrest point of embryonic-lethal mutations upshifted after embryogenesis, might also be particularly sensitive). *ts* mutations are particularly prone to be hypomorphic: since they must be able to retain a reasonable amount of activity at the permissive temperature, they may not represent complete nulls at the restrictive temperature.

Kemphues et al. (1988a) tried to avoid the problems associated with *ts* hypomorphic alleles by isolating nonconditional maternal-effect lethal mutations. These are much more likely to be null alleles, but most still turned out to be hypomorphs of zygotically active genes. In a screen for strict maternal-effect lethals on chromosome II, mutations in 17 genes fell into two descrete categories typified by their mutation rates. Thirteen of the genes were each represented by only one or two alleles with an average mutation frequency of 6×10^{-5} mutations/gene/chromosome. This is surprisingly low since a number of studies have found that the frequency of loss-of-function alleles in *C. elegans* is consistently near 5×10^{-4} after using standard mutagenesis protocols (Brenner 1974; Anderson and Brenner 1984; Park and Horvitz 1986; Rogalski and Riddle 1988; Howell and Rose 1990; Kirby et al. 1990; Mains et al. 1990b). Kemphues et al. (1988a) proposed that the maternal-effect mutations that occurred at the low frequency were likely to be rare hypomorphic alleles of zygotically active genes; that is, these strict maternal-effect mutations *did not* represent pure maternal genes. In fact, two of the genes in the low frequency category had been previously identified by zygotic-lethal alleles by other workers. In contrast, the remaining four genes isolated by Kemphues et al. (1988a) were each represented by four, five or six alleles and occurred at an expected high frequency, 4×10^{-4}. These were clearly not rare alleles of the genes and so their strict maternal-effect lethality reflected the true null phenotype.

Even though rare maternal-effect lethal alleles of housekeeping genes occur at a low frequency, they still present a problem since a very large number of genes can give rise to them. Estimates of the total number of genes required during *C. elegans* embryogenesis range from 200–1000 (Cassada et al. 1981; Wilkins 1986). In contrast, the number of genes whose null phenotype is strict maternal-effect lethality is probably quite small. There were only four such loci found on chromosome II; and since each of these was represented by four or more alleles, it was unlikely that additional genes were missed by chance. As there are six linkage groups in *C. elegans*, all of about equal size, there may be as few as 24 genes required only during embryogenesis in *C. elegans* (Kemphues et al. 1988a). This number may be even smaller since some genes must be required for more mundane functions such as meiosis and eggshell formation and some genes may serve nonessential functions at other points in the life cycle. Two of the genes on chromosome II fell into this category, so the number of pure maternal genes may be as few as 12 (Kemphues et al. 1988a).

Apparently, a substantial fraction of the pure maternal genes has been identified. Among the genes described in Section 5.2, the null phenotypes of *par-1*, *par-2*, *par-3*, *par-4*, *zyg-9*, *zyg-11*, *mei-1* and *cib-1* are strict maternal-effect lethality. This is also true of two genes not discussed earlier, *zyg-1* and *him-14* (Kemphues et al. 1988a). Not all are pure maternal genes since *mei-1* and *him-14* function during meiosis and *zyg-1* has nonessential roles at other points in the life cycle. Other identified loci may represent pure maternal genes, but it is not yet known if their strict maternal-effect lethality represents their null phenotype.

We are therefore left in a quandary when searching for genes that control development in pregastrulation embryos. Fewer than 30 zygotically active genes are transcribed preferentially prior to gastrulation (Schauer and Wood 1990), but these are unlikely to result in visible defects until much later (Storfer 1990). Mutations in genes specifically affecting pregastrulation lineages will likely show maternal effects, but these are obscured by the large number of housekeeping genes that can also mutate to maternal-effect phenotypes. Housekeeping genes can be avoided by considering only loci whose null phenotypes show strict maternal effects, but this simplifying assumption leaves fewer genes than might be expected, perhaps as few as a dozen. How can this dilemma be explained? There are several possibilities, which are not mutually exclusive.

It is certainly possible that the estimate of 12 pure maternal genes is low. It assumed that the genes of interest will be at least roughly equally mutable and that they are randomly distributed over the genome. Although these assumptions seem valid, it would not be surprising if the actual number of pure maternal genes is manifold higher.

Another possibility is that relatively few genes are actually required for the early lineage pattern. In *Drosophila*, there are probably only 30–40 pure maternal loci required for early embryonic patterning (Nüsslein-Volhard et al. 1987). Both the relative concentrations of specific gene products and the combinatorial use of different genes convey information. Therefore, only a limited number of genes is necessary for early development since a unique gene is not required to specify each structure or event. Genes may also be used efficiently in *C. elegans*. For example, the different cell cycle times in each founder cell lineage may result from partitioning different concentrations of the same oscillator to each founder cell rather than requiring unique components in each lineage (Sect. 3.1.2; Schierenberg and Wood 1985). Another example is *cib-1*, which apparently is used in P_1, P_2 and P_3 to carry out the stem-cell division pattern (Sect. 5.2; Schnabel and Schnabel 1990a).

Genetic functions are often redundant (see Brenner et al. 1990 for a recent discussion). Eliminating one gene product has little or no effect if another gene can replace the missing activity. There are estimates that half of the loci in yeast (Goebl and Petes 1986), *C. elegans* (Park and Horvitz 1986) and mice (Dove 1987) may be redundant or otherwise dispensable. In this respect, a number of gene disruption experiments in mice have resulted in very mild or no phenotypes even though the genes were chosen with the thought that they would play critical roles during embryogenesis. Redundant genes are better identified by dominant gain-of-function "poison" mutations rather than by recessive loss-of-function alleles (Suzuki 1970; Park and Horvitz 1986; Kusch and Edgar 1986), but this possibility has been explored only to a limited extent during early *C. elegans* embryogenesis (Mains et al. 1990a). In addition, genes could be redundant during one part of the life cycle but not another. As mentioned earlier (Sect. 5.2), *zyg-11* is required maternally and only for early embryogenesis, but its transcription is not restricted to oogenesis (Carter et al. 1990). Similarly, null mutations of *lin-10* affect only the vulva, yet the product of this gene is also present elsewhere, including early embryos (Kim and Horvitz 1990).

Finally, it is likely that the inclination toward genes that function *only* during early embryogenesis is in error. A case in point is *glp-1*, which would have been overlooked using this criteria. The null phenotype of this gene is a zygotic defect in germ-cell proliferation, but leaky (hypomorphic) or *ts* alleles allow animals to escape this block so that the maternal effects, a block to early cell-cell signaling, can be seen. The homologous gene *lin-12* is an example of a gene used multiple times during postembryonic development. It recently has been shown that *C. elegans* has genes similar to mammalian *ras* and epidermal growth factor receptor tyrosine kinase, and these genes both function during embryogenesis and vulval induction (Aroian et al. 1991; Beitel et al. 1991; Han and Sternberg 1991). In *Drosophila*, there appear to be many genes that have specific developmental roles but function multiple times during the life cycle. Perrimon et al. (1989) made germ-line mosaics (clones of *m/m* germ cells in *m/+* females) for a large number of zygotic, larval or pupal lethal mutations and found that a number displayed maternal effects leading to embryonic patterning defects. As described by Hooper and Scott (this Vol.), several segment polarity genes can be proided either maternally or zygotically and some function multiple times during development. Therefore, even though they may be difficult to identify against the background of housekeeping functions, genes with critical roles during early *C. elegans* development may be used for several related but different roles during the life cycle. It will be a challenge for *C. elegans* researchers to identify these genes, but it should be worth the effort.

Acknowledgements. I would like to thank James McGhee and Thomas Clandinin for comments on the manuscript. P.E.M. is supported by a grant from the Alberta Heritage Foundation for Medical Research.

References

Aamodt EJ, Chung MA, McGhee JD (1991) Spatial control of gut-specific gene expression during *Caenorhabditis elegans* development. Science 252:579–582

Albertson DG (1984) Formation of the first cleavage spindle in nematode embryos. Dev Biol 101:61–72

Albertson DG, Thomson JN (1982) The kinetochores of *Caenorhabditis elegans*. Chromosoma 86:409–428

Anderson P, Brenner S (1984) A selection for myosin heavy-chain mutants in the nematode *Caenorhabditis elegans*. Proc Natl Acad Sci USA 81:4470–4474

Aroian RV, Koga M, Mendel JE, Ohshima Y, Sternberg, P (1991) The *let-23* gene necessary for *Caenorhabditis elegans* vulval induction encodes a tyrosine kinase of the EGF receptor subfamily. Nature (London) 348:693–699

Austin J, Kimble J (1987) *glp-1* is required in the germ line for regulation of the decision between mitosis and meiosis in *Caenorhabditis elegans*. Cell 51:589–599

Austin J, Kimble J (1989) Transcript analysis of *glp-1* and *lin-12*, homologous genes required for cell interactions during development of *C. elegans*. Cell 58:565–571

Babu P (1974) Biochemical genetics of *Caenorhabditis elegans*. Mol Gen Genet 135:39–44

Beitel GJ, Clark SG, Horvitz, HR (1991) *Caenorhabditis elegans ras* gene *let-60* acts as a switch in the pathway of vulval induction. Nature (London) 348:503–509

Bennett KL, Ward S (1986) Neither a germ line-specific nor several somatically expressed genes are lost during embryonic chromatin diminution in the nematode *Ascaris lumbricoides*. Dev Biol 118:141–147

Boveri T (1910) Über die Teilung centrifugierter Eier von *Ascaris megalocephala*. Wilhelm Roux's Arch Entwicklungsmech Org 30:101–125

Brenner S (1974) The genetics of *Caenorhabditis elegans*. Genetics 77:71–94

Brenner S (1988) Forward. In: Wood WB (ed) The nematode *Caenorhabditis elegans*. Cold Spring Harbor Press, Cold Spring Harbor, New York

Brenner S, Dove W, Herskowitz I, Thomas R (1990) Genes and development: molecular and logical themes. Genetics 126:479–486

Carter PW, Roos JM, Kemphues KJ (1990) Molecular analysis of *zyg-11*, a maternal-effect gene required for early embryogenesis of *Caenorhabditis elegans*. Mol Gen Genet 221:72–80

Cassada R, Isnenghi E, Culotti M, von Ehrenstein G (1981) Genetic analysis of temperature-sensitive embryogenesis mutants in *Caenorhabditis elegans*. Dev Biol 84:193–205

Cleavinger PJ, McDowell JW, Bennett KL (1989) Transcription in nematodes: early *Ascaris* embryos are transcriptionally active. Dev Biol 133:600–604

Collins J, Saari B, Anderson P (1987) Activation of a transposable element in the germ line but not the soma of *Caenorhabditis elegans*. Nature (London) 328: 726–728

Coulson A, Sulston J, Brenner S, Karn J (1986) Toward a physical map of the genome of the nematode *Caenorhabditis elegans*. Proc Natl Acad Sci USA 83:7821–7825

Coulson A, Waterston R, Kiff J, Sulston J, Kohara Y (1988) Genome linking with yeast artificial chromosomes. Nature (London) 335:184–186

Cowan AE, McIntosh JR (1985) Mapping the distribution of differentiation potential for intestine, muscle, and hypodermis during early development in *Caenorhabditis elegans*. Cell 41:923–932

Davidson EH (1986) Gene activity in early development. Academic Press, New York

Davidson EH (1990) How embryos work: a comparative view of diverse modes of cell fate specification. Development 108:365–389

Denich KTR, Schierenberg E, Isnenghi E, Cassada R (1984) Cell-lineage and developmental defects of temperature-sensitive embryonic arrest mutants of the nematode *Caenorhabditis elegans*. Wilhelm Roux's Arch Dev Biol 193:164–179

Deppe U, Schierenberg E, Cole T, Krieg C, Schmitt D, Yoder B, von Ehrenstein G (1978) Cell lineages of the embryo of the nematode *Caenorhabditis elegans*. Proc Natl Acad Sci USA 75:376–380

Dove WF (1987) Molecular genetics of *Mus musculus*: point mutagenesis and millimorgans. Genetics 116:6–8

Driever W, Nüsslein-Vohard C (1988a) A gradient of *biciod* protein in *Drosophila* embryos. Cell 54:83–93

Driever W, Nüsslein-Vohard C (1988b) The *bicoid* protein determines position in the *Drosophila* embryo in a concentration-dependent manner. Cell 54:95–104

Eddy EM (1975) Germ plasm and the differentiation of the germ line. Int Rev Cytol 43:229–280

Edgar LG, McGhee JD (1986) Embryonic expression of a gut-specific esterase in *Caenorhabditis elegans*. Dev Biol 114:109–118

Edgar LG, McGhee JD (1988) DNA synthesis and the control of embryonic gene expression in *Caenorhabditis elegans*. Cell 53:589–599

Edgley ML, Riddle DL (1990) The nematode *Caenorhabditis elegans*. Genetic Maps 5:3.111–3.133

Emmons SW (1988) The genome, In: Wood WB (ed) The nematode *Caenorhabditis elegans*. Cold Spring Harbor Press, Cold Spring Harbor, New York

Emmons SW, Klass MR, Hirsh D (1979) Analysis of the constancy of DNA sequences during development and evolution of the nematode *Caenorhabditis elegans*. Proc Natl Acad Sci USA 76:1333–1337

Fire A (1986) Integrative transformation of *Caenorhabditis elegans*. EMBO J 5:2673–2680

Goebl MG, Petes TD (1986) Most of the yeast genomic sequences are not essential for cell growth and division. Cell 46:983–992

Greenwald I (1985) *lin-12*, a nematode homeotic gene, is homologous to a set of mammalian proteins that includes epidermal growth factor. Cell 43:583–590

Han M, Sternberg PW (1991) *let-60*, a gene that specifies cell fate during *C. elegans* vulval induction, encodes a *ras* protein. Cell 63:921–931

Hecht RM, Gossett LA, Jeffery WR (1981) Ontogeny of maternal and newly transcribed mRNA analyzed by in situ hybridization during development of *Caenorhabditis elegans*. Dev Biol 83:374–379

Hecht RM, Berg-Zabelshansky M, Rao PN, Davis FM (1987) Conditional absence of mitosis-specific antigens in a temperature-sensitive embryonic arrest mutant of *Caenorhabditis elegans*. J Cell Sci 87:305–314

Herman RK (1978) Crossover suppressors and balanced recessive lethal in *Caenorhabditis elegans*. Genetics 88:49–65

Herman RK (1988) Genetics. In: Wood WB (ed) The nematode *Caenorhabditis elegans*. Cold Spring Harbor Press, Cold Spring Harbor, New York

Herman RK, Hedgecock EM (1990) Limitations of the size of the vulval primordium of *Caenorhabditis elegans* by *lin-15* expression in surrounding hypodermis. Nature (London) 348:169–171

Hill DP, Strome S (1988) An analysis of the role of microfilaments in the establishment and maintenance of asymmetry in *Caenorhabditis elegans* zygotes. Dev Biol 125:75–84

Hill DP, Strome S (1990) Brief cytochalasin-induced disruption of microfilaments during a critical interval in 1-cell *C. elegans* embryos alters the partitioning of developmental instructions to the 2-cell embryo. Development 108:159–172

Hill DP, Shakes DC, Ward S, Strome S (1989) A sperm-supplied product essential for initiation of normal embryogenesis in *Caenorhabditis elegans* is encoded by the paternal-effect embryonic-lethal gene, *spe-11*. Dev Biol 136:154–166

Hirsh D, Vanderslice R (1976) Temperature-sensitive developmental mutants of *Caenorhabditis elegans*. Dev Biol 49:220–235

Hirsh D, Kemphues KJ, Stinchcomb DT, Jefferson R (1985) Genes affecting early development in *Caenorhabditis elegans*. Cold Spring Harbor Symp Quant Biol 50:69–78

Hirsh D, Oppenheim D, Klass M (1976) Development of the reproductive system of *C. elegans*. Dev Biol 49:200–219

Hodgkin J, Horvitz HR, Brenner S (1979) Nondisjunction mutants of the nematode *C. elegans*. Genetics 91:67–94

Horvitz, HR (1988) Genetics of cell lineage. In: Wood WB (ed) The nematode *Caenorhabditis elegans*. Cold Spring Harbor Press, Cold Spring Harbor, New York

Howell AM, Rose AM (1990) Essential genes in the *hDf6* region of chromosome I of *Caenorhabditis elegans*. Genetics 126:583–592

Hyman AA (1989) Centrosome movement in the early divisions of *Caenorhabditis elegans*: a cortical site determining centrosome position. J Cell Biol 109:1185–1193

Hyman AA, White JG (1987) Determination of cell division axes in the early embryogenesis of *Caenorhabditis elegans*. J Cell Biol 105:2123–2135

Illmensee K, Mahowald AP (1974) Transplantation of posterior polar plasm in *Drosophila*. Induction of germ cells at the anterior pole of the egg. Proc Natl Acad Sci USA 71:1016–1020

Isnenghi E, Cassada R, Smith K, Denich K, Radnia K, von Ehrenstein G (1983) Maternal effects and temperature-sensitive period of mutations affecting embryogenesis in *Caenorhabditis elegans*. Dev Biol 98:465–480

Ito K, McGhee JD (1987) Parental DNA strands segregate randomly doring embryonic development of *Caenorhabditis elegans*. Cell 49:329–336

Kemphues KJ (1989) *Caenorhabditis*. In: Glover DM, Hames BD (eds) Genes and embryos. IRL Press, Oxford

Kemphues KJ, Kusch M, Wolf N (1988a) Maternal-effect lethal mutations on linkage group II of *Caenorhabditis elegans*. Genetics 120:977–986

Kemphues KJ, Priess JR, Morton DG, Cheng N (1988b) Identification of genes required for cytoplasmic localization in early *Caenorhabditis elegans* embryos. Cell 52:311–320

Kemphues KJ, Wolf N, Wood WB, Hirsh D (1986) Two loci required for cytoplasmic organization in early embryos of *Caenorhabditis elegans*. Dev Biol 113:449–460

Kenyon C (1986) A gene involved in the development of the posterior body region of *C. elegans*. Cell 46:477–487

Kim SK, Horvitz HR (1990) The *Caenorhabditis elegans* gene *lin-10* is broadly expressed while required specifically for the determination of vulval cell fates. Genes Dev 4:357–371

Kimble J, Hirsh D (1979) Post-embryonic cell lineages of the hermaphrodite and male gonads in *Caenorhabditis elegans*. Dev Biol 70:396–417

Kimble J, Sharrock WJ (1983) Tissue-specific synthesis of yolk proteins in *Caenorhabditis elegans*. Dev Biol 96:189–196

Kimble JE, White JG (1981) On the control of germ cell development in *Caenorhabditis elegans*. Dev Biol 81:208–219

Kirby C, Kusch M, Kemphues K (1990) Mutations in *par* genes of *Caenorhabditis elegans* affect cytoplasmic reorganization during the first cell cycle. Dev Biol 142:203–215

Kusch M, Edgar RS (1986) Genetic studies of unusual loci that affect body shape of the nematode *Caenorhabditis elegans* and may code for cuticle structural proteins. Genetics 113:621–639

Laufer JS, Bazzicalupo P, Wood WB (1980) Segregation of developmental potential in early embryos of *Caenorhabditis elegans*. Cell 19:569–577

L'Hernault SW, Shakes DC, Ward S (1988) Developmental genetics of chromosome I spermatogenesis-defective mutants in the nematode *Caenorhabditis elegans*. Genetics 120:435–452

Mains PE, Sulston IA, Wood WB (1990a) Dominant maternal-effect mutations causing embryonic lethality in *Caenorhabditis elegans*. Genetics 125:351–369

Mains PE, Kemphues KJ, Sprunger SA, Sulston IA, Wood WB (1990b) Mutations affecting the meiotic and mitotic divisions of the early *Caenorhabditis elegans* embryo. Genetics 126:593–605

McGhee JD (1992) Gut esterase expression in the nematode *Caenorhabditis elegans*. Adv Dev Biochem 1:169–210

Meneely PM, Herman RK (1979) Lethal, steriles and deficiencies in a region of the X chromosome of *Caenorhabditis elegans*. Genetics 92:99–115

Merrill P, Sweeton D, Wieschaus E (1988) Requirements for autosomal gene activity during precellular stages of *Drosophila melanogaster*. Development 104:495–509

Miwa J, Schierenberg E, Miwa S, von Ehrenstein G (1980) Genetics and mode of expression of temperature-sensitive mutations arresting embryonic development in *Caenorhabditis elegans*. Dev Biol 76:160–174

Moerman DG, Benian GM, Waterston RH (1986) Molecular cloning of the muscle gene *unc-22* in *Caenorhabditis elegans* by Tc1 transposon tagging. Proc Natl Acad Sci USA 83:2579–2583

Monk M (1990) Variation in epigenetic inheritance. Trends Genet 6:110–114

Mori I, Moerman DG, Waterston RH (1988) Analysis of a mutator activity necessary for germline transposition and excision of Tc1 transposable elements in *Caenorhabditis elegans*. Genetics 120:397–407

Nigon V, Guerrier P, Monin H (1960) L'Architecture polaire de l'oeuf et les movements des constituants cellulaires au cour des premieres etapes du developpement chez quelques nematodes. Bull Biol Fr Belg 94:132–201

Nüsslein-Volhard C, Wieschaus E (1980) Mutations affecting segment number and polarity in Drosophila. Nature (London) 287:795–801

Nüsslein-Volhard C, Frohnhofer H, Lehmann R (1987) Determination of anterioposterior polarity in Drosophila. Science 238:1675–1681

Park E-C, Horvitz HR (1986) Mutations with dominant effects on the behavior and morphology of the nematode Caenorhabditis elegans. Genetics 113:821–852

Perrimon N, Mohler D, Engstrom L, Mahowald AP (1986) X-linked female-sterile loci in Drosophila melanogaster. Genetics 113:695–712

Perrimon N, Engstrom L, Mahowald AP (1989) Zygotic lethals with specific maternal effect phenotypes in Drosophila melanogaster. I. Loci on the X chromosome. Genetics 121:333–352

Pimpinelli S, Goday C (1989) Unusual kinetochores and chromatin diminution in Parascaris. Trends Genet 5:310–315

Priess JR, Hirsh DI (1986) Caenorhabditis elegans morphogenesis: the role of the cytoskeleton in elongation of the embryo. Dev Biol 117:156–173

Priess JR, Thomson JN (1987) Cellular interactions in early Caenorhabditis elegans embryos. Cell 48:241–250

Priess JR, Schnabel H, Schnable R (1987) The glp-1 locus and cellular interactions in early Caenorhabditis elegans embryos. Cell 51:601–611

Roberts L (1990) The worm project. Science 248:1310–1313

Rogalski TM, Riddle DL (1988) A Caenorhabditis elegans RNA polymerase II gene, ama-1 IV, and nearby essential genes. Genetics 118:61–74

Rogalski TM, Moerman DG, Baillie DL (1982) Essential genes and deficiencies in the unc-22 IV region of Caenorhabditis elegans. Genetics 102:725–736

Rosenbluth RE, Rogalski TM, Johnsen RC, Addison LM, Baillie DL (1988) Genomic organization in Caenorhabditis elegans: deficiency mapping on linkage group V (left). Genet Res 52:105–118

Schauer IE, Wood WB (1990) Early C. elegans embryos are transcriptionally active. Development 110:1303–1317

Schierenberg E (1985) Cell determination during early embryogenesis of the nematode Caenorhabditis elegans. Cold Spring Harbor Symp Quant Biol 50: 59–68

Schierenberg E (1986) Developmental strategies during early embryogenesis of Caenorhabditis elegans. J Embryol Exp Morphol 97s:31–44

Schierenberg E (1987) Reversal of cellular polarity and early cell-cell interaction in the embryo of Caenorhabditis elegans. Dev Biol 122:452–463

Schierenberg E (1988) Localization and segregation of lineage-specific cleavage potential in embryos of Caenorhabditis elegans. Wilhelm Roux's Arch Dev Biol 197:282–293

Schierenberg E (1989) Cytoplasmic determination and distribution of developmental potential in the embryo of Caenorhabditis elegans. Bioessays 10:99–104

Schierenberg E, Wood WB (1985) Control of cell-cycle timing in early embryos of Caenorhabditis elegans. Dev Biol 107:337–354

Schierenberg E, Miwa J, von Ehrenstein G (1980) Cell lineages and developmental defects of temperature-sensitive embryonic arrest mutants in Caenorhabditis elegans. Dev Biol 76:141–159

Schnabel R, Schnabel H (1990a) Early determination in the Caenorhabditis elegans embryo: a gene, cib-1, required to specify a set of stem-cell-like blastomeres. Development 108:107–119

Schnabel H, Schnabel R (1990b) An organ-specific differentiation gene, pha-1, from Caenorhabditis elegans. Science 250:686–688

Seydoux G, Greenwald I (1989) Cell autonomy of lin-12 function in a cell fate decision in Caenorhabditis elegans. Cell 57:1237–1245

Sigurdson DC, Spanier GJ, Herman RK (1984) Caenorhabditis elegans deficiency mapping. Genetics 108:331–345

Simpson VJ, Johnson TE, Hammen RF (1986) *Caenorhabditis elegans* DNA does not contain 5-methylcytosine at any time during development or aging. Nucl Acid Res 14:6711–6717

Sternberg PW, Horvitz HR (1984) The genetic control of cell lineage during nematode development. Annu Rev Genet 18:489–524

Sternberg PW, Horvitz HR (1989) The combined action of two intercellular signalling pathways specifies three cell fates during vulval induction in *Caenorhabditis elegans*. Cell 58:679–693

Stinchcomb DT, Shaw JE, Carr SH, Hirsh D (1985) Extrachromosomal DNA transformation of *Caenorhabditis elegans*. Mol Cell Biol 5:3484–3496

Storfer FA (1990) Contribution of embryonic gene expression to early embryogenesis in the nematode *Caenorhabditis elegans*: a genetic analysis. PhD Thesis, University of Colorado, Boulder, Colorado

Strome S (1986a) Fluorescence visualization of the distribution of microfilaments in gonads and early embryos of the nematode *Caenorhabditis elegans*. J Cell Biol 103:2241–2252

Strome S (1986b) Asymmetric movements of cytoplasmic components in *Caenorhabditis elegans* zygotes. J Embryol Exp Morphol 97s:15–29

Strome S (1989) Generation of cell diversity during early embryogenesis in the nematode *Caenorhabditis elegans*. Int Rev Cytol 114:81–124

Strome S, Wood WB (1982) Immunofluorescence visualization of germ-line-specific cytoplasmic granules in embryos, larvae, and adults of *Caenorhabditis elegans*. Proc Natl Acad Sci USA 79:1558–1562

Strome S, Wood WB (1983) Generation of asymmetry and segregation of germ-line granules in early *Caenorhabditis elegans* embryos. Cell 35:15–25

Sulston JE (1988) Cell lineage. In: Wood WB (ed) The nematode *Caenorhabditis elegans*. Cold Spring Harbor Press, Cold Spring Harbor, New York

Sulston JE, Brenner S (1974) The DNA of *Caenorhabditis elegans*. Genetics 77: 95–104

Sulston JE, Horvitz HR (1977) Post-embryonic cell lineages of the nematode, *Caenorhabditis elegans*. Dev Biol 56:110–156

Sulston JE, Schierenberg E, White JG, Thomson JN (1983) The embryonic cell lineage of the nematode *Caenorhabditis elegans*. Dev Biol 100:64–119

Suzuki D (1970) Temperature-sensitive mutations in *Drosophila melanogaster*. Science 170:695–706

Vanderslice R, Hirsh D (1976) Temperature-sensitive zygote defective mutants of *Caenorhabditis elegans*. Dev Biol 49:236–249

Vavra SH, Carroll SB (1989) The zygotic control of *Drosophila* pair-rule gene expression I. A search for new pair-rule regulatory loci. Development 107: 663–672

Ward S, Carrel JS (1979) Fertilization and sperm competition in the nematode *Caenorhabditis elegans*. Dev Biol 73:304–321

Waring DA, Kenyon C (1990) Selective silencing of cell communication influences anteroposterior pattern formation in *Caenorhabditis elegans*. Cell 60:123–131

Wieschaus E, Sweeton, D (1988) Requirements for X-linked zygotic gene activity during cellularization of early *Drosophila* embryos. Development 104:483–493

Wieschaus E, Nüsslein-Volhard C, Jürgens G (1984) Mutations affecting the pattern of the larval cuticle in *Drosophila melanogaster*. Zygotic loci on the X-chromosome and fourth chromosome. Wilhelm Roux's Arch Dev Biol 193: 296–307

Wilkins AS (1986) Genetic analysis of animal development. John Wiley & Sons, New York

Wolf N, Priess J, Hirsh D (1983) Segregation of germline granules in early embryos of *Caenorhabditis elegans*: an electron microscopic analysis. J Embryol Exp Morphol 73:297–306

Wood WB (ed) (1988a) The nematode *Caenorhabditis elegans*. Cold Spring Harbor Press, Cold Spring Harbor, New York

Wood WB (1988b) Embryology. In: Wood WB (ed) The nematode *Caenorhabditis elegans*. Cold Spring Harbor Press, Cold Spring Harbor, New York

Wood WB (1991) Reversal of handedness in *C. elegans* embryos: new evidence for extensive early cell interactions that determine cell fates. Nature (London) 349:536–538

Wood WB, Hecht R, Carr S, Vanderslice R, Wolf N, Hirsh D (1980) Parental effects and phenotypic characterization of mutations that affect early development in *Caenorhabditis elegans*. Dev Biol 74:446–469

Wood WB, Schierenberg E, Strome S (1984) Localization and determination in early embryos of *Caenorhabditis elegans*. UCLA Symp Mol Biol 19:37–49

Yamaguchi Y, Murakami K, Furusawa M, Miwa J (1983) Germline-specific antigens identified by monoclonal antibodies in the nematode *Caenorhabditis elegans*. Dev Growth Differ 25:121–131

Yochem J, Greenwald I (1989) *glp-1* and *lin-12*, genes implicated in distinct cell-cell interactions in *Caenorhabditis elegans*, encode similar transmembrane proteins. Cell 58:553–563

Note Added in Proof

The reader is directed to the following recent publications. The worm genome sequencing project has reported the first 120 kb of sequence, and there is evidence that the number genes in *C. elegans* is at least several-fold higher than previous estimates [Sulston et al. (1992) Nature (London) 356:37–41]. A gene that may be responsible for the establishment of the EMS cell fate has been described by Bowerman et al. [(1992) Cell 68:1061–1075] and this gene is also required at other times in the life cycle. Goldstein [(1992) Nature (London) 357:255–257] has report additional evidence that gut differentiation is not autonomous.

3 Analysis of Early Development in the Zebrafish Embryo

Eric S. Weinberg

1 Introduction: Some Advantages of the System

The zebrafish (*Brachydanio rerio*), a small fresh water fish native to rivers of northeast India, has long been a favorite of tropical fish fanciers. Within the past few years, it has also become an organism of great interest to vertebrate embryologists. The potential of the zebrafish as an effective experimental system can be traced to the work of G. Streisinger who recognized that the organism was highly suited to genetic analysis (Streisinger et al. 1981). After his untimely death, the development of the system was continued at the University of Oregon, and more recently, at many laboratories in the United States and Europe. The result has been a productive investigation of the embryology, genetics, neurobiology, and molecular biology of the zebrafish embryo (reviewed most recently in Kimmel and Warga 1988; Kimmel 1989; Ekker and Akimenko 1991; Fulwiler and Gilbert 1991). In this review, I will attempt to summarize what is known about the early development, genetics, and molecular biology of the zebrafish. Development of the nervous system has been more intensively investigated than any other aspect of embryogenesis in this organism. This subject has been reviewed recently (Eisen 1991) and will not be covered in detail here.

The advantages of the zebrafish, for the developmental biologist, are numerous and were recognized by Streisinger in his first zebrafish paper (Streisinger et al. 1981). The fish are easy to raise, reaching sexual maturity in as little as 2–3 months if reared at 28 °C. Their small size (adults are usually 2–3 cm) enables one to maintain the fish at a reasonable density in high quality water (one fish per two l is not difficult, and systems with good filtration can support much higher densities of fish). The fish can be fed with easily obtained aquarium food flakes, trout pellets, *Drosophila* larvae, and *Artemia*. Raising fish from fertilized eggs is straightforward, provided that the fry are fed paramecia, and then newly hatched *Artemia*, on a regular basis. Information on raising and maintaining zebrafish can be found in the excellent manual, *The Zebrafish Book* (Westerfield 1989).

The most powerful advantage of the zebrafish is the potential to develop an effective genetic system with this species. Since a generation can be

Department of Biology, University of Pennsylvania, Philadelphia, PA 19104, USA

Results and Problems in Cell Differentiation 18
W. Hennig (Ed.)
Early Embryonic Development of Animals
© Springer-Verlag Berlin Heidelberg 1992

raised in 2–3 months, and pairwise matings are very easy to do, there is no difficulty in breeding four generations per year. Gamma-ray (Chakrabati et al. 1983; Walker and Streisinger 1983; Streisinger 1984), chemical (Streisinger 1984; Grunwald and Streisinger 1991a), and ultraviolet light (Grunwald and Streisinger 1991b) mutagenesis have been used to obtain mutants. Mutant screens are practical, especially if facilities to house large numbers of fish are available. Since hatched fry have many properties of mature fish, mutant screening done several days after fertilization should identify genes essential to various physiological functions in addition to those involved in development. Several methods allow the production of haploids, which live for several days, or of viable homozygous diploids, enabling one to screen embryos directly for mutant phenotypes which would be masked as recessives in normal diploid fish (see below). In this way, inbreeding to produce fish homozygous for a recessive mutant allele is not necessary for mutant screens.

The second most important advantage is that the embryos are a delight to work with. The eggs are laid by the females and immediately fertilized by sperm released by males. Since this occurs (mostly) after the onset of a light period (the fish are kept on a strict light cycle – usually 10 h dark, 14 h light), one can obtain large quantities of eggs and embryos at the same stage. A good mating can produce several hundred eggs, so that if many pairs are mated (fish can be left as pairs the evening before eggs are desired), it is possible to obtain thousands of eggs in a morning. Embryogenesis is rapid – the body plan is laid down within the first 12 h, pairs of somites appear at intervals of approximately 30 min between 10 and 24 h, and by this time the optic cups with lens, otic placodes, brain, and heart have formed, and muscular twitches occur readily (outlined in Westerfield 1989). The joy of working with these embryos is that because the embryo is transparent and large, all of these events are easy to observe with a good quality dissecting microscope. The clarity of the embryo has made it possible to analyze the development of the nervous system at a detailed cellular level (reviewed in Eisen 1991). In these respects, the advantages of the fish embryo over dark murky amphibian embryos are impressive. Moreover, an obvious advantage for developmental studies of the fish embryo (shared with amphibians) over mammalian embryos is that one does not have to recover uterine implants to study key early developmental stages.

Experimental manipulations of the zebrafish embryo have been successful, and analyses of lineage and embryonic fate have been extremely revealing (see below). Transplantation, ablation, and embryo reconstitutions, however, are still in their infancy. Also encouraging has been the success in producing transgenic zebrafish. Several groups have been able to show stable transmission of injected DNA through the germline (Stuart et al. 1988; Stuart et al. 1990; Culp et al. 1991). At present, next to the mouse, the zebrafish would seem to the best species for vertebrate transgenic studies. An important step in working out methods to create genetically engineered fish is the recent demonstration that chimeric zebrafish can be produced efficiently (Lin et al. 1992).

The molecular events in the formation of the vertebrate body plan appear to be highly conserved, despite the considerable variation among the vertebrate classes in both the form of gastrulation and early cleavage, and in the degree of fixation of lineage during early development. Kimmel (1989) has pointed out that the most important common features of vertebrate development will emerge from comparative studies on the major vertebrate developmental systems. Already, work on the expression of particular genes important in development (e.g., *Brachyury*, *Pax*, and *Hox* genes; see Sect. 6) have pointed to the unity of the molecular mechanisms underlying the formation of the vertebrate body plan. Two attributes of the zebrafish, however, are not particularly advantageous for genetic and molecular studies: a haploid chromosome number of 25 (Endo and Ingalls 1968), and a genome size of approximately 2×10^9 (Hinegardner and Rosen 1972).

2 Early Embryogenesis

2.1 Cleavage

Zebrafish embryonic stages were first defined by Hisaoka and Battle (1958), based in part on earlier descriptions (Roosen-Runge 1936, 1938, 1939) and the availability of a time lapse film (Lewis and Roosen-Runge 1944). The development of the blastula has been documented by scanning electron microscopy (Beams et al. 1985). The timing of developmental stages noted in this review will be for embryos grown at 28.5 °C (Westerfield 1989). The time required to reach late blastula is a fairly constant 3.5 h at temperatures between 24 and 34 °C, but later stages are considerably more temperature-dependent (Schirone and Gross 1968). Figure 1 illustrates a number of the stages described below. Up to 500 eggs, spawned by an oviparous female during courtship with a single male, are immediately fertilized externally. This usually takes place within minutes of the onset of a light cycle, but can also occur several hours into the daylight period. The eggs float free within a proteinaceous membrane termed the chorion. The percentage of fertilized eggs can vary considerably, but it is not unusual to obtain fertilization levels greater than 90%. It is difficult to distinguish between fertilized and unfertilized eggs until cell division starts. The sperm enters the egg through a micropile directly at the animal pole. Within minutes of fertilization, a clear cytoplasmic region forms at the animal pole, and there are extensive streaming movements within the bulk of the egg cytoplasm. The first cell division takes place about 35–40 min post fertilization.

The early cell division pattern of the zebrafish embryo is the typical teleost meroblastic (incomplete) cleavage (Wilson 1891). The most prominant feature is that the cells divide at the animal pole of the egg, with the yolk sequestered in a single large cell making up the bulk of the embryo volume. The first five cleavages are vertical, alternating at right angles, resulting in a 32-cell embryo (the yolk cell itself is not counted, and

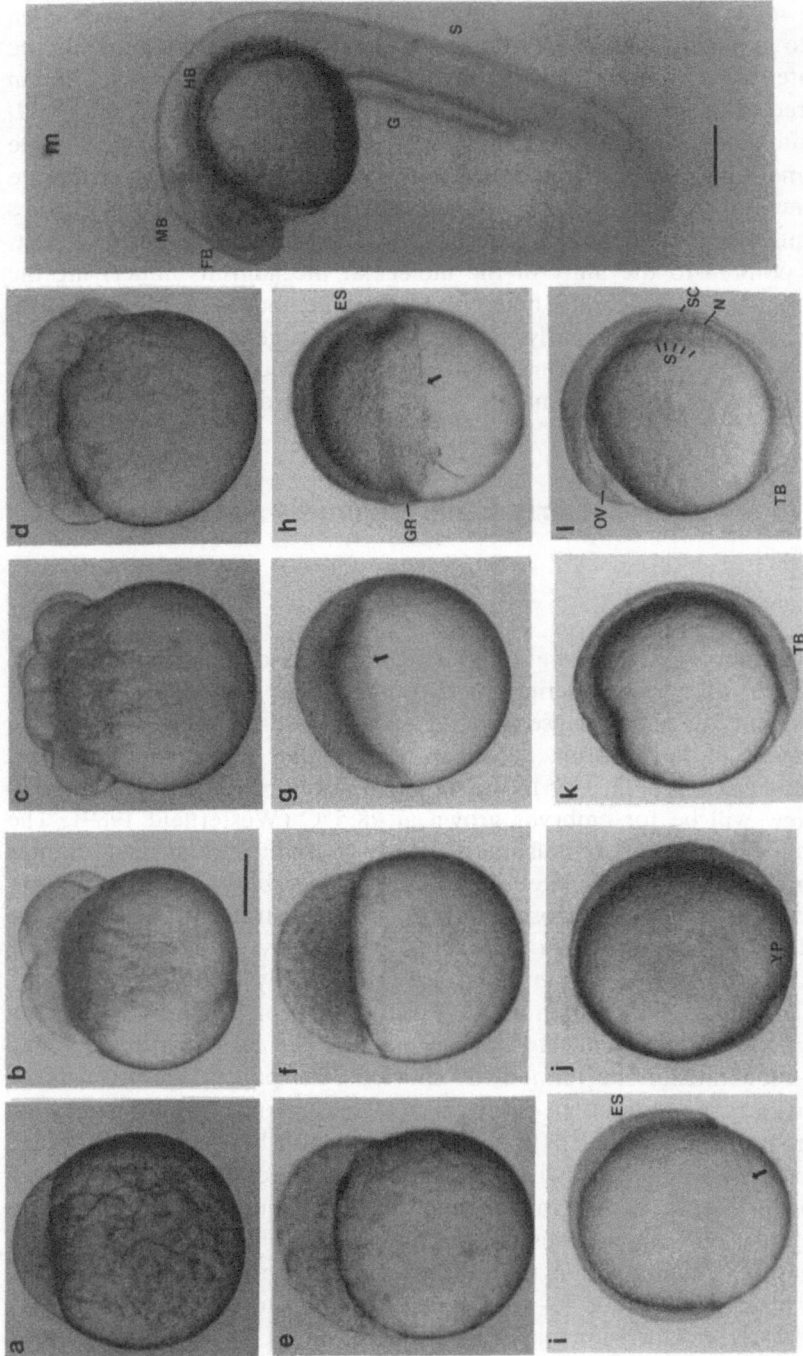

Fig. 1a–m

apparently does not have a nucleus at this stage) arranged in four rows of 8 cells (Hisoaka and Firlit 1960; Kimmel and Law 1985a). Up to the 16-cell stage, all blastomeres are part of a syncitium with the yolk cell, due to a lack of membrane formation between the yolk and the blastomeres. Cytoplasmic connections also are maintained between the individual blastomeres at the 16-cell stage, as demonstrated by the flow of fluorescent dextran from single blastomeres into adjacent blastomeres as well as to the underlying yolk cell (Kimmel and Law 1985a). At the 32-cell stage, the central 8–12 non-marginal cells not only are completely delimited by cell membrane and separated from from the yolk by a small blastocoel, but also lose cytoplasmic bridges to the more marginal cells (Kimmel and Law 1985a). The sixth cleavage is vertical, resulting in two layers of 32 cells. A fairly synchronous doubling of cells continues until the 512-cell stage (approximately 2.8 h) (Kimmel and Law 1985a). During these divisions, the marginal blastomeres adjacent to the yolk remain connected to the yolk cell by cytoplasmic bridges (Kimmel and Law 1985a). The more internal cells ("deep cells") in the blastoderm do not seem to be connected.

Some variation of the early cleavage pattern has been observed, with the most common difference being horizontal divisions at the fifth cleavage in some embryos (Beams and Kessel 1976; Kimmel and Law 1985a). The 7th and 8th cleavages are also somewhat variable. Although usually horizontal, these divisions are vertical in approximately 10% of the embryos (Kimmel and Warga 1987a). Subsequent divisions are even more variable, but none of these differences seem to affect the outcome of embryonic development (Kimmel and Warga 1987a). Relevant to these findings is my observation that treatment of embryos at 16 °C for the first hour of development often results in irregular cleavage patterns and the sloughing off of blastomeres (as many as one-third of the cells can detach from the blastoderm); nevertheless, these embryos usually develop normally.

An important finding is that there does not appear to be a fixed relationship between the plane of first cleavage and the dorsal-ventral axis of the embryo (Kimmel and Law 1985c; Kimmel and Warga 1987a). The position of cells in the blastula can be easily correlated with the first cleavage plane. Clones of blastula cells marked with fluorescent dextran were followed until the embryonic shield (the first visible sign of the future dorsal midline, see below) was evident. No relationship was found between the position of the

◄ **Fig. 1a–m.** The first day of zebrafish embryogenesis. **a** 1-cell; **b** 4-cell; **c** 8-cell; **d** 64-cell; **e** 128-cell; **f** "high" blastula; **g** "doming" and beginning of epiboly; **h** 6-h embryo, 55% epiboly: the germ ring and embryonic shield have formed; **i** 75% epiboly; **j** 90% epiboly; **k** 10-h embryo, just prior to somite formation; **l** 13-h embryo, 5-somite stage; **m** 24-h embryo. All embryos are photographed at the same magnification: *scale bars* in **b** and **m** are 200 μ. Embryos are aligned with the animal pole of the egg upwards, and in those stages where the dorsal-ventral polarity can be determined (**h-m**), with the dorsal side to the *right*. *Arrows* in **g**, **h**, and **i** indicate the position of the front of migrating cells. *GR* Germ ring; *ES* embryonic shield; *YP* yolk plug; *TB* tail bud; *OV* optic vesicle; *S* somites; *N* notochord; *SC* spinal cord; *FB* forebrain; *MB* midbrain; *HB* hindbrain; *G* gut

clone in the blastula, relative to the plane of first cleavage, and the location of the clone with respect to the embryonic shield in the later embryo.

2.2 The Blastula Stage

The blastula stage lasts about 2 h, from the 128-cell embryo (at just over 2 h postfertilization) to an embryo with somewhat over 3000 cells (Westerfield 1989). The blastula cap of cells is sometimes called the blastoderm or blastodisc. These terms are also occasionally used to describe the embryonic cell mass in later embryonic stages as well. During blastulation, regular cleavage planes are not seen, and three distinct cell (or nuclear) populations are formed. The outer cells form a thin single squamous epithelial cell layer called the enveloping layer (EVL). Work on *Fundulus* has shown that this layer, containing numerous junctional complexes, is responsible for establishment of a high electrical resistance and a barrier to the transport of small ions (Betchaku and Trinkaus 1978; Keller and Trinkaus 1987). Studies of this sort have not yet been done on the zebrafish, but morphologically, the EVLs of the two species look quite similar. The EVL cells eventually give rise to the periderm overlying the embryo, and do not appear to contribute to any true embryonic tissue (although this has not been proven for the zebrafish). A second specialized region is the yolk syncitial layer (YSL). This syncitium, underlying the blastoderm, forms in the zebrafish embryo after the 9th or 10th division as a ring at the blastoderm margin (Kimmel and Law 1985b). The mode of its formation essentially was described for the sea bass embryo 100 years ago (Wilson 1889). The marginal cells of the blastoderm, which are evidently still connected by cytoplasmic bridges to the yolk cell, collapse into a thick, non-yolky cytoplasmic layer adjacent to the yolk. After formation of the YSL ring, cytoplasmic bridges to the blastomeres cease to exist (there is no transfer of injected fluorescent dextran); on the other hand, diffusion of low molecular weight substances from the yolk cell to the blastomeres occurs until the beginning of gastrulation, a conclusion based on the transfer of injected lucifer yellow (Kimmel and Law 1985b). The third group of cells in the blastodisc are the deep cell blastomeres, or deep layer (DEL) cells, which are the most numerous cells at this stage, and from which the whole embryo proper is derived.

At about the same time the YSL is formed, there is a midblastula transition in the zebrafish embryo (work of D.A. Kane, cited in Kimmel 1989). The rate of cell division slows and the cycles become asynchronous, accompanied by an increase in RNA synthesis. In the zebrafish, the midblastula transition is later in haploid, and earlier in triploid embryos, a result consistent with the premature transition induced in polyspermic *Xenopus* embryos (Newport and Kirschner 1982). The blastodisc then goes through a series of distinctive shape changes, first as a sort of crown of cells perched high above the yolk cell, later as a smooth, flattened mass of cells embedded into the yolk cell so that the whole embryo is spherical.

2.3 Epiboly

The first of three major cell movements, epiboly, begins at 4.3 h of development. The upper portion of the yolk bulges up in the direction of the animal pole, forming a so-called dome, and the blastoderm cells start spreading over the yolk cell (Warga and Kimmel 1990). In teleosts, epiboly is the process by which cells migrate down over the surface of the yolk cell, eventually completely internalizing the yolk in a thin shell of cells (Trinkaus 1984a,b). In the zebrafish, mixing and scattering of the DEL cells accompanies the expansion (Kimmel and Law 1985c; Kimmel and Warga 1986), with radial intercalations between cells located in the deeper and more superficially located regions of the blastula (Warga and Kimmel 1990). DEL cells near the animal pole do not tend to spread, whereas cells closer to the margin migrate extensively toward the vegetal pole (Kimmel and Warga 1987a). The EVL cell monolayer also spreads downwards towards the vegetal pole, but does not mix with the DEL cells (Kimmel and Warga 1987a; Warga and Kimmel 1990). About 1 h after these movements begin in the zebrafish embryo (5.2 h), 50% of the yolk cell is covered by these migrating cells (the "50% epiboly stage"). Epiboly then continues until the entire yolk surface is covered with cells, but is simultaneous with two additional cell movements.

2.4 Involution: The Onset of Gastrulation

At 5.2 h, the second morphogenetic movement, involution, begins (Warga and Kimmel 1990). Below the thin sheet of EVL cells, the leading edge of the cell mass now consists of two layers – an outer epiblast and an inner hypoblast. The view that the teleost hypoblast is formed by involution of cells at the margin of the blastoderm (Wilson 1889; Morgan 1895; Pasteels 1936) has recently been shown to be correct for the rosy barb (Wood and Timmermans 1988) and the zebrafish (Warga and Kimmel 1990). An alternative conclusion that the hypoblast was derived from deep lying blastoderm cells from points closer to the animal pole (Ballard 1966a,b,c; summarized in Trinkaus 1984b) was based on cell-marking procedures done with embryos of *Salmo* and other large teleosts. Although it is conceivable that the mechanism of hypoblast formation may vary in different fish species, it is more likely that the cell-marking methods used on *Salmo* were not sensitive enough to detect involution (Warga and Kimmel 1990).

At the front margin of the cells undergoing epiboly in the rosy barb and the zebrafish, there is clearly a movement of cells inward and then underneath the originating layer, a result which conforms to earlier observations (Pasteels 1936). Within 15 min of the beginning of involution, a distinct thickened band (the "germ ring") forms around the circumference of the epiboly front. Careful observation of events at the germ ring of the rosy barb embryo with Nomarski optics identified the site of involuting cells

(Wood and Timmermans 1988). So-called tension striae seem to link the YSL to the yolk cell at the margin. Above this, the DEL cells extend inward towards the YSL and are replaced by neighboring cells at the margin. This involution of DEL cells, only at the edge of the germ ring, seems to take place simultaneously around the circumference of the margin. The involuting cells then form the hypoblast layer which extends back up under the front of downward migrating cells, and there is no obvious mixing with the upper epiblast layer.

Similar observations of DEL cell involution were made in the zebrafish embryo by following cells which had been injected with fluorescent dextran (Warga and Kimmel 1990). Again, only DEL cells located at the blastoderm margin migrate inwards, and all of the involuting cells wind up in the hypoblast. The hypoblast eventually spreads from the margin and extends to the animal pole so that the whole egg cell is covered by the two cell layers (and the third outer EVL layer, as well). Not all of the hypoblast cells have a net upward movement with respect to the yolk. The later involuting cells may actually continue a net downward movement toward the vegetal pole, but move upwards with respect to the epiblast cells, which continue their epibolic movement downwards. The process of involution can be formally thought of as gastrulation (Kimmel 1989; Warga and Kimmel 1990), being analogous to the involution at the blastopore in amphibians.

2.5 Convergent Extension

The third type of embryonic cell movement in teleost embryos is termed convergent extension (or convergence and extension). Shortly after the beginning of involution, in addition to the downward epibolic movement of the blastoderm, and the internalization of the hypoblast (which in many cells involves upward movement toward the animal pole), there is a movement of all DEL cells and their derivatives (both epiblast and hypoblast) toward the future dorsal side of the embryo (Ballard 1973; Kimmel and Warga 1986, 1987a; Warga and Kimmel 1990). This results in a thinning of the cell layers on the ventral side, and a thickening of layers at the dorsal side. The embryonic axis forms at the dorsal side with extensive mediolateral inter-calations of the migrating cells. The formation of a thickening of the germ ring on the dorsal side, the "embryonic shield", (occurring at 6 h in the zebrafish) was first described by Oppenheimer (1937) for *Fundulus*, and was recognized in the zebrafish embryo by Hisaoka and Battle (1958). During convergence in the zebrafish embryo, there is extensive mixing within the hypoblast and epiblast layers, but no mixing between the two layers (Warga and Kimmel 1990). Mixing of these cells at this stage occurs only within the primordia of specific tissues, and not between cells which will become different tissues (Warga and Kimmel 1990), since lineages are already tissue-restricted (Kimmel and Warga 1986; Kimmel et al. 1990). In contrast to the DEL cells, EVL cells do not converge, and thus there are no ventral to dorsal movements of these cells as they undergo epiboly (Kimmel and Warga 1987a).

2.6 Somitogenesis and Segmentation

At 9–10 h 100% epiboly is reached. At this time, dorsal convergence has resulted in the formation of a mass of cells at the dorsal side of the embryo, while the ventral side of the embryo contains only the outer EVL layer, and perhaps a one-cell thick inner layer. The cells at the animal pole form a rounded head rudiment, the cells at the vegetal pole form a "tail bud" with a distinct angular shape. Somitogenesis starts approximately 1 h later and is easily observed in the clear embryo. The rostrocaudal sequence of somite formation is similar to that in other vertebrates, although in the zebrafish, the somites are added at a rate of two per hour and can be seen clearly through the transparent embryo (Hanneman and Westerfield 1989). A set of two to six cells in each of the first ten somites synchronously undergo shape changes and develop long parallel actin bundles by 14.5 h postfertilization (Felsenfeld et al. 1991). These cells, termed muscle pioneer cells or MPs, are located alongside the notochord and are the first myotomal muscle cells to differentiate. Since these cells form synchronously in a set of somites, they should be useful for studies of muscle development. The MPs are probably the first muscle cells to express myosin heavy chain (Van Raamsdonk et al. 1978, 1982) and acetylcholine esterase (Hanneman and Westerfield 1989), and also express high levels of *engrailed* homeoprotein (Hatta et al. 1991a) (see Sect. 6.2). The zn-5 monoclonal antibody, which recognizes cell surfaces of a number of particular cell types in the zebrafish embryo (Trevarrow et al. 1990), specifically stains the somites in the MP cells (Hatta et al. 1991a).

Primary neurons form in the embryo as early as 10–12 h after fertilization, even prior to the formation of the neural tube (Kimmel and Westerfield 1990). The birthdays of the earliest neurons occur at 8–10 h after fertilization, during gastrulation (Mendelson 1986a), but they do not initiate axons until 18 h of development (Mendelson 1986b). Axons grow from neurons at specific locations after the appearance of somites (reviewed with references in Kimmel et al. 1991). By the end of the first day of development, there is a small collection of highly characteristic axonal bundles in the embryo (Hanneman et al. 1988; Trevarrow et al. 1990).

Somitogenesis is one of four types of segmentation that can be recognized in the zebrafish embryo during the second 12 h of development (Kimmel et al. 1988). Mesodermal segmentation also occurs in the head region some hours later to give rise to the pharyngeal arches, structures which are composed of cells from all three germ layers and the neural crest (shown for the sea lamprey, Langille and Hall 1988). There is also segmental organization of the spinal cord and the hindbrain (Myers et al. 1986; Westerfield et al. 1986; Kimmel et al. 1988; Trevarrow et al. 1990). The spinal cord segments correspond to a myotomal segment; a single cluster of motoneurons present on each side of each segment can be visualized with the neuron-specific monoclonal antibody zn-1 (Kimmel et al. 1988). Each cluster of motoneurons leads directly to the middle of each myotome (Myers et al. 1986). Segmentation is also seen in the hindbrain of all vertebrates; the individual segments are termed rhombomeres or hindbrain neuromeres. Seven to nine segments have been recognized in the zebrafish beginning at 15–16 h of development (Kimmel et al. 1988; Trevarrow et al. 1990): the neuronal components of each segment have been identified, and the patterns of innervation have in part been traced to the pharyngeal arches.

As was first shown in frogs (Elsdale et al. 1976) and then in chicks (Veini and Bellairs 1985) and the Mexican axolotl (Armstrong and Graveson 1988), the development of myotomes from somites in the zebrafish is transiently sensitive to brief heat shock (Kimmel et al. 1988). The period of heat shock sensitivity of a particular somite moves rostrocaudally at the same rate (two somites per hour) as the formation of the somites themselves, but precedes somite furrowing by 2–2.5 h. Spinal cord segmentation is affected at the same point in time at which myotome pattern was affected. The finding that both mesodermal and neural segmentation is affected at the same region by heat shock suggests that there is some sort of coordinate patterning going on between these two tissue elements (Kimmel et al. 1988).

3 Cell Lineage and Embryonic Fate

It is now clear that the zebrafish embryo has no fixed cell lineage during cleavage and blastulation. By the beginning of gastrulation at 5.2 h, however, a fate map can be drawn on the cup of cells overlying the animal pole half of the yolk. The evidence for this conclusion comes mainly from the tracer dye injection experiments of Kimmel and colleagues (Kimmel and Law 1985c; Kimmel and Warga 1986, 1987a,b; Warga and Kimmel 1990; Kimmel et al. 1990; Hatta et al. 1990; reviewed in part in Kimmel and Warga 1988). Figure 2 illustrates the methods which have been used in zebrafish lineage analysis, and presents the fate map of the early gastrula.

3.1 Genetic Mosaic Analysis of the Pigmented Retina

The first insight into the lack of a fixed cell lineage in the zebrafish cleavage and blastula stages came from a classic clonal analysis of genetic mosaics in the retina generated by gamma irradiation (Streisinger 1984; Streisinger et al. 1989). The overall conclusion was that a single tissue, the pigmented retina (PRE), was formed from descendents of almost every blastomere present at the 32-cell stage. [Another application of radiation-induced mosaicism to the development of muscle (Felsenfeld et al. 1990), will be discussed in Sect. 5.4.] The retina experiments depend on the phenotype of mutants of the *gol-1* ("golden-1," see Table 1) gene. Homozygous *gol-1* fish lack pigmented melanocytes for the first 76 h of development, whereas the wild-type homozygotes and *gol-1/+* heterozygotes become pigmented at 48 h (Streisinger et al. 1981). The retina epithelium is an excellent site for mosaic analysis since it is a highly pigmented tissue easy to score for *gol-1* phenotype in the 2-3-day-old embryo. The strategy of irradiation-induced mosaicism, used routinely for cell lineage analysis in *Drosophila* (reviewed in Wieschaus 1978), was applied to the zebrafish embryo. Specifically, *gol-1/+* heterozygotes were irradiated with gamma rays at various times after fertilization (Streisinger et al. 1989). Mosaic eyes could be identified at 76 h, with the frequency of such eyes increasing with the dose of radiation (Streisinger 1984). At a dose which produces 10% mosaic eyes, the vast majority of unpigmented patches should be clonal. Mosaicism might result from somatic crossing over, loss of the chromosome containing the wild-type gene, or a new mutation (Streisinger 1984); no information is available on which of these events actually takes place, although by analogy to other systems, somatic recombination is by far the most likely explanation. A supposition for this work is that golden and non-pigmented cells should not divide at different rates and should both contribute in the same way to the development of the pigmented retina. This was by and large borne out (Streisinger et al. 1989) since the unirradiated and irradiated embryo retinas (wild-type or golden) at 76 h had about the same number (550) of cells (except for embryos irradiated at the two-cell stage, which had slightly fewer cells).

The number of ancestral PRE cells can be calculated simply by expressing the inverse of the mean fraction of golden cells in the retina, and the results indicate that each cell of the first eight (and probably first 32) blastomeres give rise to retinal precursors (Streisinger et al. 1989). A descendent of a single 2-, 4-, or 8-cell blastomere cannot make up the whole PRE since no retina was found to be entirely golden when the embryo had been irradiated at these early stages (golden cells composed 27, 13, and 7% of the PRE cells in these cases). Irradiation of embryos at the 64-cell stage or later resulted in a smaller percentage (2.4%) of unpigmented PRE cells in each retina and the overall percentage of golden cells decreased only moderately in embryos irradiated at later stages. Therefore, the number of ancestral cells present from 64 cells to over 3000 cells probably does not change. The overall estimate is that 38–76 ancestral cells are present during this entire time period. After the 5-h 3000-cell stage, it is not known whether or not these cells divide, so the actual retinal primordium could be composed of more than 38–76 cells. At the 1000-cell stage, the average contribution of an ancestral cell is 7 cells, and only rarely does a precursor give rise to as many as 30 cells. The conclusion is that most of the ancestral cells do contribute a small number of cells to the PRE. In some cases, only a single golden cell was seen in the PRE – some of the ancestral cells may

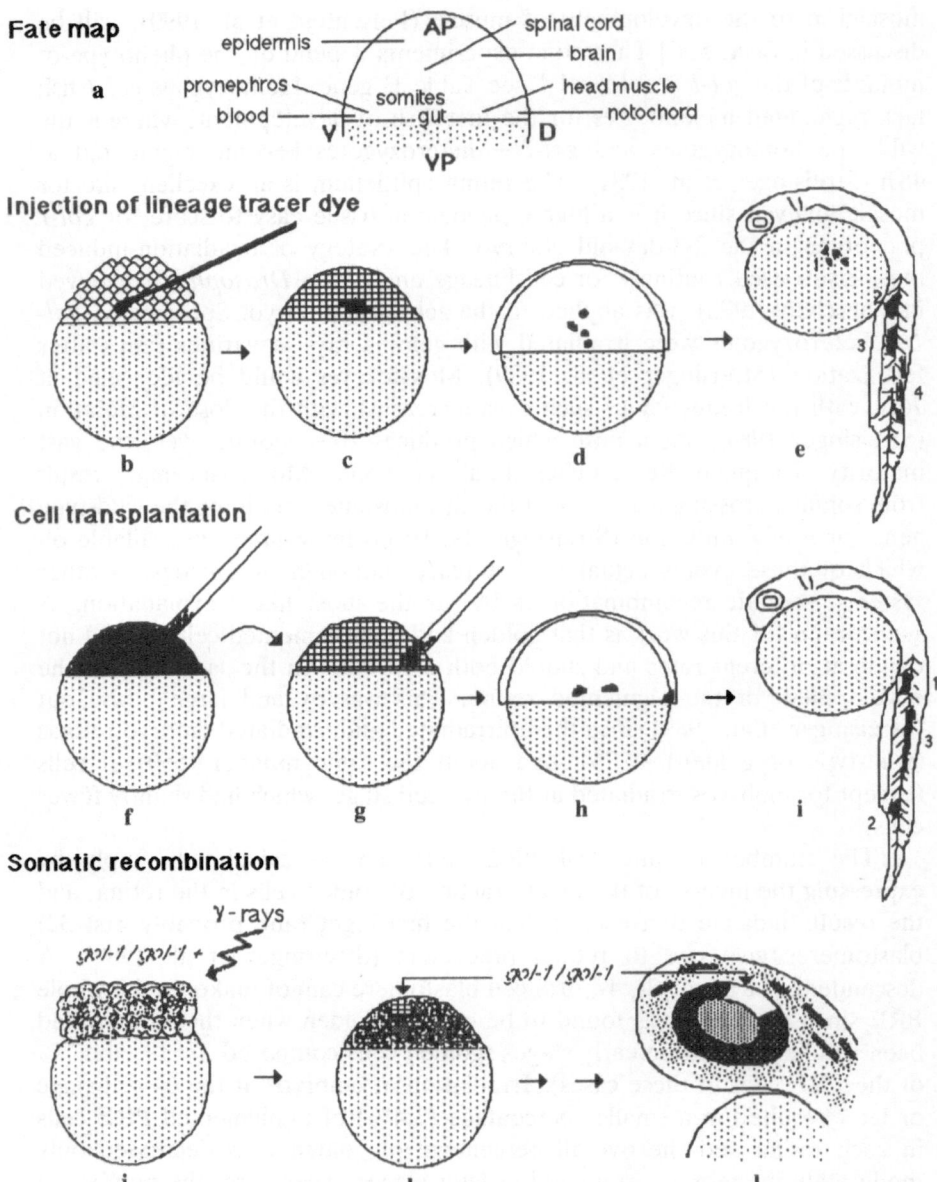

Fig. 2. Analysis of lineage in the zebrafish. **a** The fate map of the 50% epiboly stage (after Kimmel et al. 1990). **b-e** Scheme of a typical lineage tracer injection experiment: **b** a fluorescent dye is injeced into a single blastomere in the early blastula (the dye is indicated in *black);* **c** a patch of cells can be seen in the 1000–2000 cell embryo; **d** in the 50% epiboly embryo, four distinct clones of cells can be seen; **e** in the 24-h embryo, these clones have given rise to periderm (*1*), somitic mesoderm (*2* and *3*), and neural ectoderm (*4*). The clones of cells can be visually followed during the period between **d** and **e**, allowing an assignment of fate to various latitudes and longitudes on the cup of cells in the 50% epiboly embryo. **f-g** Scheme of a cell transplantation experiment: **f** cells are removed with a pipette from

have a reduced, inherent capacity to replicate, or some irradiated cells may have a lower replication rate due to the treatment. At later stages, Streisinger et al. (1989) found no evidence for a pool of cells that contributes to the PRE of both eyes; at earlier stages, when most, if not all, cells are ancestral cells, it was not surprising that there is such a pool.

These results with irradiated late cleavage and blastula embryos are somewhat surprising in the light of lineage tracer experiments which showed that there are no tissue-restricted lineages up until the late blastula (Kimmel and Warga 1986, 1987a; see below). In contrast, from the mosaic analysis of the retina, a restricted lineage would be expected starting at the 64-cell stage, since from this point on, not every cell contributes to the retina. One explanation offered by Streisinger et al. (1989) is that although only about 40 cells contribute descendents to the PRE, perhaps all cells retain the potential to become PRE cells. Certain cells might be excluded from contributing to the PRE precursor line merely because of location in the embryo. A second possibility discussed is that the 40 precursor cells might not proliferate in the blastula, and only later divide to give rise to the actual PRE founder cells. The supposition here is that such non-proliferating blastula cells might have been underrepresented in experiments in which Kimmel and Warga injected tracer into 1000-cell blastulae (see Sect. 3.2), and would not have been scored. A third possibility, not discussed by Streisinger et al. (1989), is that all cells in the blastula could initially have the potential to form retinal precursors, but that at a particular point there is a stochastic commitment accompanied by negative feedback on the non-committed cells.

3.2 Lineage Analysis by Injection of Tracer Dyes

Injection of dyes into embryonic cells was first used to trace blastomere lineages in the leech embryo (Weisblat et al. 1978, 1980). The Kimmel group applied this method to the zebrafish embryo by injecting fluorescent

---

the animal pole region of a donor 1000-cell embryo, marked earlier by injection of a tracer dye at the one-cell stage (cells are *black*), and transferred (**g**) to the marginal region of a recipient embryo at the same stage; later, at the 50% epiboly stage (**h**), the transplanted cells can be seen as three patches near the margin; in this hypothetical experiment, these patches give rise in the 24-h embryo (**i**) to cells in the somitic mesoderm (*1* and *2*) and notochord (*3*). **j-l** An example of the application of somatic recombination, in this case using heterozygous *gol-1/gol-1* + embryos (Streisinger et al. 1989): **j** γ-rays are used to induce somatic recombination in cleavage stage embryos, resulting in some *gol-1/gol-1* cells; **k** the homozygous *gol-1* cells give rise to clones of cells in the late blastula (the clone is *white* and the rest of the embryo is *shaded* – the phenotypes of these cells cannot be distinguished at this stage, however); **l** in the 48-h embryo, unpigmented sectors can be seen in the otherwise pigmented retina – these patches are derived from the *gol-1* clones (the rest of the embryo is mostly unpigmented, although shown here in *gray* to clearly distinguish the unpigmented sector in the retina)

Eric S. Weinberg

Table 1. Mutants of developmental interest

Gene[a]	Common name	Mutant alleles	Map distance to centomere (cm)	Embryonic phenotype	Earliest time of expression	Cell autonomy	References
gol-1[b]	Golden-1	b1, b7	44.5–47.5	At 2 days: no body melanophores, no pigment in PRE	Distinguished from wild type at 2 days	Autonomous in PRE[c]	Walker and Streisinger (1983); Streisinger et al. (1986, 1989)
gol-2[b]	Golden-2	b2, b3	28.5	At 2 days: large body melanophores (light-brown); PRE is brown	Distinguished from wild type at 2 days	NR	Streisinger et al. (1986)
alb-1[b]	Albino	b4	18	At 2 days: no body melanophores, no pigment in PRE	Distinguished from wild type at 2 days	Autonomous in PRE[d]	Streisinger et al. (1986); Lin et al. (1991)
spa-1[b]	Sparse	b5	12	At 2 days; fewer body melanophores than wild type	Difficult to distinguish from wild type	NR	Streisinger et al. (1986)
spt-1	Spadetail	b104, b141	NR	Trunk somites do not form, other somites normal; hooked tail; lethal at 7 days	Gastrulation: incorrect cell movements	Auton. for trunk mesoderm[e] non-auton. for neural defects	Kimmel et al. (1989); Ho and Kane (1990); Kimmel et al. (1991a)
cyc-1	Cyclops	b16	28	FP does not form; cyclopia; curving of body axis; lethal within 4 days	Flattening of midline at 9–10h	Autonomous for FP specification, not differentiaton[e]	Hatta et al. (1991b)
ned-1	Neural degeneration	b39	~0	Massive cell degeneration in CNS; some CNS cells normal; lethal at day 6	Degenerating CNS cells noted at 24h	NR	Grunwald et al. (1988)

fub-1	Unbundled fibers	b45, b126	24, 29	Myofibrillar filament bundles disordered; lack of muscle movements	Lack of ordered fibrils in MP cells – 13–15 h	Autonomous within individual myotubes[c]	Felsenfeld et al. (1990, 1991)
nic-1	Nicotinic acetylcholine receptor	b107	17.9	Lacks functional acetylcholine receptor; trunk muscles paralyzed, heart beats normally	Spontaneous body contractions not seen at 18 h	NR	Westerfield et al. (1990)
ntl-1	No notochord/ no tail	NR	NR	Notochord does not form, somite MP cells absent, no nuclear T-gene protein	NR	NR	M. Halpern, cited in Kimmel et al. (1991); Schulte-Merker, cited in Ruiz i Altaba (1991); M. Halpern and C. Kimmel, personal communic.

[a] Zebrafish genes have been generally named with a three-letter apellation based on phenotype, followed by a number, according to the conventions used for C. elegans (Horvitz et al. 1979). Alleles of the gene are designated in brackets by a b followed by a number. In this convention gol-1[b1] and gol-1[b7] are alleles of the same gene, and gol-2[b2] is an allele of a different gene, with a similar phenotype. However, at a March 1992 meeting held in Ringberg (Tegernsee), Germany, the consensus among zebrafish researchers was to use only a three letter name, and to drop the use of the number which followed. Except for the case of gol-2, a new name for the gene will have to be found. gol-1 will probably be named gol. Alleles will still be denoted by numbers within brackets, following the three letter gene name.
[b] Information adapted from Table 1 in Streisinger et al. (1986).
[c] Demonstrated by gamma ray-induced mosaicism.
[d] Demonstrated by formation of chimeras.
[e] Demonstrated by cell transplantation.
Abbreviations: NR = not reported; auton. = autonomous; PRE = pigmented retina; FP = floor plate; MP = muscle pioneer cells; CNS = central nervous system.

conjugated dextran into blastomeres at the cleavage or blastula stages, and then following the fate of these cells during later embryogenesis. Embryos so injected can be repeatedly viewed by using a silicon-intensified target videocamera. Cells injected at the 32- or 64-cell stage gave rise to labeled clones which were coherent in the blastula, but which dispersed and intermixed during epiboly and gastrulation (Kimmel and Law 1985c; Kimmel and Warga 1986). Kimmel and Warga (1987a) followed the fate of a single particular blastomere ("L1211") injected at the 64-cell stage in ten different embryos. At 24–30 h each embryo had a unique distribution of labeled tissues and a scattered spatial pattern. The results showed that the descendents of one particular blastomere from the 64-cell stage embryo could form virtually any embryonic tissue in any part of the embryo, and moreover, that exactly which cell types were derived from this blastomere varied greatly from embryo to embryo. Generally, the probability of a particular tissue containing labeled cells was proportional to the number of cells in that tissue at 24–30 h of development. One reason for the variability in position of the derived cells in the later embryo is that the cleavage planes of the blastomeres and their derivatives are quite variable from the 7th cleavage on, ruling out a strictly defined relationship between the cell lineage and the eventual position of the cells in the blastula and in later stages (Kimmel and Warga 1987). Another reason for positional variation in later stages is that there is no fixed relationship between the early cleavage planes and dorsoventral polarity of the later embryo (Kimmel and Law 1985c; Kimmel and Warga 1987a).

When cells were injected at the 1000-cell blastula (the advantage here is that smaller clones are produced which are less scattered) and the progeny cells followed, a general correlation was observed between the position of the injected blastomere in the blastula, and the position of the labeled descendent cells in the gastrula (Kimmel and Warga 1987a). Clones derived from a cell at the blastula animal pole remained at the animal pole in the 50% epiboly early gastrula, and eventually, descendents wound up in the head region of the 24–30 h embryo. Clones at the blastula margin tended to be located at or near the germ ring of the 50% epiboly embryo, and then were often dispersed along the longitudinal axis of the embryo at later stages. Those clones most distant from the embryonic shield (dorsal side) appeared to move the furthest, winding up in the tail region. In these clones, cells originally at the ventral side of the 50% epiboly embryo migrated all the way to the dorsal midline. In all of these cases, independent of the final position of the cells, the clone developed into a variety of tissue types which included EVL cells as well as other embryonic tissues. EVL cells, however, did not participate in convergence movements, tending to stay at the same relative dorsoventral position in the embryo as at the time of formation of the embryonic shield at the 50% epiboly stage. The major difference between cells marked at the 64-cell stage and cells marked at the 1000-cell blastula was that the eventual position in the embryo (e.g., tail or head) of the clonal descendents was much more predictable in the latter case. In both series of experiments, however, a variety of tissue types was produced from a single clone, indicating that even at the 1000-cell stage, cell lineage was unfixed.

The one exception to the lack of lineage restriction during cleavage and blastula stages demonstrated by lineage tracer experiments is the formation of the YSL. A subset of the original 20 marginal cells of the 64-cell embryo was shown to give rise to the YSL three or four division cycles later (Kimmel and Law 1985b). The lineage could be quite variable: observation of embryos with Nomarski optics revealed that a particular blastomere at the 64-cell stage sometimes gave rise to YSL nuclei, in other embryos it gave rise only to blastomeres, and in other cases it formed both YSL nuclei and blastomeres. The restriction found was that YSL nuclei are never derived from the four internal non-marginal blastomeres of the 16-cell embryo. This may merely be due to the fact that the descendents of these four blastomeres give rise only to the non-marginal cells, which unlike the marginal cells at this stage, do form membrane between the cell cytoplasm and the yolk cell. The descendents of these cells are also probably non-syncitial, and therefore would not be able to contribute nuclei to the YSL by a collapse of cell contents into the yolk cell.

In contrast to the indeterminate lineage in the blastula, cells in the 5.2 h early gastrula have sharply restricted lineages (Kimmel and Warga 1986; Kimmel et al. 1990). In other words, a fate map can be first drawn at the early gastrula stage (Fig. 2). EVL or DEL cells injected at the 1000-cell stage yield clones that give rise to several tissues at separate locations in the 24-h embryo (Kimmel and Warga 1986; Kimmel et al. 1990). Dispersed cells are typical of the history of clones derived from the blastomeres of the 1000-cell embryo. Small clones of cells at gastrula (derived from the injected blastomeres), however, were shown to give rise to cells of a single tissue type. EVL cells at gastrula gave rise exclusively to periderm in the 24–30 h embryo. In the first study of gastrula clones, DEL cells gave rise to either neural cells, muscle, kidney, or epidermis (Kimmel and Warga 1986). In only five cases (out of 373 clones which were followed) did descendents of a gastrula cell populate more than one tissue type, and in two of these cases, the lineage suggested that the cells arose from the neural crest. With rare exceptions, therefore, the lineage from gastrula on was fixed to a particular tissue type. The clonally related cells, however, need not wind up in the same region of the embryo (Kimmel and Warga 1986, 1987b). This was clearly shown for neuroepithelial cells extending along the anterior-posterior axis, neurons in different hindbrain segments, and muscle cells in different somites. Sibling hindbrain neurons (Kimmel and Warga 1986), muscle cells (Kimmel and Warga 1987b), or sibling cells can even cross the midline to reside in the left and right sides of the embryo. Kimmel and Warga (1986) postulate that the first cell determination may be that of tissue-specific lineage, at the time of early gastrula. Subsequent divisions produce cells that will populate different segments or regions of the embryo. Lastly, for the central nervous system (CNS) and somites, divisions sometimes occur which will partition cells to the right and left sides of the midline (restricted to the tail segments for myotomes).

108 Eric S. Weinberg

In a more extensive study of the timing of lineage fixation, Kimmel et al. (1990) demonstrated that the lineage restriction of EVL cells to periderm precedes, by approximately 1 h, the restriction of clones of DEL cells to particular embryonic tissues. Prior to 3 h, a particular EVL cell can be shown to give rise to both EVL and DEL cells (Kimmel and Law 1985c; Kimmel and Warga 1987; Kimmel et al. 1990), but DEL cells do not become EVL cells (Kimmel et al. 1990; Warga and Kimmel 1990). Between 3 and 4 h, an increasing number of EVL cells give rise only to EVL cells, and eventually periderm, and after 4 h, lineage restriction is complete. The lineage restrictions for a DEL cell to form one particular tissue occur more gradually, between 3 and 5 h. The type of tissue to which a DEL cell will be restricted cannot be predicted by its depth in the blastoderm. There is a strong relationship between latitude of a DEL cell at the 50% epiboly stage, and the eventual germ layer to which it will contribute. Ectoderm is derived from most of the blastoderm, but not from cells near the margin. Endoderm is derived from cells at, or very near, the margin; mesoderm from intermediate positions. At the 50% epiboly stage, the derived fate map for the ectoderm shows that the animal pole cells become brain, if from the dorsal side, and epidermis, if from the ventral side. The forebrain and nasal and eye tissues come from the animal pole region, the midbrain and hindbrain come from cells located more toward the margin. Some cells forming the spinal cord and neural crest come from the ventral side. For mesodermal tissues, head mesoderm maps to the dorsal side of the ring of cells which will become mesoderm ("cirumferential ring"); myotomal muscles of trunk and tail map to intermediate and ventral locations, notochord and hatching gland (a tissue of as yet undefined embryonic layer) to the dorsal region, blood to the ventral region, and pronephros to an intermediate region. Boundaries between these regions overlap to some extent, but this might reflect the imprecision of following small clones of cells rather than single cells in the gastrula (the clones were labeled by injection of cells at midblastula). Lineage tracing can be used to determine the origins of very specific tissues. For example, particular jaw muscles (see Sect. 6.2) were found to be derived from paraxial mesoderm, and not neural crest, by following the derivatives of marginal cells labeled with dye at the 1000-cell stage (Hatta et al. 1990).

An interesting point raised by Kimmel et al. (1990) is whether lineage restrictions to a single germ layer occur prior to restrictions to a tissue type. The answer appears to be negative since in cases of cells which contributed descendents to more than one tissue, about as many cases were found in which two derived tissues were from different germ layers (e.g., nerve and muscle) as from the same germ layer (e.g., epidermis and nerve). The lineage hierarchies which might occur after tissue specification, suggested earlier (to particular embryonic regions or segments, and then to embryonic sides; Kimmel and Warga 1986), need to be investigated more extensively. To summarize the major points: embryonic layer restrictions of DEL clones map roughly to latitude with respect to the animal pole; particular tissue restrictions map with respect to the eventual dorsal/ventral orientation of the embryo (a polarity which is only obvious on formation of the embryonic shield). The fate map, therefore, can be constructed before the first visible signs of dorsal/ventral polarity, and just prior to the onset of involution.

What is the relationship of the fate map to the involution and convergence movements? Lineages during gastrulation and later stages were determined for cases in which involution of the cells was actually observed (Warga and Kimmel 1990). Without exception, involuting cells formed mesoderm and endoderm, and cells which did not involute formed ectoderm. The epiblast is thus the precursor of the ectoderm, and the hypoblast is the precursor of the mesoderm and the endoderm. From the results of Kimmel et al. (1990), and the lineage results of Warga and Kimmel (1990), we can infer that the epiblast is derived from the upper two-thirds of the hemisphere of cells present at the 50% blastula stage, and the hypoblast consisting of involuted cells is derived from the cells at the margin and the lower third of the hemisphere. In almost every case observed (Warga and Kimmel 1990), the earliest cells to undergo involution became endoderm, and later involuting cells would up as mesoderm. Within a germ layer, the earlier involuting cells returned furthest toward the animal pole and eventually populated the anteriormost embryo structures. The last cells to involute generally appeared in the posterior trunk and tail regions. Since convergence movements occur in both the hypoblast and epiblast, but mediolateral intercalations at the dorsal midline occur separately within each of these layers, there is a spreading apart of the clones along the anterior-posterior embryonic axis, but no mixing of hypoblast and epiblast cells.

The fate map of the zebrafish (Kimmel et al. 1990) resembles that constructed for several other teleost species (Ballard 1973, 1981, 1982, 1986), but the zebrafish map has fewer overlap regions. The major difference is that the zebrafish map has head mesoderm deriving from the dorsal blastoderm margin; Ballard's maps show that the head mesoderm is derived from deep regions at the animal pole (this region produces ectoderm in the zebrafish). As pointed out by Kimmel et al. (1990) and Kimmel (1989), the zebrafish fate map resembles strikingly that of *Xenopus* (Dale and Slack 1987a) or ascidians (Nishida 1987), but instead of a vegetal pole, there is a vegetal margin, and the yolk cell seems to be inserted into the vegetal pole, compressing the mass of cells into an overlying cup. A fate map can be made much earlier for *Xenopus* (Dale and Slack 1987b; Moody 1987) than for the zebrafish, and this is undoubtedly because there is only limited mixing of cells during cleavage and blastula stages in the amphibian species. In the lack of fixed lineage in the blastula and cleavage stages, the zebrafish (and presumably other teleost) embryos most resemble mammalian embryos (Rossant 1985; Beddington 1986; Lawson et al. 1986; Sanes et al. 1986; Soriano and Jaenisch 1986). Genetically marked single cells (marked at the 8-64-cell stage) will contribute to most somatic tissues (Gardner et al. 1985; Soriano and Jaenisch 1986), and blastomeres intermix during embryogenesis (McLaren 1976; Kelly 1977; Rossant 1985).

An important consideration is whether the coherent regions identified with particular fates at the 50% epiboly stage are the result of cell determination (i.e., autonomous restriction of potential), or merely derive from the position of these cells in the embryo. Experiments which perturb the system such as cell transplantation (see the following section), cell ablation, or genetic manipulation would shed light on this issue (as pointed out in Marcey and Nüsslein-Volhard 1986). Kimmel et al. (1990) have noted that by 5.2 h, the time in which the fate map can be constructed, the most extensive radial intercalations which mix blastomeres during epiboly have been completed. Further cell movements (continuation of epiboly, involution, and convergence), although dramatic, do not destroy the coherence of many of the regions identified on the fate map at 5.2 h. The lineage that can be worked out, therefore, conceivably could be merely a result of the position of clones in the cell hemisphere in the early gastrula, and not the result of cell determination or commitment at this developmental time.

3.3 Cell Transplantation Experiments

Cell transplantations can be performed to distinguish between cell commitment and determination, on the one hand, and a more passive fixation of cell lineage such as location of particular cells with respect to a morphogenetic signal, on the other. Transplantations of gastrula cells in *Salmo* (Luther 1936) and *Fundulus* (Oppenheimer 1936, 1938) indicate that in these species, gastrula cells are not necessarily committed to a specific cell fate. Cell transplantations have recently been performed in the zebrafish with cells derived from late blastula/early gastrula (Ho and Kane 1990; Hatta et al. 1991b) and midblastula (Lin et al. 1992) embryos. Experiments involving transplants of cells from embryos at least 5.2 h of age would be most relevant for the present discussion, since many cells from younger embryos would not be expected to be committed from the lineage tracer experiments. The use of transplantation to test for cell autonomous action of particular genes and to produce chimeric fish which can genetically transmit the genotype of the transplanted cells will be covered in Section 4.5. At this point, the discussion will be restricted to tests of cell determination.

Experiments of Ho and Kane (1990) and Hatta et al. (1991b) were designed to test the autonomy of cells homozygous for the *spt-1* and *cyc-1* mutations (see Sects. 5.1 and 5.2), respectively. In the Ho and Kane experiments the transplantation of wild-type cells as controls was also illuminating for tests of cell commitment. Wild-type cells from the mid anterior-posterior position in the blastoderm of 5-h embryos (but selected randomly with respect to the dorsal-ventral axis, since the earliest marker of dorsal-ventral polarity, the embryonic shield, had not yet formed) were placed in the lateral margin, the dorsal marginal zone, or the non-marginal region of host embryos. The donor cells could be distinguished from the host cells because they had been marked with fluorescent dye. If they had been committed to a cell fate at 5 h, the donor cells, derived from a position in the donor embryo in most cases coincident with the somitic muscle area of the fate map, should have given rise mostly to somitic muscle, irrespective of their position in the host embryo. When their position in the host was at the lateral margin, the donor cells did form mainly somitic muscle, but when placed in the dorsal margin or the lateral non-marginal zone, the transplanted cells would up in the notochord or in the spinal cord, respectively. This indicates that at 5 h, at least, these cells were not committed to a somitic muscle fate. Unfortunately, the description of the donor sites of the cells was scanty, and the timing of the transfer was a bit earlier than the time at which the fate map was determined (Kimmel et al. 1990). Nevertheless, these results point to the regulative nature of the cells which would normally give rise to somitic muscle in the unperturbed embryo. Using a similar approach, Hatta et al. (1991b) transplanted cells from within or near the embryonic shield of the early gastrula to similar positions in host embryos. The wild-type donor cells formed floor plate, notochord, and other tissues, indicating that the system has potential for analyzing the commitment of the transferred cells. No information, however, was presented on transfers of cells to other regions in the host embryo. Although definitive experiments on the extent of cell commitment in the zebrafish gastrula have yet to be done, the experiments described here indicate that the methodology is now in place for a detailed analysis of the problem.

4 Genetic Manipulations

4.1 Obtaining Haploid Embryos and Fry:
An Advantage for Mutant Screens

Fertilization of eggs (obtained by squeezing anesthetized female fish) with UV light-treated sperm activates the egg, but the paternal genome is genetically destroyed (Streisinger et al. 1981). The resulting egg develops into a haploid embryo and then a fry, eventually dying several days later (survivors, with occur 0.03% of the time were shown to be diploid and probably arose through non-disjunction). The contribution of the UV-treated sperm to the embryo was assayed by using the recessive *gol-1* mutation. As mentioned above, homozygous *gol-1* fish lack pigmented melanocytes for the first 76 h of development, whereas the wild-type and *gol-1/gol-1* + heterozygotes become pigmented at 48 h. Streisinger et al. (1981) used wild-type (*gol-1* +/*gol-1* +) males as a source for the UV-treated sperm and *gol-1/gol-1* females for the eggs. No pigment was observed at times when *gol-1* heterozygotes would normally make pigment; thus, the UV treatment is

112   Eric S. Weinberg

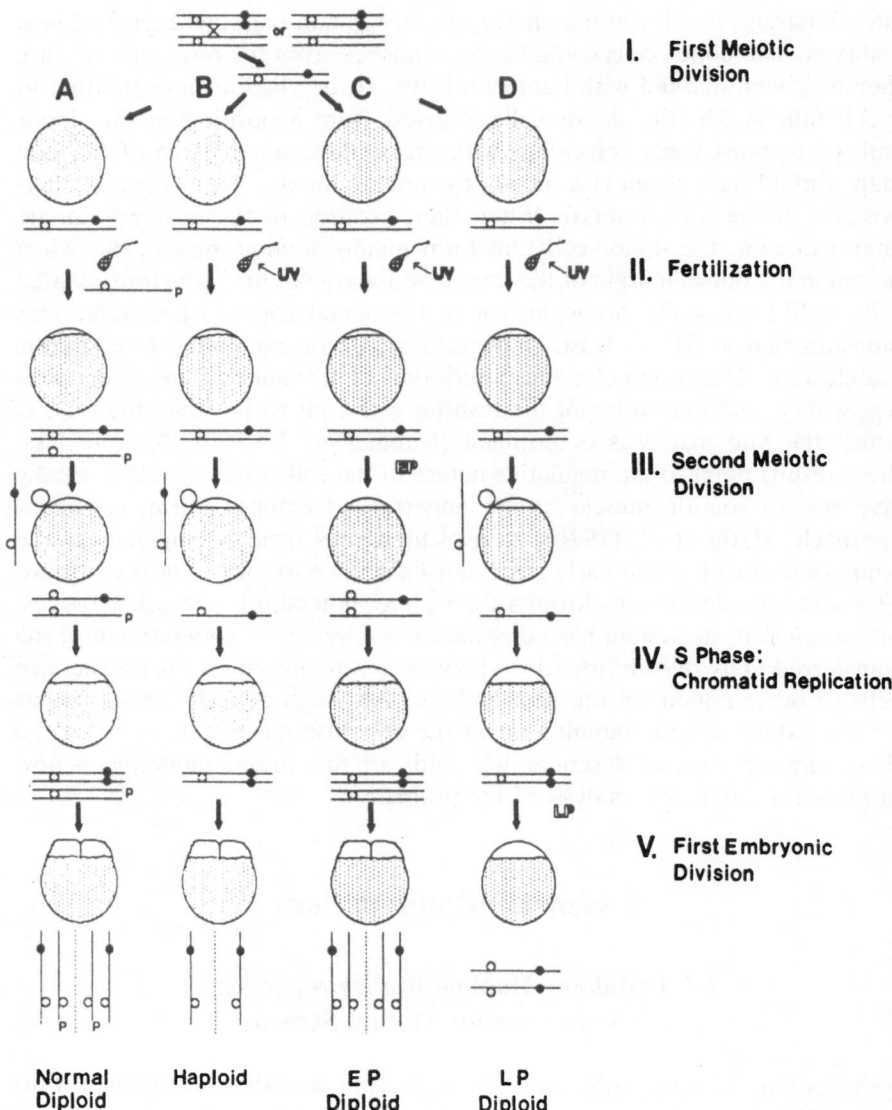

Fig. 3. Obtaining haploid and parthenogenetic diploid embryos. The *four columns* represent different modes of producing embryos: **A** normal diploid embryos resulting from fertilization of eggs with untreated sperm; **B** haploid embryos produced by fertilization of eggs with UV-irradiated sperm; **C** parthenogenetic diploid embryos resulting from fertilization of eggs with UV-irradiated sperm and pressure treatment during the first few minutes after fertilization; **D** parthenogenetic diploid embryos produced by fertilization of eggs with UV-irradiated sperm and pressure treatment during the second half of the one-cell stage. Chromosomes are marked with a *black dot* to indicate a mutant allele of a gene, some distance to the right of the centromere. During the meiotic prophase, crossing over may occur between the gene and the centromere. Only the chromosomes derived from the recombinant half-tetrads are diagrammed below each egg or embryo. The non-recombinant tetrads are

effective in eliminating the male genetic contribution in these experiments. Once an interesting phenotype is recognized in the haploid embryos of a particular cross, the mutagenized parent can be set aside and bred to recover the mutant. The ability to produce haploid embryos provides a powerful tool for mutant screens. However, it is sometimes difficult to distinguish abnormal embryonic development from the background of variations in growth which appears in activated haploid embryos. The more conventional method of inbreeding the progeny of mutagenized fish to obtain homozygotes eliminates this problem and bypasses the somewhat time-consuming procedure of squeezing to obtain eggs. However, if mutants are recognized in a haploid screen, the 3-month wait for the inbred generation can be avoided. The production of haploid embryos and other genetic manipulations is shown in Fig. 3.

4.2 Production of Clonal Strains Using "Late Pressure" and Heat Shock

Clonal strains of the zebrafish can be readily produced (Streisinger et al. 1981). Two methods, hydrostatic pressure and heat shock, were used to produce viable homozygous fish from these haploid zygotes. In the hydrostatic pressure experiments, the activated eggs were subjected to 8000 p.s.i. for 7 min in a French press beginning at 22 min after fertilization ("late pressure"). This treatment prevents the first embryonic cell division and results in the diploidization of the haploid egg by retention of both chromatids of each chromosome. In contrast to "early pressure" treatment, which prevents the second meiotic division (see below), the diploid zygotes formed after late pressure treatment should be homozygous (see Fig. 3). The initial experiment resulted in the normal appearance of 29% of the embryos, and 20% of these survived to adults. These adults were crossed with each other

---

present in a percentage of the eggs (the percentage is a function of the distance of the locus from the centromere), but are not diagrammed. At meiosis I (*I*), half-tetrads segregate and eggs shed from the female are arrested at this stage. Eggs are fertilized (*II*) with untreated (**A**) or UV-irradiated (**B, C, D**) wild-type sperm. The paternal chromosome is denoted with a *P* at one telomere. UV irradiation results in destruction of functional genes; there is no chromosomal contribution to the embryo. Fertilization results in the activation of meiosis II (*III*), resulting in the partition of the chromatids of the maternal chromosomes and the extrusion of the second polar body. In the early pressure (*EP*) embryos (column **C**), this division is prevented and the daughter chromatids both remain in the zygote. Note that in this case, the two chromatids may be heterozygous for the locus in question (some zygotes containing non-recombinant half-tetrads – not diagrammed – will be homozygous for either the wild-type or mutant allele). The chromosomes then replicate (*IV*) so that each chromosome contains two identical chromatids at the end of S-phase. The first embryonic division then occurs (*V*), resulting in a two-cell embryo. However, if pressure is applied during the latter part of the first cell cycle (*LP*, column **D**), cell division is prevented and the chromatids will not be partitioned. In contrast to *EP* treatment, *LP* results in obligate homozygosity. (After Kimmel 1989)

and the progeny were used to obtain eggs for another round of activation with UV light-treated sperm and hydrostatic pressure. Of these diploidized eggs, 20% developed into normal-looking 24-h embryos and 68% of these survived to adulthood. Pairs of these adults were once again crossed and their progeny again used to obtain eggs for late pressure treatment. The success rate in this third round was 38% normal 24-h embryos, 69% of which survived to adults. Clones derived from these fish (e.g., C32, C29) are now used as "wild-type" strains for mutant screens and breeding in several laboratories. These clonal strains, however, do not seem as hardy as uncloned lines. Streisinger et al. (1981) suggested crossing two clonal lines to obtain a hardier strain. The clonal lines also often have strange sex ratios, with an overwhelming predominance of either males or females. Females can be converted to males by treating embryos with 17-α-methyl-testosterone for 2 weeks starting at 1-day postfertilization (Yamamoto 1969; Streisinger et al. 1981). Despite the difficulty of working with clonal lines, they are useful, perhaps essential, to provide a stable and constant background for mutagenesis studies.

Late pressure treatment was originally shown to produce homozygous embryos by assaying for isozymes of esterase (Streisinger et al. 1981). Three alleles of the *est-3* gene were identified which encoded electrophoretically distinguishable forms of the enzyme. Eggs from females heterozygous for the locus were late pressure-treated and shown always to be homozygous. This test ruled out a contribution of a chromosome from non-disjunction at the first meiotic division, since the resulting embryo would have been heterozygous for the esterase.

The second method used by Streisinger et al. (1981) to generate homozygous diploids was a heat shock treatment of the haploid zygotes (41.4 °C for 2 min, applied 13 min after fertilization). The effectiveness of heat shock was somewhat better than the pressure method in these original experiments – 25–33% appeared normal at 24 h and 38–61% of these survived to adulthood. Subsequent use of heat shock for making homozygous embryos resulted in initial survival of only 19–23% (Grunwald and Streisinger 1992a,b).

4.3 "Early Pressure" Treatment and Mapping Genes

If eggs are subjected to hydrostatic pressure for the period between 1.5 and 6 min ("early pressure") after activation with UV-irradiated sperm, the second meiotic division is blocked (Streisinger 1981). The spawned zebrafish egg is blocked at meiosis II, and the extrusion of the second polar body normally occurs only after fertilization. Early pressure treatment results in the failure to partition the sister chromatids of each chromosome at this time, yielding a diploid chromosome number. As shown in Fig. 3, if a crossover event had occurred during oogenesis between a particular locus and the centromere in the tetrad, the half tetrad in the shed egg would be heterozygous. The early pressure-treated embryos are thus heterozygous for many loci, the frequency of heterozygosity for a particular locus increasing with the distance of the locus from the centromere. On the other hand, late pressure treatment results in homozygosity because the sister chromatids of the half tetrad are already partitioned (into the egg and the polar body) and the diploidization results from lack of partition of the daughters of single chromatids after a round of replication. Eggs derived from *est-3* heterozygote mothers were used to test whether early pressure treatment was effective for mapping the distance from the centromere (Streisinger et al. 1981). Of the activated embryos surviving early pressure treatment, 14% were heterozygous for the esterase enzymes (as mentioned above, late pressure treatment yielded no heterozygous embryos). The *est-3* locus, therefore, is located 7 (uncorrected) map units from the centromere.

Streisinger et al. (1986) applied this method to map four pigment pattern mutants, *gol-1* ("golden-1"), *gol-2* ("golden-2"), *alb-1* ("albino"), and *spa-1* ("sparse") (see footnote to Table 1 explaining the system used to name mutants and gene loci). These mutants exhibit distinct phenotypes (see Table 1), and map to distinct loci which show no genetic linkage. For each locus, eggs from heterozygous females (e.g., *gol-1/gol-1* +) were activated with UV-irradiated sperm and subjected to early pressure. Since heterozygotes and wild-type homozygotes cannot be distinguished, the frequency of half tetrads carrying chromatids recombinant between the locus and the centromere $(1-2 m)$ could be established from the frequency of embryos with mutant phenotype (m). The values obtained for $1-2 m$ were: *gol-1*, 0.89–0.95; *gol-2*, 0.57; *alb-1*, 0.36; and *spa-1*, 0.24. The high value for *gol-1* is interesting in that it is greater than the theoretical maximum of 0.67 and must, therefore, indicate a high level of chiasma interference (Perkins 1955) (it will be interesting to learn whether such interference is a general property of zebrafish meiosis, or whether it is restricted to the *gol-1* chromosome). If there is complete interference, corrections need not be made for additional crossovers, and the recombination frequency is just half the value of the fraction of half tetrads carrying a recombinant chromatid, $(1-2 m)/2$. Streisinger et al. (1986) also calculated gene-centromere distances assuming no interference at all, using the Haldane equation (Haldane 1919). The values for *gol-1* for the two extremes were 44.5 (complete interference) and 110 (no interference). In contrast, a much tighter linkage was found between the *spa-1* locus and the centromere, with distances of 12 (complete interference) and 14 (no interference). In later work, at least one locus (*ned-1*) has been positioned very close to the centromere, since recombinants could not be detected after early pressure treatment (Grunwald et al. 1988). Segregation analysis was also done by the half tetrad method, crossing parents each homozygous for one of the above-mentioned pigment mutations (Streisinger 1984). The results are exactly what one would expect for unlinked genes (However, ratios in this analysis are far from the 1:1:1:1 distribution because of the extensive distances of the loci from the centromeres).

4.4 Mutagenesis

Mutagenesis of the germ line has been induced in the zebrafish by gamma rays (Chakrabarti et al. 1983; Walker and Streisinger 1983), ethylnitrosourea (ENU) (Grunwald and Streisinger 1992b), and ultraviolet light (Grunwald and Streisinger 1992a). As will be discussed below, gamma irradiation probably results in large deficiencies or rearrangements, whereas ENU causes base substitutions (Richardson et al. 1987; Pastnik et al. 1989) and ultraviolet light mostly induces point mutations and frameshifts (Drake 1969). The latter two approaches might be more useful in the generation of mutations at specific loci.

4.4.1 Gamma Ray Mutagenesis

Gamma ray mutagenesis has been effective in the zebrafish in generating new alleles of previously recognized genes, and creating lesions in new loci. Chakrabarti et al. (1983) and Walker and Streisinger (1983) measured specific locus mutations as well as recessive lethal frequencies. Three specific, unlinked pigment pattern genes were studied: *gol-1*, *gol-2*, and *alb-1*. The frequency of gamma ray-induced mutations was scored at each of these loci by irradiating eggs (immediately after fertilization) from wild-type females fertilized with sperm from homozygous mutant males, and observing pigmentation in the embryo (Chakrabarti et al. 1983). Mutations, identified by a lack of pigmentation at the *gol-1* and *alb-1* loci were induced at the same rate (0.26–2.3% of the embryos, depending on the dose); mutants at the *gol-2* locus were induced at half the rate. In the case of the *gol-1* locus, approximately the same range of mutagenesis was seen when sperm (from wild-type males) was irradiated and used to fertilize eggs from homozygous mutant famales. Mosaic golden phenotypes were observed, at a rate of one-fourth that of the frequency of all golden mutants, after irradiation of either egg or sperm. Gamma irradiation of embryos at the cleavage and blastula stages was also performed, and the efficiency of mutagenesis tested for the *gol-1* locus (Walker and Streisinger 1983). Adults derived from the mutagenized blastulae were crossed with *gol-1/gol-1* fish of the opposite sex and approximately 3.5% of the fish were shown to carry new *gol-1* alleles. Not surprisingly, the germ lines in the adults grown from the mutagenized embryos were mosaic, with transmission frequencies ranging from 3 to 52%. The overall frequency of induced *gol-1* mutations in blastulae (1.2×10^{-5} per roentgen, normalized per haploid chromosome set of the supposed pregonial cells; see below) was very close to the rate produced by irradiating fertilized eggs or sperm (4×10^{-5} per roentgen; Chakrabarti et al. 1983).

Lethal mutants were induced as well. Eggs were irradiated with gamma rays, activated with UV-irradiated sperm, subjected to heat shock to form homozygous diploids, and then scored for survival (Chakrabarti et al. 1983). Lethal mutations were induced at a rate of 6×10^{-3} per roentgen, of which recessive lethals made up two-thirds, a rate only 100-fold over that observed for mutations in a specific locus such as *gol-1*. These rate measurements are, however, somewhat inexact, both because of the high degree of lethality in control, unirradiated but heat-shocked embryos, and because of cellular damage without genetic consequences due to the gamma irradiation. Lethal mutations were also scored in experiments in which blastulae were gamma irradiated (Walker and Streisinger 1983). The rate of 1.2×10^{-3} per roentgen is based on few actual, recovered lethals, but is quite close to the rates determined by irradiation of eggs. Increased gamma irradiation doses appeared to induce maleness in the population (Walker and Streisinger 1983). A surprising result was that some of the males grown from the irradiated blastulae, crossed with the *gol-1* test females, gave rise to mosaic F1 individuals. A possible explanation is that there is delayed fertilization with the *gol-1* sperm from the male, resulting in a mixed haploid/diploid individual (Walker and Streisinger 1983).

The comparison of the rates of mutagenesis for lethal and for site-specific mutants led to the inference that gamma mutagenesis induces large deficiencies or structural alterations (Chakrabarti et al. 1983; Walker and Streisinger 1983). If most of the gamma ray-induced lesions were small, the rate of "lethal" mutations should be much greater than 100-fold that of mutagenesis at the *gol-1* locus. The suppositions were that the *gol-1* locus and the two other specific loci tested are not particularly more mutable than the average locus, and that one would expect far in excess of 100 genes essential for embryonic viability. Chakrabarti et al. (1983) did consider explanations other than the formation of large lesions: (1) that the three specific loci tested are in fact much more mutable than the average locus (not unreasonable since these mutations are among the few spontaneous mutations recognized so far and may have been detected because of high spontaneous rates); (2) that there may be gene redundancy in the zebrafish, due perhaps to tetraploidy or to the presence of most essential genes present in multiple copies; and (3) that the inducibility of repair by the UV-irradiated sperm may reduce the frequency of lethal mutations (lethals are assayed in heat-shocked eggs activated with UV-irradiated sperm, the specific locus mutations are assayed in normally fertilized eggs). Consistent with the idea that gamma rays induce large deficiencies was the observation that the new gamma ray-induced *gol-1* mutations were also recessive lethals (Chakrabarti et al. 1983). Also supporting the idea is the finding that ENU and ultraviolet light-induced *gol-1* mutations and recessive lethals appear in a ratio of 5000 to 1 rather than 100 to 1 (Grunwald and Streisinger et al. 1992a,b). These chemical mutagens most likely induce point mutations or small lesions, in contrast to what is believed for gamma ray mutagenesis. On the other hand, there was no significant difference in the gene-centromere distances determined for the spontaneous standard *gol-1(b1)* allele and the gamma ray-induced *gol-1(b7)* allele, when measured in the same genetic background (Streisinger et al. 1986) (admittedly, a large deletion encompassing the *gol-1* gene in *b7* might extend away from the centromere, and therefore would not result in a different gene-centromere map distance). Whether or not the gamma ray-induced mutations are in fact actually large deletions must await the molecular characterization of mutant alleles of cloned genes and/or systematic molecular genomic analysis and mapping.

Many interesting gamma ray-induced mutations were recognized by screening haploid or parthenogenetic diploid offspring of fish raised from irradiated embryos. Of a total of 225 mutagenized mothers screened in this way, 75 individual mutations were identified by observing embryonic phenotype (Felsenfeld et al. 1990). Among these, seven were found with non-motile phenotypes (Felsenfeld et al. 1990; Westerfield et al. 1990), several with degeneration of the nervous system (Grunwald et al. 1988), and some with tissue-specific defects (Hatta et al. 1991b). (The characteristics of some of these mutants will be discussed below in Sect. 5; see Table 1.)

Gamma ray mutagenesis was also used to estimate the number of pregonial cells present in the blastula embryo (Walker and Streisinger 1983). Irradiation of embryos at cleavage and early blastula produced fish (both males and females) which transmitted the gol-1 phenotype to only a fraction of their progeny (when mated with gol-1/gol-1 fish). If only one pregonial cell was present in the embryo at the time of irradiation, and a gamma-induced lesion occurred in one of the two wild-type genes in this cell, one would expect the individual eventually to transmit the gol-1 phenotype to 25–50% of its progeny (assuming that gamma radiation causes lesions in both strands of a duplex). Transmission frequencies of 0.40–0.51 were in fact observed for a number of the irradiated fish. The more typical values were transmission frequencies of 0.03–0.13, arguing for perhaps five pregonial cells in the blastula. A puzzling aspect of these results, not commented on, is the bimodal distribution of transmission frequencies. Nevertheless, the results are interesting in light of the indeterminate lineage of the pre-gastrula embryo. Apparently, only a small number of cells in the early blastula contribute to the germ line, in contrast to the much larger pool of precursor cells in the cleavage and blastula stages which give rise to the pigmented retina (Walker and Streisinger 1983). One possible explanation for the apparant low number is that most of the blastomeres with potential to give rise to pregonial cells might divide only rarely and only a few such cells would normally give rise to the germ line. Another reason might be that cells undergoing somatic recombination are somehow less likely to divide and/or provide pregonial progenitors. However, the contrast between the number of blastula cells contributing descendents to the pigmented retina and to the germ line is also reflected in the much higher percentage of embryos showing retinal mosaicism than germ line mosaicism for the gol-1 phenotype in parallel experiments with ENU-induced mutations (Grunwald and Streisinger 1992b; see below). I interpret the results of recent chimeric fish experiments (Lin et al. 1992; see Sect. 4.5 below), on the other hand, as indicating that the number of pregonial precursors at the blastula may indeed be higher than the five or so suggested by Walker and Streisinger (1983).

4.4.2 Chemical Mutagenesis

The potential of chemical mutagenesis in zebrafish was first suggested by Streisinger (1984). EMS (ethyl methanesulfonate) was used to induce mosaicism for the *gol-1* gene in the retina, and ENU was shown to induce recessive lethals in the germ line. ENU at concentrations of 1–2 mM appears to be very effective in inducing recessive lethals when applied to sperm (Grunwald et al. 1992b). Three approaches to measuring the average number of independent lethal mutations in fish produced from ENU-treated sperm ("G0-mutagenized fish") suggested that the value may be as high as ten per germ line. Because of mosaiscism in the germ line resulting from direct mutagenesis of sperm, the average number of lethal mutations transmitted *per gamete* from these fish is about one-fourth this level. The level of lethals was estimated by determining the viability of heat shock, homozygous offspring of these fish, by scoring clusters of these embryos (from single G0-mutagenized fish) which showed abnormal morphology, and by outcrossing the G0-mutagenized fish.

ENU-treated sperm also gave rise to *gol-1* mutants (Grunwald and Streisinger 1992b). These were identified by fertilizing *gol-1* eggs with ENU-treated sperm, and looking for mosaic pigmented retinas in these embryos. At 2 mM ENU, 5% of the embryos showed gol-1 mosaicism in their retinas (a similar value was obtained for the *alb-1* locus), but only 2 of the 15 fish with this property were able to transmit the gol-1 phenotype to the next generation. If embryos are mosaic for *gol-1* in the cleavage stage, the mutation may easily escape entering the germ line since germ line precursors are thought to be few in number (Walker and Streisinger 1983), in contrast to the number of cells providing descendents to the pigmented retina (Streisinger et al. 1989). The Grunwald and Streisinger (1992b) data are thus consistent with the earlier work. The ratio between fish showing *gol-1* expression in the retina to those transmitting the trait through the germ line (~10:1), and the proportion of mutant embryos which survive, compared to control embryos, (~0.4) led Grunwald and Streisinger (1992b) to estimate that 1 in 500 G0-mutagenized fish would show a heritable mutation at any one specific locus of interest, a value very close to that reported for the mouse (Russell et al. 1979; Johnson and Lewis 1981). They also estimated that there are approximately 5000 genes required for normal embryogenesis, based on the ratio of ENU-induced lethals to ENU-induced *gol-1* mutants (10 per germ line vs 0.002 per germ line).

4.4.3 Ultraviolet Light Mutagenesis

Ultraviolet light was also effective in inducing new mutations in sperm (Grunwald and Streisinger 1992a). Embryos (the G0-mutagenized generation), raised from *gol-1* or *alb-1* eggs fertilized with mutagenized sperm, showed mosaicism in their retinas at a frequency dependent on the dose of UV radiation. The incidence of mosaicism at any one particular dose was similar for the two loci. The new mutations appeared to be fixed in the embryos by the 2- or 4-cell stage since the pigmented retinas of most of the mosaic embryos were composed of approximately one-fourth or one-half unpigmented cells. Recessive lethals were assayed by observing development and viability of homozygous embryos derived from heat-shocked eggs of G0-mutagenized fish. Grunwald and Streisinger (1992a) noted that the lethal mutations were observed during late embryogenesis and larval growth – in most cases the embryos carrying the lethal allele could not be distinguished from control embryos in the blastula and gastrula stages. This observation was taken to indicate that relatively few zygotic genes have a role in the establishment of the embryo body plan. Grunwald and Streisinger (1992a) also recognized 13 clones of mutants with specific phenotypes from 29 G0-mutagenized fish. The germ lines of the G0-mutagenized fish were mosaic in all cases, with the mutant clones representing approximately one-eighth, one-fourth, or one-half of the germ line, as indicated by the fraction of the homozygous embryos which displayed a particular developmental phenotype. The phenotypes of the mutants were quite interesting: six involved degeneration of the nervous system, three had spinal abnormalities, one lacked spontaneous movements, one had small eyes, one lacked eyes completely, and one lacked a brain and a tail. Based on approaches very much like those used for ENU mutagenesis, an estimate was made that the germ line of fish grown from eggs fertilized with sperm treated with ultraviolet light (at 292 $erg\,mm^{-2}$) carried an average of at least 2.3 recessive lethal mutations. This value is in the same range as that resulting from ENU mutagenesis. Comparison of the recessive lethal frequency and the specific locus mutation frequency yields a ratio of 3000:1 for UV mutagenesis.

4.5 Chimeric Individuals

In a recent exciting contribution, chimeric adult fish have been produced by injection of 20–100 blastomeres form wild-type 1000–2000 cell blastulae into the blastoderm of recipient albino *alb-1/alb-1* embryos of the same stage (Lin et al. 1992). Of the 70 fish which grew to adults from the chimeric embryos, 23 were pigmented and were easily distinguished from the albino strain. The pigmentation appeared in transverse bands, the narrowest of which was about 1/21 the length of the fish. The bands occurred independently on either side of the fish; each band containing the transverse stripes of dark color which would normally appear in that region. These patterns are similar to those seen by B. Mintz in her landmark chimeric mice experiments (Mintz 1967, 1971). By analogy to the mouse, Lin et al. (1992) conclude that the pigmented cells of the zebrafish midflank stripes are derived clonally from at least 21 pairs of precursor cells.

The results of these experiments not only showed that pigmentation can be used as a dominant, visible marker to assay the success of cell transplantation, but also that the descendents of the transplanted cells can contribute to-the germ line of the chimeric animal (Lin et al. 1992). Of 28 chimeric fish which were mated, 5 yielded pigmented offspring. One of these five chimeric fish was originally produced with donor cells derived from a transgenic fish heterozygous for a RSV-*lacZ* DNA fragment (Culp et al. 1991). When this chimeric fish was mated to an albino fish, about half of the pigmented offspring, but none of the albino offspring, inherited the transgene. This proves that the pigmented offspring of the chimeric fish were genetically derived from the transplanted donor cells.

The germ lines of the five chimeric individuals which transmitted the pigmentation were mosaic since the frequency of pigmented offspring was between 1 and 43%. Especially noteworthy were three of the fish which transmitted the pigmentation to 1, 7, and 14% of their progeny, respectively. These values are lower than the minimum of 20% predicted if the germ line were formed from only five pregonial cells, as estimated from gamma ray mutagenesis experiments (Walker and Streisinger 1983). The overall fraction of chimeric fish which transmitted the wild-type pigmentation allele through their germ lines (5/28) also argues for a larger pool of pregonial precursors than was suggested by the mutagenesis work. The chimeras were formed by transplanting 1–10% of the cells from one blastula to another. If there were only five pregonial precursors at this time, it is unlikely that such a high percentage of the chimeric fish would have established germ lines with donor genotype cell components.

5 Developmental Mutants

5.1 Spadetail (*spt-1*)

Table 1 lists a number of zebrafish mutants useful for developmental and genetic studies. The "spadetail" mutant (Kimmel et al. 1989; Ho and Kane 1990) has severe deficiencies in embryonic trunk mesodermal tissues (e.g., lacking trunk somites). The phenotype is recessive, segregates as a single, fully penetrant locus, and results in lethality within 1 week of fertilization. Two mutant alleles of the gene have been identified: the original mutation of unknown origin (*b104*) whose phenotype is well characterized (Kimmel et al. 1989), and a non-complementing gamma ray mutation (*b141*) which has not yet been studied. Mutant *b104* embryos look normal until 7–8 h when a swelling can be seen at the junction of prospective head and trunk regions at the dorsal side of the embryo. At 10 h a second swelling develops in the mutant at the anus near the trunk-tail boundary, and the excess cells then enter and enlarge the tail bud. Trunk somites fail to form in these embryos, although tail somites appear on schedule. Some other trunk mesodermal structures such as pronephric kidneys, pectoral fins, blood cells, and the anus are missing in some embryos. The notochord is present, but is bent and kinked. Anterior structures such as the brain, heart, and hatching gland are normal, as is the tail. In older embryos there is a regulative repopulation of the trunk with cells that form somites. The origin of these cells is not known.

The abnormal swelling in the tail is formed by an excess of cells in this region. This surfeit of tail cells, along with the deficiency of paraxial mesoderm cells in the trunk, suggested that abnormal cell migration during convergent extension might be involved in the mutant phenotype (Kimmel et al. 1989). The idea was first tested by tracing the lineage of cells in normal and mutant embryos with fluorescent dyes. Instead of converging dorsally towards the prospective trunk area and giving rise to trunk somite tissue, cells from the lateral portion of the germ ring of mutant early gastrulae move to a posterior position and enter the tail bud. These inappropriately positioned cells develop into tail muscle, or surprisingly, notochord. The *spt-1* mutation not only results in inappropriate cell movements, but also in some cases to a switch in the tissue fate of the cells.

Cell transplantation experiments revealed that the mutation acts autonomously in somitic mesodermal precursors (Ho and Kane 1990). Wild-type and *spt-1* blastula cells, each marked with different tracer dyes, were cotransplanted into the lateral marginal zone of either wild-type or *spt-1* blastulae. In both types of hosts, the transplanted wild-type cells migrated towards the dorsal side of the embryo and gave rise to myotomal muscle as in control wild-type embryos. In contrast, the transplanted *spt-1* cells moved to the posterior end of the embryo, entered the tail bud, and formed tail mesenchyme. Wild-type host cells failed to influence the behavior of the mutant cells in these transplants. If donor cells of the two genotypes were transplanted into the dorsal marginal zone (the area normally fated to give rise to notochord), both types of cells, along with the host dorsal marginal derivatives, formed notochord. Similarly, if both types of cells are placed in the lateral non-marginal zone of host embryos, they converged together towards the dorsal side and both formed spinal cord. The *spt-1* product is therefore necessary for correct convergence of lateral mesoderm cells and their expression as somitic muscle, but not for convergence and differentiation of cells forming the notochord or spinal cord.

A secondary effect of *spt-1* is the perturbation of primary motoneurons in the 1-day-old embryo (Eisen and Pike 1991). The neurons are not segmentally organized and are fewer in number than in the wild-type embryo. In some cases neuron-like cells appear to be abnormally located outside the spinal cord. Single motoneurons transplanted from mutant to wild-type embryos and vice versa take on characteristics of the host and not donor embryos (Eisen and Pike 1991; Kimmel et al. 1991). The action of the mutant is non-autonomous on neural cells, with the phenotype probably resulting from primary defects in the mesoderm.

5.2 Cyclops (cyc-1)

The *cyc-1 (b16)* mutation, or "cyclops," selected in a screen of gamma ray mutagenized fish, has recently been described (Hatta et al. 1991b). It is a recessive lethal, inherited in Mendelian fashion, with embryos surviving less than 4 days. The phenotype is pleiotropic with the most notable abnormalities being cyclopia and body axis curvature. Fusion of the two eyes is apparently the result of an absence of ventral forebrain which tends to spearate the two eyes during normal development. The floor plate, formed from a row of single cells dorsal to the notochord and ventral to the midbrain, hindbrain, and spinal cord, is absent in the mutant embryos. The effect of the mutation can first be seen at 9–10 h, at the end of gastrulation, as a flattening in the midline of the presumptive head. Mesoderm, including the notochord, develops normally in the mutant embryos.

A mosaic analysis, similar to that described for the spadetail mutant, has been performed by cell transplantation (Hatta et al. 1991b). Between 10 and 100 cells were removed from the embryonic shield area of tracer-marked, donor, early gastrulae (either mutant or wild-type), and introduced into the same area of wild-type or mutant host late blastulae or early gastrulae. Wild-type ectodermal cells could differentiate into floor plate in mutant hosts, showing that the mutation acts in an autonomous fashion in ectoderm. In some transplants, wild-type donor mesoderm cells formed notochord in host embryos, but did not induce floor plate in mutant hosts. Inductive effects from wild-type notochord cells, therefore, are not sufficient to induce floor plate in the mutants. However, when wild-type ectodermal donor cells did form floor plate in mutant hosts, some adjacent mutant cells were recruited to form floor plate. The *cyc-1* mutation, therefore, does not block floor plate differentiation. Hatta et al. (1991b) postulate that there are two specification pathways: induction from the notochord, which is blocked in the *cyc-1* mutant, and induction between floor plate precursors ("homeogenetic induction"), which can still occur in the cyclops embryos. The Hatta et al. (1991b) experiments also reveal non-autonomous effects on axonal growth. Axons are scattered in the mutant spinal cord, but can be properly organized if they are near donor wild-type floor plate cells.

5.3 Neural Degeneration (*ned*) Mutants

One of the most common phenotypes recognized in screens of gamma ray or ultraviolet light mutagenesis is the degeneration of the CNS, especially the brain (Grunwald et al. 1988; Grunwald and Streisinger 1992a). One of these mutants, *ned-1(b39rl)* has been studied in detail (Grunwald et al. 1988). This mutation was induced by gamma irradiation and was uncovered in a screen of early pressure-treated homozygous embryos. It is inherited as a recessive lethal at a single locus, with a map position very close to the centromere. The mutant phenotype, first obvious at 36 h by the opacity of the CNS, is the result of massive cell death in the brain and spinal cord. By the end of the second day, the embryonic body is thinner than normal, and the tail turns upward. Death occurs by the sixth day after fertilization. Microscopic inspection of the embryos revealed that degenerating cells were found in the CNS as early as 24 h after fertilization. The most severely affected regions were in the dorsal tissue of the midbrain and hindbrain. There was extensive cell death in the central retina but not in the stem cells at the marginal zone, indicating that the mutation probably affects only post mitotically developing cells. Cell degeneration was not encountered in non-neuronal tissues at this time, and neural tissue outside the CNS was also normal. The most striking aspect of the specificity of the mutation was that even within the CNS, specific neurons such as sensory neurons, hindbrain interneurons (including the Mauthner cell), and spinal primary motoneurons and Rohon-Beard neurons, were normal in the day 2 embryo. These neurons make up a class of cells termed "primary neurons," whose distinguishing characteristic is that they begin to grow neurites prior to completion of the first day of embryogenesis. Despite the massive cell degeneration in the CNS, the *ned-1* mutation did not prevent early muscle contractions, presumably because they are dependent only on the primary neurons.

5.4 Motility Mutants

Screening early pressure parthenogenetic diploid offspring or activated haploid offspring of 225 gamma ray mutagenized females resulted in the isolation of seven non-motile mutant lines (Felsenfeld et al. 1990). Two of these seven are alleles (*b45*) and (*b126*) of the *fub-1* (fibrils unbundled) gene (Felsenfeld et al. 1990), one (*nic-1*) is deficient in functional acetylcholine receptors (Westerfield et al. 1990), and four remain to be characterized. In all seven strains, lack of motility was the only apparent phenotype. The *fub-1* mutants were initially characterized by Felsenfeld et al. (1990). The *b45* and *b126* mutants appeared to be allelic since they did not complement one another. However, they did map at slightly different distances from the centromere, a result which might be due to genetic background, to different sizes of the gamma ray-induced lesions, or to the two mutants being located in closely linked genes that code for interacting protein products. The initial indication of mutant phenotype was seen at about 18 h as a failure to exhibit spontaneous body movements. The *b45* homozygous individuals could twitch only slightly at 24 h, but not at all at 36 h. Individuals homozygous for the weaker *b126* allele showed some response to tactile stimuli at this time. At 36 h, mutants did not have the usual muscle striations but the ratio of thin to thick filaments was about the same as in wild-type embryos. The defect (more severe in *b45* than in *b126*) appeared to be in the ordering and assembly of the myofibrils and not in the absence of a major myofibrillar component. In this regard, two-dimensional gel analysis of proteins from mutant and wild-type embryos did not reveal and differences. Muscle cells began to die at about 4 days and the embryos did not survive past 7 days. All skeletal muscle appeared to be affected and the heart did not pump blood. More recent electron microscope evidence indicates that the first indication of the *fub-1(b45)* mutation can be seen as early as 14.5 h. The formation and ordering of myofibrillar bundles, which occur at this time in muscle pioneer cells in wild-type embryos, do not take place in these cells in the mutant (Felsenfeld et al. 1991).

Gamma ray-induced mosaic embryos were created to test for cell autonomy of the *fub-1* mutation (Felsenfeld et al. 1990). Two-cell heterozygous embryos irradiated with gamma rays sometimes developed into embryos with mutant and wild-type cells. Such embryos had a scattered distribution of the two-cell types in their striated muscle, as expected from the multi-cell origin of somites deduced from lineage tracer studies (Kimmel and Warga 1987). Since mutant cells in these mosaic muscles could be completely surrounded by wild-type cells, the action of the *fub-1* gene is most likely cell autonomous, and the defect does not result from a lack of a diffusible factor or a lack of innervation.

The *nic-1* mutation was also identified by non-motility of 1-day-old embryos in a screen of gamma ray-induced mutants (Westerfield et al. 1990). The one allele identified at this locus, *b107*, segregates as a single Mendelian trait and has a recessive lethal phenotype. No contractile movements were seen in mutant embryos after tactile or electrical stimulation of the spinal cord. Direct stimulation of muscles, however, produced contractions. The defect in the mutant is thought to be an absence of acetylcholine receptors since the muscle of mutant individuals fails to stain with α-bungarotoxin or with monoclonal antibodies directed against the receptor, and there is a lack of spontaneous or evoked electrical activity.

5.5 No notochord/no tail (*ntl-1*)

As of this writing, there is little published information about this very interesting mutant which resembles the phenotype of mouse *Brachyury* ("*T*") locus mutants (C. Kimmel, cited in Ruiz i Altaba 1991). The notochord is not formed in the *ntl-1* mutant, and somites are abnormal. One defect in the somites is an absence of muscle pioneer cells in the myotomes and it has been suggested that the notochord may induce the formation of muscle pioneer cells in the adjacent somitic mesoderm (M. Halpern and C. Kimmel, personal communication). The *ntl-1* mutant should be informative in studying other interactions between the notochord and surrounding tissues. Moreover, because the zebrafish *T* gene has been cloned and its expression found to be localized in the presumptive notochord and tail regions (Schulte-Merker 1992; see Sect. 6.5), it will be of great interest to see if a lesion in the zebrafish *T* gene is found in *ntl-1* mutants. In this regard, it is noteworthy that no nuclear *T*-gene protein is found in *ntl-1* mutants (S. Schulte-Merker, C. Kimmel, cited in Ruiz i Altaba 1991).

6 Gene Expression

6.1 *Hox* Genes

The homeobox, a 183-bp segment endoding a DNA-binding domain of 61 amino acids first recognized in the *Drosophila* homeotic genes (McGinnis et al. 1984; Scott and Weiner 1984), is also found in the genes of vertebrates (reviewed in Kessel and Gruss 1990; Boncinelli et al. 1991). Those vertebrate genes with homeoboxes of the *Drosophila Antennapedia* type are termed *Hox* genes. In the mouse and human genomes, there are over 30 of these genes, arranged in four main clusters which are probably the result of ancient duplications of chromosome segments (Kappen et al. 1989; Kessel and Gruss 1990). Within each cluster the *Hox* genes not only have the same order in mouse and man, but they can also be aligned with the *Antennapedia* and *bithorax* clusters in *Drosophila* (Duboule and Dolle 1989; Graham et al. 1989; Kessel and Gruss 1990; Boncinelli et al. 1991) and with homeobox genes in *Caenorhabditis elegans* (Kenyon and Wang 1991). In mammals there are at least 13 gene subfamilies, with paralogous members of each subfamily occupying the same respective positions in the different clusters. Expression patterns of the mammalian homeobox genes and their role in specification of positional information have been reviewed recently (Reid 1990; Kessel and Gruss 1990).

Seven different zebrafish *Hox* (*Antennapedia*-like homeobox) gene sequences have now been isolated and partially characterized (Eiken et al. 1987; Njølstad et al. 1988a,b,c, 1990; Runstadler and Kocher 1991). The various genes are listed in Table 2, classified with respect to homologies with mouse *Hox* genes. Three of the cloned sequences, ZF-22 (Njølsted et al. 1990), ZF-21 (Njølstad et al. 1988b,c), and ZF-13 (Njølstad et al. 1988a) appear to be orthologous to the *Hox 2.2*, *Hox 2.1*, and *Hox 2.6* genes in the mouse *Hox 2* cluster. Two of the sequences, ZF-61 (Njølstad et al. 1990) and ZF-25 (Eiken et al. 1987), are homologous to the mouse *Hox 3.3* and *Hox 3.4* genes (originally named *Hox 6.1* and *Hox 6.2*) in the *Hox 3* cluster. Not yet matched with a specific homologue, ZF-54 (Njølstad et al. 1988c) appears to be most similar to mouse *Hox 2.1*, and ZF-26 (a PCR fragment, Runstadler and Kocher 1991) is most similar to mouse *Hox 2.6*.

Transcripts from four of these genes (ZF-25, ZF-13, ZF-21, and ZF-22) have been analyzed on Northern blots. In all cases the transcripts were first detected only after the appearance of somites in the embryo, and then reached fairly constant levels from 28 to 48 h (Eiken et al. 1987; Njølstad et al. 1988a,c, 1990). In situ hybridizations have been done with probes for three of the genes. At 56 h, the ZF-13 (*Hox 2.6*-like) gene is strongly expressed throughout the CNS, and more weakly, in the gut and in trunk muscle (Njølstad et al. 1988a). The ZF-21 and ZF-22 genes (*Hox 2.1* and *Hox 2.2*) are both expressed in the 56-h embryo at a high level in the hindbrain. There is a sharp rostral border of expression in the posterior part of the

Table 2. Zebrafish genes of developmental interest

Gene	Class	Mouse homologue	Transcript size (kb)	Time of expression (h) first	Time of expression (h) peak	Location of expression[a]	References
ZF-13[b,c]	Antp	Hox-2.6	2.5, 4.6	15	28–48	Brain and spinal cord, gut, muscle	Njølstad et al. (1988a)
ZF-21[b,c]	Antp	Hox-2.1	2.3	12.5	15–48	Posterior hindbrain, gradient in spinal cord	Njølstad et al. (1988b,c); Njølstad and Fjose (1988)
ZF-22[b]	Antp	Hox-2.2	1.4	12.5	NR	Posterior hindbrain, gradient in spinal cord	Njølstad et al. (1990)
ZF-61[b]	Antp	Hox-3.3	NR	NR	NR	Somites 5–7; notochord; some neurons in the CNS; PF	Njølstad et al. (1990); Molven et al. (1990)
ZF-25[b,c]	Antp	Hox-3.4	1.4, 2.1	15	15–48	NR	Eiken et al. (1987)
ZF-54[b,c]	Antp	?	NR	NR	NR	NR	Njølstad et al. (1988c)
ZF-26[d,c]	Antp	?	NR	NR	NR	NR	Runstadler and Kocher (1991)
eng-2[b,d]	engrailed	En-2	3.8	16	16–48	Midbrain/hindbrain stripe; somites: MPs and other cells; jaw muscles; CNS; PF	Fjosse et al. (1988); Njølstad and Fjosse (1988); Patel et al. (1989a); Hatta et al. (1990, 1991a); Holland and Williams (1990)
En-1[d,c]	engrailed	?	NR	NR	NR	NR	Holland and Williams (1990)
mshA,B,C[d,c]	msh	?	NR	NR	NR	NR	Holland (1991)
Pax[zf-a][c]	pax	Pax-6	3.0	9–12	?	Spec. regions of brain, spinal cord; eye	Krauss et al. (1991b; Puschel et al. 1992)
Pax[zf-b][c]	pax	Pax-2	NR	NR	NR	Spec. regions of brain, spinal cord; NP	Krauss et al. (1991a)
wnt-1[b]	wnt	Wnt-1	3.9, 4.9	12–13	16–42	Midbrain/hindbrain stripe and other CNS areas	Molven et al. (1991)
Brachyury "T"[c]	Brachyury "T"	Brachyury "T"	2.5	3–4	5.2–9.5	Germ ring, prospective and differentiated notochord, tail bud	Schulte-Merker et al. (1992)

[a] Sites listed may not show expression at the same developmental stage. See text for details. [b] Sequence derived from genomic clone. [d] Sequence derived from PCR fragment. [c] Sequence derived from cDNA clone. [e] Sequence information limited to homeobox or paired box.

Abbreviations: NR = not reported; CNS = central nervous system; PF = pectoral fin; MPs = muscle pioneer cells; spec. = specific; NP. = nephritic primordia

hindbrain, and the level of transcripts decreases more gradually towards the spinal cord (Njølstad and Fjose 1988; Njølstad et al. 1990). The mouse *Hox-2.1* and *Hox-2.2* genes are also expressed in a similar region of the CNS (Krumlauf et al. 1987; Holland and Hogan 1988; Schughart et al. 1988; Graham et al. 1989). Unfortunately, patterns of in situ hybridization are not yet reported for earlier stages of zebrafish embryos.

Thus far, the spatial expression of zebrafish *Hox* genes has been most effectively investigated by staining with specific antibodies rather than by in situ hybridization. Molven et al. (1990) used polyclonal antibody directed against part of the *Xenopus* XlHbox1 protein (Oliver et al. 1988a). Since this *Xenopus* protein is most likely the homologue of the mouse *Hox-3.3* (and zebrafish ZF-61) product, the staining pattern is probably reflective of expression of the zebrafish *Hox-3.3* gene. Determination of just which gene product is being detected, however, awaits a more complete characterization of the antibody specificity and of the zebrafish gene. Expression was first detected at 14 h in epithelial cells lining somites 5 and 6. At 16 h, staining was seen in somites 5, 6, and 7, with a sharp cut-off between somites 4 and 5. Weaker staining was exhibited by the notochord and by some cells in the CNS. At 22 h, staining was stronger in the same somites and in the spinal cord, but absent from the notochord. At 48 h, the somites no longer stained, but now many (but not all) neurons did. The homeodomain protein recognized by this antibody thus has a very specific pattern of appearance. It is not specific for a particular tissue type, but rather delineates a particular group of segments or cells within a tissue. Particularly interesting is the pattern of expression in the pectoral fin, the probable precursor to the tetrapod forelimb. A gradient of antibody staining in the fin was seen at 48 h, very much like a gradient seen in the forelimb bud of mouse and frog embryos, using the same antibody (Oliver et al. 1989b). Other mouse *Hox* genes are also expressed in gradients in mouse limb buds (Dolle et al. 1989; Oliver et al. 1989).

6.2 *Engrailed* Genes

Engrailed genes encode proteins with homeodomains considerably diverged from the *Antennapedia*-type sequence. The *Drosophila engrailed* gene (Fjose et al. 1985; Poole et al. 1985) has multiple essential developmental functions including roles in segmentation, imaginal disk compartmentation, and neurogenesis (references in Ingham 1988; Patel et al. 1989a). The *Drosophila* genome also contains a second closely linked gene of this class, *invected* (Coleman et al. 1987). Two homologues, *En-1* and *En-2*, have been identified in the mouse (Joyner et al. 1985; Joyner and Martin 1987). Deletion of the *En-2* homeobox by homologous recombination results in defects in cerebellar development (Joyner et al. 1991). Although two different *engrailed* genes have been identified in the zebrafish genome (Fjose et al. 1988; Holland and Williams 1990), only one of them, the homologue of *En-2*, has been sequenced outside of the homeobox region (Fjose et al. 1988). This zebrafish gene, denoted as *eng-2* (Hatta et al. 1991a), encodes a transcript with similar size (and timing of appearance) to that of the mouse *En-2* RNA (Joyner and Martin 1987). In situ hybridization of embryos also revealed similarities in expression of the genes in the two species – a narrow stripe of hybridization in the 56-h zebrafish embryo brain is restricted to the caudal part of the midbrain and rostral part of the hindbrain (Njølstad and Fjose 1988), very similar to a major site of mouse *En-2* expression (Davidson et al. 1988; Davis and Joyner 1988; Davis et al. 1988, 1991). The stripe of *eng-2* expression is contained within the region of zebrafish *Hox 2.1* and *Hox 2.2* expression discussed above, but the *Hox* gene transcripts are found in the spinal cord as well (Njølstad and Fjose 1988; Njølstad et al. 1990).

A more detailed picture of zebrafish *engrailed* gene expression has been obtained by staining tissue sections (Patel et al. 1989a; Hatta et al. 1990, 1991a) with the 4D9 monoclonal antibody raised against the *Drosophila invected* homeodomain (Patel et al. 1989a) and the αEnhb-1 polyclonal antibody raised against the mouse *En-2* homeodomain (Davis et al. 1991). Since engrailed proteins in a variety of organisms have been identified with these antibodies (Gardner et al. 1988; Hemmati-Brivanlou and Harland 1989; Patel et al. 1989a; Davis et al. 1991), it is likely that more than one *engrailed* gene product may be recognized in the zebrafish tissue sections. The earliest appearance of engrailed antigen was in a stripe of cells in the brain primordium, at 10–11 h postfertilization at the time of neural tube formation (Patel et al. 1989a; Hatta et al. 1991a). At 20 h, the furrow separating the midbrain from the hindbrain forms in the middle of the stained area (Hatta et al. 1991a). Labeling of this restricted region of the CNS persists through larval formation, and eventually is restricted to parts of the cerebellum, optic tectum, and particular midbrain tegmental nuclei (Hatta et al. 1991a). The same hindbrain/midbrain region identified at 56 h by in situ hybridization with the *eng-2* gene probe (Njølstad and Fjose 1988) is recognized by antibody staining.

The antibodies also stain a subset of 2–6 cells in each somite (Patel et al. 1989a; Hatta et al. 1991a). Except for the first few somites, in which there is simultaneous appearance of antigen at 13–14 h, the staining occurs in an anterior to posterior wave, lagging behind formation of the somites by 1–2 h. Hatta et al. (1991a) have identified these stained cells as muscle pioneer cells (see Sect. 2.6). Beginning at about 20 h, additional somite cells are recognized by the antibody (Hatta et al. 1991a), with staining appearing as a second wave lagging the first wave by about ten somites (Patel et al. 1989a). Hatta et al. (1991a) point out that in both the brain and somites, the sites of *engrailed* expression mark sites of division of tissue into subdivisions: the stripe in the brain is just where the furrow separating the hindbrain and midbrain develops, and the muscle pioneer cells are located at the position of the division of the myotome into dorsal and ventral muscle regions.

A very special region of staining is found in the cells which develop to form two specialized jaw muscles, the levator arcus palatini (LAP) and the dilator operculi (DO) (Hatta et al. 1990). A scattered group of 15 cells just posterior to the eyes stains strongly at 24–26 h. The engrailed-positive cells eventually form a single cluster in the mandibular arch by 48 h, and then develop into the LAP and DO, the only head muscles which stain with the antibody. Whereas the hindbrain/midbrain stripe, somite cells, and LAP and DO jaw muscles were stained with both the 4D9 and αEnhb-1 antibodies, other regions (cells in the pectoral girdle, pectoral fin epithelium, teeth, otic vesicle, hindbrain, and spinal cord) were stained with only one or the other antibody (Hatta et al. 1991a). This finding suggests that there are at least two engrailed proteins in the embryo, but it is not known at present whether they are encoded by different genes. Although zebrafish *engrailed* expression is to some extent segmental in the CNS and somites, it is also expressed in particular cells and tissues which do not appear related to segmental organization (Hatta et al. 1991a). The expression in the pectoral fin is in the ventral-anterior half of the epidermis, similar to the ventral-dorsal gradient of expression of En-1 protein found in the limb bud of the mouse and chick (Davis et al. 1991). As mentioned above, at least one Hox protein in the zebrafish is also expressed as a gradient in the pectoral fin (Molven et al. 1990), also similar to the pattern of *Hox* genes in the mouse forelimb (Dolle et al. 1989; Oliver et al. 1991), the most probable anatomical homologue. The engrailed protein obviously has multiple functions in vertebrate development, many of which are highly conserved through evolution.

6.3 *Pax* Genes

The *Pax* gene family encodes proteins which contain the paired-box (Bopp et al. 1986), a DNA binding domain (Goulding et al. 1991; Treisman et al. 1991). *Pax* genes in *Drosophila* are involved in segmentation and in neurogenesis at later embryonic stages (Bopp et al. 1986, 1989; Baumgartner et al. 1987; Patel et al. 1989b). Eight *Pax* genes have been identified in vertebrates thus far (Dressler et al. 1988; Burri et al. 1989; Kessel and Gruss 1991; Walther et al. 1991). Three of the mouse genes, *Pax-3*, *Pax-6*, and *Pax-7* contain a homeobox in addition to the paired-box (Kessel and Gruss 1990; Walther et al. 1991), a combination also found in the *Drosophila paired* gene (Bopp et al. 1986). In the mouse, various *Pax* genes are expressed in segmented mesoderm, in specific regions of the CNS and the eye, and in embryonic kidney (reviewed in Kessel and Gruss 1990; Krauss et al. 1991b), and mutations in one gene, *Pax-1*, are probably responsible for the *un* (undulated) phenotype in mice (Balling et al. 1988).

Sequences of two zebrafish *Pax* genes have recently been isolated from a cDNA library (Krauss et al. 1991a,b; Püschel et al. 1992). Both genes have homeoboxes in addition to the paired-box sequences. One of the genes, *Pax[zf-b]* (Krauss et al. 1991a) is probably a homologue of mouse *Pax-2* (Dressler et al. 1990); the other, *Pax[zf-a]* (Krauss et al. 1991b; Püschel et al. 1992), is a homologue of the murine *Pax-6* gene (Walther et al. 1991).

In situ hybridization revealed a highly localized expression pattern of the *Pax-6* gene during brain development (Krauss et al. 1991b; Püschel et al. 1992), quite similar to that of the mouse homologue. Transcripts first appear in 9–12 h embryos in the neural tube: in the diencephalon and in the hindbrain. In the 17-h embryo, the signal was more intense, especially in a particular area of the diencephalon surrounding the presumptive thalamus, and in a restricted longitudinal column throughout the hindbrain and spinal cord. A region between the midbrain and the diencephalon, however, did not express the transcript. The gene also appears to be strongly expressed in the eye. The pattern of localization of the *Pax-2* homologue RNA is distinct from that of *Pax-6* (Krauss et al. 1991a). In the 9–10 h embryo, the gene is expressed in two transverse stripes of cells in the rostral ⅓ of the embryo. At 10–12 h, transcripts are found in the otic placode, the optic stalk, and the Wolffian duct. At 14–15 h, single cells in the spinal cord begin to express the gene. It appears that the regions of *Pax[zf-a]* expression are quite distinct from the stripe of *engrailed* expression (Patel et al. 1989a; Hatta et al. 1990, 1991a). The region of *Pax[zf-b]* expression in the midbrain, however, might overlap with the *engrailed* stripe.

6.4 *wnt-1*

The zebrafish *wnt-1* gene was isolated by screening a genomic library with a fragment from the human homologue, *Wnt-1* (Molven et al. 1991). The first gene of this class to be isolated, mouse *Wnt-1* (formerly named *int-1*, reviewed in Nusse 1988; McMahon and Moon 1989), is a site of frequent insertion by the mouse mammary tumor virus. In the mouse embryo, *Wnt-1* expression is found only in the CNS (Shackleford and Varmus 1987; Wilkinson et al. 1987), restricted to cells at the dorsal midline of the neural tube and in the lateral walls of the midbrain (Wilkinson et al. 1987). Recent experiments show that when both mouse alleles are mutated by targeted recombination, the dorsal midbrain and hindbrain do not develop normally (McMahon and Bradley 1990; Thomas and Capecchi 1990). In contrast to the other genes discussed in this section, *Wnt-1* has none of the characteristics of a DNA binding protein. It is secreted by cells and is found on the cell surface (Papkoff et al. 1987; Bradley and Brown 1990; Papkoff and Schryver 1990).

   The sequence of the zebrafish *wnt-1* gene is more similar to mouse *Wnt-1* than to the other reported *Wnt-1*-related genes, and for this reason is considered to be the true homologue (Molven et al. 1991). However, Southern blotting revealed a potential family of related sequences in the zebrafish genome (Molven et al. 1991). The first *wnt-1* RNA signal seen in Northern blots was at 12–13 h of development, and expression of the two or three transcripts increased to a level which remained constant from 16–42 h. Molven et al. (1991) also found by in situ hybridization that the *wnt-1* RNA was restricted to the CNS. The localization was highly specific: along the

dorsal midline of the spinal cord and hindbrain, ending in a fork at the
rostral part of the midbrain; in a transverse band at the midbrain-hindbrain
junction; and in the dorsal and lateral midbrain walls. This pattern is almost
identical to that reported for *Wnt-1* in the embryonic mouse brain (Wilkinson
et al. 1987; Davis and Joyner 1988). In both mouse and zebrafish, the *Wnt-1*
and *engrailed* genes are expressed in the same midbrain-hindbrain transverse
stripe (Wilkinson et al. 1987; Davidson et al. 1988; Davis and Joyner 1988;
Davis et al. 1988, 1991; Patel et al. 1989; Hatta et al. 1991a; Molven et al.
1991), suggesting that the two genes interact during development (Molven et
al. 1991), perhaps with particular *Hox* and *Pax* gene proteins, to specify the
regionalization of the hindbrain and midbrain.

6.5 *Brachyury* (*T* gene)

The *Brachyury* or "*T*" gene was first isolated from the mouse genome by a
tour de force of positional cloning (Herrmann et al. 1990). The loss of function phenotype of the mouse *T* gene is a severe disturbance in mesoderm
formation, resulting in lack of notochord development from the notochordal
plate, complete absence of the posterior region of the embryo, failure to
form the allantois, and embryonic death at 10 days of gestation (Chesley
1935; Glueksohn-Schoenheimer 1944; Grüneberg 1958; Yanagisawa et al.
1981). The expression of the *T* gene is correlated with the tissues affected
in mutant strains. Transcripts are found first in mesoderm and primitive
ectoderm near the primitive streak, and later become increasingly restricted
in location so that by the end of gastrulation, only the notochord contains
T-gene transcripts (Wilkinson et al. 1990).

Mouse *T* gene cDNA probes were used to isolate cDNA clones for the
zebrafish (Schulte-Merker et al. 1992) and *Xenopus* (Smith et al. 1991)
homologues. Much as in the mouse embryo, the pattern of expression of the
T gene in these organisms is observed in presumptive mesodermal cells
and then in the notochord. The following is a description of the results
of Schulte-Merker et al. (1992), who carefully studied the expression of
the zebrafish *T* gene. A single 2.5-kb zebrafish *T* gene transcript is present at
3–4 h, even before the beginning of epiboly, and peaks during gastrulation
between 5.2 and 9.5 h of development. The *T* gene RNA was first detected
by whole mount in situ hybridization in the 4.2 h doming embryo in a ring-
like area at the margin of the blastoderm. At the time of formation of the
embryonic shield, the site of RNA is restricted to the whole germ ring,
with expression in both epiblast and hypoblast cells. After involution, most
of the hypoblast cells no longer express the gene. A subset of hypoblast
cells along the dorsal midline continues to be stained strongly with the
gene probe. At 10 h of development, the *T* gene RNA is located in chorda-
mesoderm cells in the axis and in the tail bud. At 36 h of development
the expression is restricted to cells in the notochord. The location of
T gene protein in the embryo, detected with antiserum prepared against

recombinant *T* gene protein, paralleled the distribution of the transcripts. The pattern of *Zbra* expression thus clearly becomes more restricted as development proceeds, a situation similar to that in the mouse embryo. The most startling example was the demonstration of *T* gene protein in EVL cells in the germinal ring, cells which will not contribute at all to the tissues of the embryo.

Schulte-Merker et al. (1992) also demonstrated that *T* gene protein could be induced in blastula animal caps by treatment with activin A. These results, similar to activin A and FGF induction of the *Xenopus T* gene (Smith et al. 1991), indicate that the *T* gene is part of sequential pathway of responses specifying the formation of mesoderm. To find out whether all zebrafish blastula cells could potentially respond to the inductive signal, marked animal pole cells were transplanted into the marginal zone or into the animal pole area of host embryos at the late blastula/early blastula stage (Schulte-Merker et al. 1992). The donor animal pole cells were able to produce *T* gene protein if placed at the marginal zone, but not if transplanted into the animal pole. Therefore, blastula cells which would not normally make the protein were able to respond to an inductive signal if placed in an appropriate position in the embryo.

7 Transgenic Fish

7.1 The Zebrafish Is Only One of Many Fish Species Subject to Transgenic Studies

There are now three published accounts of the production of transgenic zebrafish in which stable germ line transmission of a foreign gene has been demonstrated (Stuart et al. 1988, 1990; Culp et al. 1991). The zebrafish is by no means the only fish species to have been subject to this approach (reviewed in Dunham 1990; Houdebine and Chourrout 1991) – DNA has been introduced into the eggs of at least ten other species. In many of these other species, the foreign DNA was shown to be present (presumably integrated) in fry or adults derived from injected eggs, but in only three cases has germ line transmission to the F1 generation been shown unequivocally (Guyomard et al. 1989; Inoue et al. 1990; Zhang et al. 1990). Expression of the injected gene has been detected in a number of cases, but thus far, only for fish which developed from the injected eggs. The mouse metallothionein promoter was shown to drive expression of the human growth hormone gene in Atlantic salmon (Rokkones et al. 1989) and carp (reported in Dunham 1990) and the *E. coli* β-galactosidase gene in Atlantic salmon (McEvoy et al. 1988). Other examples include the chloramphenicol transacetylase (CAT) gene expressed from the SV40 early promoter in tilapia (Indiq and Moav 1988) and medaka (Chong and Vielkind 1989), the neomycin resistance gene in goldfish (Yoon et al. 1990), and the bovine growth hormone gene in northern pike (Schneider et al. 1989), all expressed from the Rous sarcoma virus long terminal repeat (RSV-LTR); and the chick delta-crystallin gene, presumably expressed from its own promoter, in medaka (Ozato et al. 1986). Of all of these fish species, only medaka matches the zebrafish in having both a short generation time and easily obtainable eggs. The zebrafish should therefore be an excellent species to work out methods useful for the genetic engineering of commercially important fish. The rest of this section will deal only with the zebrafish studies.

7.2 Methods of Introducing DNA into Zebrafish

Injection of DNA into cytoplasm of the one-cell or two-cell embryo has until now been the method of choice for production of transgenic zebrafish (Stuart et al. 1988, 1990; Culp et al. 1991). Injection solutions generally contain phenol red (0.5–2%) to facilitate visualization of the DNA solution after injection into the cell. In all three reports, approximately 300 pl of DNA solution was injected. Stuart et al. (1988) showed that increasing amounts of DNA were toxic to the embryo. Survival values at 10 days were: 43% when 4.5 pg DNA was introduced per embryo, 24% for 15 pg, 16% for 30 pg, and only 3% for 90 pg; most experiments were done at 15–30 pg per embryo (300 pl of a solution containing DNA at 0.05–0.1 mg/ml) to try to optimize for integration and expression, with some sacrifice of viability. Culp et al. (1991) achieved a survival of approximately 30% of fertilized eggs after dechorionation and injection. Both linearized (Stuart et al. 1988) and supercoiled (Stuart et al. 1990) plasmids were injected. Eggs can be placed at lower temperatures to slow division during the first hour of development (Culp et al. 1991; C. Fulwiler, pers. comm.; E. Weinberg, unpubl. results). In my experience, 1 h at 16 °C results in no loss in viability and greatly extends the time window for injection into one- and two-cell embryos.

In the published injection experiments, the chorion membrane was removed either manually (Stuart et al. 1988) or with pronase (Stuart et al. 1990; Culp et al. 1991). Although the zebrafish chorion is quite tough, I have found that it is possible to inject directly through the membrane by positioning the eggs in a slitted agarose mold. Injection through the chorion has been successful in other fish species (Dunham et al. 1987; Brem et al. 1988; other references cited in Dunham 1990) and a variety of devices to hold the eggs of these fish have been designed (Rokkones et al. 1985; Chourrout et al. 1986). For the zebrafish, the slits should be a bit wider than the eggs and have an angular shape so that the egg is immobilized against a vertical rear slit wall during injection. If the needle is thin enough, withdrawal can be achieved without the aid of forceps. I have designed a thin plastic cover with openings which fit over the slits. Even if the needle catches on the chorion on withdrawal, the egg is held back by the plastic cover and the needle can be removed smoothly without damaging the cell.

Recently, success in introducing DNA into zebrafish eggs has been claimed for electroporation (Buono and Linser 1991) and particle gun (Zelenin et al. 1991) methods. Electroporation has been used successfully with fertilized medaka eggs to introduce DNA containing the rainbow trout growth cDNA sequence fused to the mouse metallothionein promoter (Inoue et al. 1990). Approximately 75% of the embryos survived to hatching, and 4% of these were shown to contain the injected DNA sequence by Southern blotting. Germ line transmission of the transgene was demonstrated for one of the fish raised from these embryos. The results for electroporation of zebrafish eggs (Buono and Linser 1991), however, are preliminary and seem to be difficult to reproduce. It is claimed, by the criterion of dot blot hybridization, that 65% of the 6-day-old fish derived from pulsed eggs contained an RSV-CAT construct, and CAT activity was detected in assays of batches of these fish. The particle gun was used by Zelenin et al. (1991) to introduce plasmid DNA containing a neomycin resistance gene into zebrafish-fertilized eggs. Larvae raised from these eggs are claimed to have increased resistance to G418 and to contain amplified amounts of plasmid DNA. Evidence for stable integration of DNA is lacking, however, and the data on resistance to G418 are difficult to interpret.

7.3 Integration and Expression
of the Transgenes During Development

Stuart et al. (1988) injected a construct containing a hygromycin B resistance gene expressed from an SV40 early promoter (pSV-hygro). Dot blotting of DNA isolated from injected embryos showed that there was a tenfold amplification of DNA per embryo between 6–10 h of development. Southern blots showed that the DNA was converted to a high molecular weight form within 0.5 h after injection. Most of the amplified DNA was degraded at 12 h of development, but a small amount (somewhat greater than the amount originally injected) persisted in high molecular weight form. At 3 weeks, the amount of foreign DNA was estimated to be less than one copy per cell. At 4 months, 5% (28 of 547) of the fish were judged to have retained pSV-hygro sequences by dot blots of total DNA extracted from fins.

Stuart et al. (1990) used constructs containing the early SV40 promoter fused to CAT (pUSVCAT) or both the SV40 early promoter and the RSV-LTR fused to CAT {pSVeRSVCAT}. Of 55, 4-month-old pUSVCAT fish, 10 were shown to have CAT activity in their fins. In the experiments of Culp et al. (1991), the F0 fish were not directly assayed for the presence of the injected RSV-*lacZ* or SV40-*lacZ* DNA. As will be discussed in the next section, that integration of the RSV-*lacZ* (but not the SV40-*lacZ*) DNA into the germ line did occur is clear from transmission ratios of 50% from the F1 to F2 generation. Although expression of the RSV-*lacZ* construct was detected in embryos after injection, the offspring of transgenic fish did not express β-galactosidase.

7.4 Transmission of Transgenes Through the Germ Line

In the original experiments from the Westerfield group (Stuart et al. 1988), 20 fish judged to contain pSV-hygro DNA were crossed to uninjected fish, and the F1 offspring were tested for the presence of the transgene. Only one of the injected fish passed the gene to her offspring. The transmission rate from this fish was about 20%, strongly indicating a mosaic germ line. The offspring each contained about 100 copies of the transgene per cell, in high molecular weight form. Restriction enzyme analysis indicated that the DNA in each F1 fish was present in the same complex array, perhaps as a multimer of both head-to-head- and head-to-tail-arranged linear plasmid. Candidates for junction fragments were seen on Southern blots, but physical proof of actual integration was not provided. That integration did in fact take place, however, is supported by the transmission of the transgene to the F2 generation. The transmission rate in this case was 50%, indicating a completely heterozygous non-mosaic germ line, and the number of copies per cell remained at 100, suggesting the stable integration of a tandemly repeated array at a single site in the genome. Putative junction fragments could not be recognized, however, in Southern blots of DNA from several tissues of the F1 founder female. Expression of the transgene was not detected in the F1 fish – neither by resistance of the fish to hygromycin, nor by Northern blot analysis of extracted RNA.

In contrast, expression of transmitted recombinant CAT genes was observed in the second study from the Westerfield lab (Stuart et al. 1990). In this work, approximately 5% of the fish raised from injected eggs contained the foreign DNA (pUSVCAT) in the germ line. Of 87 adults tested, 4 gave rise to F1 embryos which expressed CAT. Since two of these four fish had no CAT activity in their fins, the dot blots of DNA from fin clips in the previous Stuart et al. (1988) study were probaby also not predictive of the yield of expressing transgenics. As in the previous work with pSV-hygro DNA, the germ lines of (three of the four) fish were mosaic (the F1 embryos were CAT positive in 12, 25, 36, and 54% for the four founder fish). F2 fish were obtained by breeding each of these four F1 individuals. The F1 fish appeared to have true heterozygous non-mosaic germ lines (as in the previous study) since they transmitted the CAT gene to 50% of the F2 offspring when bred with uninjected fish. The presence of DNA in fish of the F1 and F2 generations for one of the lines ("CATfishI") was demonstrated by dot blotting, and the copy number was strikingly constant (about 60 copies per cell). Putative junction fragments, which seemed to be the same in different individuals tested, were visible in Southern blots of F1 fish from the CATfishI line. The arrangement of the multiple copies is somewhat ambiguous, however, since although all copies seem to be tandemly linked, there are more than two weakly hybridizing fragments (which would have been the expectation for a single integration of a tandemly repeated array). All four transgenic lines showed CAT expression (by assay of excised tissues, and in some cases by staining with antibody against CAT) in fins, skin, heart, muscle, gills, and eyes, but not in the brain, liver, or gonad, and only weakly in the gut. The relative amount of expression in the different tissues was characteristic for each of the four lines.

A higher rate of transmission of injected DNA was found by Culp et al. (1991) with an RSV-*lacZ* construct. Fish reared from injected eggs were mated to each other or to uninjected fish and the DNA from F1 embryos was assayed for the presence of the *lacZ* gene by PCR. Three separate experiments gave positive F1 fish from 7, 19, and 25% of the fish raised from injected eggs. As in the work of Stuart et al. (1990), the founder F0 fish were mosaic; percentages of transmission for any one fish ranged from 6–24%. Individual F1 fish transmitted the transgene to 50% of the individuals in the next generation, indicating stable integration into chromosomal DNA. Southern blotting of DNA in the F1 fish indicated that tandem copies of the transgene were present, usually in single arrays. Some of the tandem arrays were made up of unit lengths that differed from the length of the plasmid DNA. Culp et al. (1991) suggest that amplification of the DNA must follow alterations or rearrangements in the original plasmid sequence. In two F0 fish, there were multiple integration events. A study of the inheritance of the DNA in one of these fish was very revealing. Of four bands on the Southern blot of F1 offspring from the F0 founder fish, two (denoted "2/4") *always* were found together in some of the F1 fish. This indicated that these bands were the product of a single integration, or two closely linked integration events. The other two bands ("1 and 3") *never* appeared together with the 2/4 bands, strongly suggesting that they were the result of integration into a different germ line precursor cell. Bands 1 and 3, however, did segregate in a Mendelian fashion to the F1 generation. Quantitation of the transmission of insertions which occur in distinct germ line precursors might be useful in determining the number of germ line precursor cells. This approach, as well as experiments with chimeric fish suggested in Section 4.5, can be potentially used to verify or contradict the conclusions of Walker and Streisinger (1983).

There were two surprising results in the Culp et al. (1991) study regarding specificity of integration and expression. First, parallel experiments in which an SV40-*lacZ* construct was injected yielded no transgenic founder fish; second, the F1 fish carrying the *lacZ* transgenes did not express β-galactosidase (even though the construct could be expressed transiently in embryos following injection). Since *lacZ* expression seems to be low or absent, Culp et al. (1991) suggest that perhaps the CAT expression assyed by Stuart et al. (1990) was low and did not truly reflect the number of fish containing the transgene. Their value of 4–5% of the injected fish which transmit the foreign DNA might in fact be an underestimate, but the DNA might not have been expressed in all the fish. C. Fulwiler has also obtained high frequencies of inheritance of injected DNAs (cited in Culp et al. 1991).

These studies most certainly represent only the beginning of a fruitful approach to the study of gene expression and gene function in the zebrafish. As pointed out by Culp et al. (1991), the frequency of transgenics may be high enough to make insertional mutagenesis practical. Moreover, if the problem of expression can be solved, the production of transgenics should enable one to study, among other things, the effects of ectopic expression of a gene, the consequences of expression of antisense transcripts, the effects of expression of mutant protein products (e.g., to test for dominant negative effects during development), the sequence requirements for promoter and enhancer function, and the use of enhancer and gene traps for gene isolation.

7.5 Transient Expression of Injected DNA

The ease in identifying specific cell types and tissues in the zebrafish embryo, and the ability to introduce DNA into the egg and obtain expression during embyogenesis, should permit a convenient assay for promoter function. Many laboratories have now achieved expression of β-galactosidase when constructs containing various viral promoters (e.g., SV40, RSV, CMV), fused to the *lacZ* gene, are injected into 1–2 cell embryos. The general result with these promoters is that expression is highly mosaic, and is not specific for any one cell type. My experience with the CMV-*lacZ* construct indicates that no more than 10–20% of the cells of the gastrula or later embryo expresses the DNA. Expression in individual cells can be seen as early as the blastula stage, but again only in a small fraction of cells. An interesting observation, however, is that there is often a high level of expression in a layer between the yolk and the embryonic cells. Since this is the location of the yolk syncytial layer, it is probable that the YSL nuclei are very active in transcribing the injected DNA.

The restriction of expression to a small number of cells might be due to a non-homogeneous distribution of injected DNA in the original embryo. Westerfield et al. (1992) observed that injected ethidium bromide-tagged DNA forms a rounded mass which is distributed to only a subset of cells. They also noted, however, that expression of the DNA was more widely distributed than the injected tagged DNA. I have found that DNA injected at the vegetal pole of the yolk, at the opposite side of the egg from the first embryonic cells, can be expressed in blastula cells. This indicates that the DNA somehow traverses the large distance across the yolk, perhaps aided by the powerful streaming events in the yolk which occur during the first divisions. I have also noted that there is little advantage of injecting DNA directly into the first cell over injecting into the yolk at the base of the first 2–4 cells. These results taken together, indicate that all cells in the blastula are likely to have access to at least some injected DNA. If this is the case, why do only relatively few cells express the injected constructs? Perhaps amplification or integration must occur as a prerequisite for expression. Alternatively, the amount of DNA which enters a particular nucleus might have to exceed a particular level to titrate out proteins with negative effects on transcription.

Although the mosaic expression of injected DNA creates some difficulties for the assay of tissue-specific transcription, one can observe many embryos and tabulate the distibution of tissue types in which a particular construct is active. This type of analysis has recently demonstrated that the mouse *Hox-1.1* and human *Hox-3.3* promoters (Westerfield et al. 1992), and the rat GAP-43 (Reinhard et al. 1991) promoter, function with correct (although somewhat imprecise) tissue specificity in the zebrafish embryo. In the Westerfield et al. (1992) experiments, both of the *Hox* promoters were active in the putative sclerotome in somites, and in the spinal cord (with the *Hox-3.3* promoter active in more dorsal regions than the *Hox-1.1* promoter). The expression of *Hox-1.1* in both somites and spinal cord was strongest in anterior segments, corresponding to the area where the gene is expressed in mice (Mahon et al. 1988; Dressler and Gruss 1989; Püschel et al. 1990). Moreover, various deletions in the *Hox-1.1* construct, previously shown to alter the expression pattern in mouse (Püschel et al. 1991), also have similar effects in the fish embryo. The specific expression of the *Hox-3.3* promoter also followed an expected pattern, in this case that of the putative endogenous gene (Molven et al. 1990). A similar assay performed with the rat GAP-43 promoter (Reinhard et al. 1991) showed specific expression in neurons of the 50-h embryo. The particular region within the central nervous system showing expression depended on the amount of 5' promoter sequence in the construct. A 5-kb segment directed expression in posterior regions as well as in the brain, whereas a 1-kb segment was active mainly in the brain. These results are extremely encouraging. They indicate a unity of regulation of key developmental and tissue-specific genes in vertebrates. Consequently, the fish embryo will be of great use for rapid assay of promoters of genes from other vertebrate classes, as well as genes from the zebrafish itself.

8 Summary and Future Prospects

It is now 10 years since Streisinger's first zebrafish publication, a period in which the perception of the zebrafish has changed from a rather esoteric curiosity to a species central in the efforts to work out mechanisms of vertebrate development. The considerable advances during the decade are due mainly to the group at the University of Oregon, which has pioneered the genetics and developmental studies on this organism. I have tried to summarize various approaches currently being used to dissect the development of the embryo. My aim was to give a comprehensive introduction to the literature for students and researchers considering working with the organism, and to provide background on aspects of development and molecular biology of the zebrafish which would be of interest to the larger community of developmental biologists.

The strengths of the zebrafish system are the ability to perform genetic analysis, the ease of visualizing all aspects of embryonic development, and the development of transgenic techniques. Mutant screens currently underway in a number of laboratories should be extremely fruitful. Mutants affecting fundamental aspects of differentiation and morphogenesis have already been reported (see Sect. 5), and other interesting phenotypes are being discovered monthly. Many of the mutants which have been studied thus far (e.g., cyclops and spadetail) have revealed aspects of embryonic control which had not been apparent from work with other species. The nature of the mutant screens (especially those designed to identify maternal genes) should identify loci which would be difficult to recognize in other vertebrates, even in the mouse. These genes might very well be those involved in setting up the polarity and body plan of the early embryo. On the other hand, many of the mutants will turn out to be in genes already identified in other organisms. The *ntl-1* mutation, for example, is a good candidate for the fish *Brachyury (T)* gene.

One should not minimize the difficulties ahead, however. Although rudimentary genetics is currently being carried out with the fish, it should be noted that there is absolutely no linkage information available at this time. The only map distances published are gene-centromere distances. Undoubtedly, the powerful techniques of linkage analysis using RFLP and simple sequence repeat polymorphisms will soon be applied to the zebrafish genome, facilitating the mapping of genes of interest – but this is still to come. Positional mapping and cloning of genes of any organism is theoretically possible, but for the zebrafish this will not be any easier than for the mouse or human (since the fish genome is of comparable size). In fact, because the zebrafish research community is small and has few of the resources available to human genome investigators, additional techniques will have to be developed to efficiently clone fish genes based just on their position in a linkage map.

An encouraging development has been the high efficiency of transgenic individuals which can be produced by DNA injection. If insertional mutagenesis turns out to occur at the same rate as in mice, this approach may be valuable for the identification and isolation of developmentally interesting genes. Screening transgenic lines for mutations is facilitated by the ability to use haploid embryos in the initial survey. However, the efficiency of experiments of this sort will be considerbly lower than screens of fish treated with the more conventional chemical and gamma ray mutagens. Using the current technology, one would probably have to screen thousands of fish raised from injected eggs to see a phenotype. The payoff, however, would be that the interrupted gene could easily be isolated. The development of enhancer and gene trap assays for insertions is currently underway in a number of laboratories and should aid in this approach to gene identification.

The unity of vertebrate embryonic mechanisms is becoming clearer, both from comparative embryology (see Kimmel 1989; Davidson 1991) and from the study of particular gene products (see Sect. 6). Isolated zebrafish genes with developmental functions will certainly be of interest to those working on mice, *Xenopus*, or chick, and vice versa. Important in this regard is the recent finding that at least some mammalian promoters function with appropriate tissue and regional specificity in the fish embryo (Reinhard et al. 1991; Westerfield et al. 1992). The ability to create chimeric mice with germ lines containing cells derived from transplanted blastomeres is also very encouraging (Lin et al. 1992). If a system analogous to the mouse embryonic stem cell could be worked out in the zebrafish, targeted mutagenesis could be applied in the fish as well. On the horizon are also approaches such as disruption of gene function by antisense methods, and overexpression and ectopic expression of genes during embryogenesis. In principle, many of the experiments being done with transgenic mice could also be done with zebrafish, more easily, and at less expense. The second decade of zebrafish biology is ahead of us. It should be a most exciting period.

Acknowledgements. Part of this review was written while on sabbatical at the Max-Planck-Institut für Entwicklungsbiologie in Tübingen. I would like to thank Dr. Christiane Nüsslein-Volhard for her hospitality, for the exceptional research environment in her laboratory, and for many insights into the strategies of studying development. Special thanks are extended to Stefan Schulte-Merker who taught me much about zebrafish, and was generous in all respects. I thank Miguel Allende, Mary Mullins, Bob Riggleman, and Mathias Hammerschmidt for lively discussions and experimental advice. Stefan Schulte-Merker, Nancy Hopkins, Chuck Kimmel, Monte Westerfield, and David Grunwald generously discussed their work in progress and provided me with manuscripts prior to publication. Nancy Hopkins, Wolfgang Driever, Carl Fulwiler, José Campos-Ortega, Chuck Kimmel, and Monte Westerfield were all kind and informative hosts during visits to their laboratories. David Bumcrot and Miguel Allende provided helpful comments on the manuscript.

References

Armstrong JB, Graveson AC (1988) Progressive patterning precedes somite segmentation in the Mexican axolotl (*Abystoma mexicanum*). Dev Biol 126:1–6

Ballard WW (1966a) The role of the cellular envelope in the morphogenetic movements of teleost embryos. J Exp Zool 161:193–200

Ballard WW (1966b) Origin of the hypoblast in *Salmo*. I. Does the blastodisc edge turn inward? J Exp Zool 161:201–210

Ballard WW (1966c) Origin of the hypoblast in *Salmo*. II. Outward movements of deep central cells. J Exp Zool 161:211–220

Ballard WW (1973) Morphogenetic movements in *Salmo gairdneri* Richardson. J Exp Zool 184:27–48

Balling R, Deutsch U, Gruss P (1988) *Undulated*, a mutation affecting the development of the mouse skeleton, has a point mutation in the paired box of Pax-1. Cell 55:531–535

Baumgartner S, Bopp D, Burri M, Noll M (1987) Structure of two genes at the *gooseberry* locus related to the *paired* gene and their spatial expression during *Drosophila* embryogenesis. Genes Dev 1:1247–1267

Beams HW, Kessel RG (1976) Cytokinesis: a comparative study of cytoplasmic division in animal cells. Am Sci 64:279–290

Beams HW, Kessel RG, Shih CY, Tung HN (1985) Scanning electron microsocopic studies on blastodisc formation in the zebrafish, *Brachydanio rerio*. J Morphol 184:41–50

Beddington R (1986) Analysis of tissue fate and prospective potency in the egg cylinder. In: Rossant J, Pedersen RA (eds) Experimental approaches to mammalian development. Cambridge Univ Press, New York

Betchaku T, Trinkaus JP (1978) Contact relations, surface activity, and cortical microfilaments of marginal cells of the enveloping layer and of the yolk syncitial and yolk cytoplasmic layers of *Fundulus* before and after epiboly. J Exp Zool 206:381–426

Boncinelli E, Simeone A, Acampora D, Mavilio F (1991) *Hox* gene activation by retinoic acid. Trends Genet 7:329–334

Bopp D, Burri M, Baumgartner S, Frigerio G, and Noll M (1986) Conservation of a large protein domain in the segmentation gene paired and in functionally related genes of *Drosophila*. Cell 47:1033–1040

Bopp D, Jamet E, Baumgartner S, Burri M, Noll M (1989) Isolation of two tissue-specific *Drosophila* paired box genes, *Pox meso* and *Pox neuro*. EMBO J 8:3447–3457

Bradley RS, Brown AMC (1990) The proto-oncogene *int-1* encodes a secreted protein associated with the extracellular matrix. EMBO J 9:1569–1990

Brem G, Brenig B, Horstgen-Schwark G, Winnacker EL (1988) Gene transfer in tilapia (*Oreochromis niloticus*). Aquaculture 68:209–219

Buono RJ, Linser PL (1991) Transgenic zebrafish by electroporation. Bio-Rad US/EG Bull 1354

Burri M, Tromvoudis Y, Bopp D, Frigerio G, Noll M (1989) Conservation of the paired domain in metazoans and its structure in three isolated human genes. EMBO J 8:1183–1190

Chakrabarti S, Streisinger G, Singer F, Walker C (1983) Frequency of gamma-ray induced specific locus and recessive lethal mutations in mature germ cells of the zebrafish, *Brachydanio rerio*. Genetics 103:109–123

Chesley P (1935) Development of the short-tailed mutant in the house mouse. J Exp Zool 70:429–435

Chong SSC, Vielkind JR (1989) Expression and fate of CAT reporter gene microinjected into fertilized medaka (*Oryzias latipes*) eggs in the form of plasmid DNA, recombinant phage particles, and its DNA. Theor Appl Genet 78:369–380

Chourrout D, Guyomard R, Houdebine L (1986) High efficiency gene transfer in rainbow trout (*Salmo gairdneri* Rich.) by microinjection in the egg cytoplasm. Aquaculture 51:143–150.

Coleman KG, Poole SJ, Weir MP, Soeller WC, Kornberg TB (1987) The *invected* gene of *Drosophila*: sequence analysis and expression studies reveal a close kinship to the *engrailed* gene. Genes Dev 1:19–28

Culp P, Nüsslein-Volhard C, Hopkins N (1991) High-frequency germ-line transmission of plasmid DNA sequences injected into fertilized zebrafish eggs. Proc Natl Acad Sci USA 88:7953–7957

Dale L, Slack JMW (1987a) Regional specification within the mesoderm of early embryos of *Xenopus laevis*. Development 100:279–295

Dale L, Slack JMW (1987b) Fate map for the 32-cell stage of *Xenopus laevis*. Development 99:527–551

Davidson D, Graham E, Sime C, Hill R (1988) A gene with sequence similarity to *Drosophila engrailed* is expressed during the development of the neural tube and vertebrae in the mouse. Development 104:305–316

Davidson EH (1991) Spatial mechanisms of gene regulation in metazoan embryos. Development 113:1–26

Davis CA, Joyner AL (1988) Expression patterns of the homeo box-containing genes *En-1* and *En-2* and the proto-oncogene *int-1* diverge during mouse development. Genes Dev 2:1736–1744

Davis CA, Noble-Topham SE, Rossant J, Joyner AL (1988) Expression of the homeobox-containing gene *EN-2* delineates a specific region of the developing mouse brain. Genes Dev 2:361–371

Davis CA, Holmyard DP, Millen KJ, Joyner A (1991) Examining pattern formation in mouse, chicken, and frog embryos with an *En*-specific antiserum. Development 111:287–298

Dollé P, Izpisua-Belmonte J-C, Falkenstein H, Renucci A, Duboule D (1989) Coordinate expression of the murine *Hox-5* complex homeobox-containing genes during limb pattern formation. Nature 342:767–772

Drake JC (1969) Mutagenic mechanisms. Annu Rev Genet 61:247–268

Dressler GR, Gruss P (1989) Anterior boundaries of *Hox* gene expression in mesoderm-derived structures correlate with the linear order along the chromosome. Differentiation 41:193–201

Dressler GR, Deutsch U, Balling R, Simon D, Guenet J-L, Gruss P (1988) Murine genes with homology to *Drosophila* segmentation genes. Development 104:181–186

Dressler GR, Deutsch U, Chowdhury K, Nornes HO, Gruss P (1990) *Pax-2*, a new murine paired-box-containing gene and its expression in the developing excretory system. Development 109:787–795

Duboule D, Dolle P (1989) The structure and functional organization of the murine *Hox* gene family resembles that of *Drosophila* homeotic genes. EMBO J 8:1497–1505

Dunham RA (1990) Genetic engineering in aquaculture AgBiotech News Info 22:401–406

Dunham RA, Eash J, Askins J, Townes TM (1987) Transfer of the metallothionein-human growth hormone fusion gene into channel catfish. Trans Am Fisheries Soc 116:87–91

Eiken HG, Njølstad PR, Molven A, Fjose A (1987) A zebrafish homeobox-containing gene with embryonic transcription. Bioch Biophys Res Comm 149:1165–1171

Eisen JS (1991) Developmental neurobiology of the zebrafish. J Neurosci 11:311–317

Eisen JS, Pike SH (1991) The *spt-1* mutation alters segmental arrangement and axonal development of identified neurons in the spinal cord of embryonic zebrafish. J Neurosci 10:34–43

Ekker M, Akimenko M-A (1991) Embryology and genetics of the zebrafish, *Brachydanio rerio*. Int J Biol Med 7:553–560

Elsdale T, Pearson M, Whitehead M (1976) Abnormalities in somite segmentation following heat shock to *Xenopus* embryos. J Embryol Exp Morphol 35:625–635

Endo A, Ingalls TH (1968) Chromosomes of the zebra fish. A model for cytogenic, embryologic, and ecologic study. J Hered 59:382–384

Felsenfeld AL, Walker C, Westerfield M, Kimmel CB, Streisinger G (1990) Mutations affecting skeletal muscle myofibril structure in the zebrafish. Development 108:443–459

Felsenfeld AL, Curry M, Kimmel CB (1991) The *fub-1* mutation blocks initial myofibril formation in zebrafish muscle pioneer cells. Dev Biol 148:23–30

Fjose A, McGinnis WJ, Gehring WJ (1985) Isolation of a homeobox-containing gene from the *engrailed* region of *Drosophila* and the spatial distribution of its transcripts. Nature 313:284–289

Fjose A, Eiken HG, Njølstad PR, Molven A, Hordvik I (1988) A zebrafish *engrailed*-like sequence expressed during embryogenesis. FEBS Lett 231: 355–360

Fulwiler C, Gilbert W (1991) Zebrafish embryology and neural development. Curr Opinions Cell Biol 3:988–991

Gardner RL, Lyon MF, Evans EP, Burtenshaw MD (1985) Clonal analysis of X-chromosome inactivation and the origin of the germline in the mouse embryo. J Embryol Exp Morphol 88:349–363

Gluecksohn-Schoenheimer S (1938) The development of normal and homozygous *Brachy (T/T)* mouse embryos in the extraembryonic coelom of the chick. Proc Natl Acad Sci USA 30:134–140

Goulding MD, Chalepakis G, Deutsch U, Erselius JR, Gruss, P (1991) Pax-3, a novel murine DNA binding protein expressed during early neurogenesis. EMBO J 10:1135–1147

Graham A, Papalopulu N, Krumlauf R (1989) The murine and *Drosophila* homeobox gene complexes have common features or organization and expression. Cell 57:367–378

Grüneberg H (1958) Genetical studies on the skeleton of the mouse XXIII: the development of *Brachyury* and *Anury*. J Embryol Exp Morphol 6:424–443

Grunwald DJ, Streisinger G (1992a) Induction of mutations in the zebrafish with ultraviolet light. Genet Res Camb 59:93–101.

Grunwald DJ, Streisinger G (1992b) Induction of recessive lethal and specific locus mutations in the zebrafish with ethyl nitrosourea. Genet Res Camb 59:103–116

Grunwald DJ, Kimmel CB, Westerfield M, Walker C, Streisinger G (1988) A neural degeneration mutation that spares primary neurons in the zebrafish. Dev Biol 126:115–128

Guyomard R, Chourrout D, Leroux C, Houdebine LM, Pourrain F (1989) Integration and germ line transmission of foreign genes micro injected into fertilized trout eggs. Biochemie 71:857–863

Haldane JBS (1919) The combination of linkage values, and the calculation of distances between linked factors. J Genet 8:299–309

Hanneman EH, Westerfield M (1989) Early expression of acetylcholinesterase activity in functionally distinct neurons of the zebrafish. J Comp Neurol 284:350–361

Hanneman E, Trevarrow B, Kimmel CB, Westerfield M (1988) Segmental pattern of development of the spinal cord and hindbrain of the zebrafish embryo. Development 103:49–58

Hatta K, Schilling TF, BreMiller RA, Kimmel CB (1990) Specification of jaw muscle in zebrafish: correlation with *engrailed*-homeoprotein expression. Science 250:802–805

Hatta K, BreMiller R, Westerfield M, Kimmel CB (1991a) Diversity of expression of *engrailed*-like antigens in zebrafish. Development 112:821–832

Hatta K, Kimmel CB, Ho RK, Walker C (1991b) The cyclops mutation blocks specification of the floor plate of the zebrafish central nervous system. Nature 350:339–341

Hemmati-Brivanlou A, Harland RM (1989) Expression of an *engrailed*-related protein is induced in the anterior neural ectoderm of early *Xenopus* embryos. Development 106:611–617

Herrmann BG, Labeit S, Poustka A, King TR, Lehrach H (1990) Cloning of the *T* gene required in mesoderm formation in the mouse. Nature 343:617–622

Hinegardner R, Rosen DE (1972) Cellular DNA content and the evolution of teleostean fishes. Am Natur 106:311–319

Hisaoka KK, Battle HL (1958) The normal developmental stages of the zebrafish, *Brachydanio rerio* (Hamilton-Buchanan). J Morphol 102:311–328

Hisaoka KK, Firlit CF (1960) Further studies on the embryonic development of the zebrafish, *Brachydanio rerio* (Hamilton-Buchanan). J Morphol 107:205–255

Ho RK, Kane DA (1990) Cell-autonomous action of zebrafish *spt-1* mutation in specific mesodermal precursors. Nature 348:728–730

Holland PWH (1991) Cloning and evolutionary analysis of *msh*-like homeobox genes from mouse, zebrafish and ascidian. Gene 98:253–257

Holland PWH, Hogan BLM (1988) Expression of homeobox genes during mouse development: a review. Genes Dev 2:773–782

Holland PWH, Williams NA (1990) Conservation of *engrailed*-like homeobox sequences during vertebrate evolution. FEBS Lett 277:250–252

Horvitz RH, Brenner S, Hodgkin J, Herman RK (1979) A uniform genetic nomenclature for the nematode, *Caenorhabditis elegans*. Mol Gen Genet 175:129–133

Houdebine LM, Chourrout D (1991) Transgenics in fish. Experientia 47:891–897

Indiq FE, Moav B (1988) A prokaryotic gene is expressed in fish cells and persists in tilapia embryos and adults following microinjection. In: Zohar Y, Breton B (eds) Reproduction in fish: basic and applied aspects of endocrinology and genetics. INRA Press, Paris, p 221

Ingham PW (1988) The molecular genetics of embryonic pattern formation in *Drosophila*. Nature 335:25–34

Inoue K, Yamashita S, Hata JI, Kabeno S, Asada S, Nagahisa F, Fujita T (1990) Electroporation as a new technique for producing transgenic fish. Cell Differ Dev 29:123–128

Johnson FM, Lewis SE (1981) Electrophoretically detected germinal mutations induced in the mouse by ethylnitrosourea. Proc Natl Acad Sci USA 78: 3138–3141

Joyner AL, Martin GR (1987) *En-1* and *En-2*, two mouse genes with sequence homology to the *Drosophila engrailed* gene: expression during embryogenesis. Genes Dev 1:29–38 (erratum, Genes Dev 1:521)

Joyner AL, Kornberg T, Coleman KG, Cox DR, Martin GR (1985) Expression during embryogenesis of a mouse gene with sequence homology to the *Drosophila engrailed* gene. Cell 43:29–37

Joyner AL, Herrup BA, Auerbach CA, Rossant J (1991) Stable cerebellar phenotype in mice homozygous for a targeted deletion of the *En-2* homeobox. Science 251:1239–1243

Kappen C, Schughart K, Ruddle FH (1989) Two steps in the evolution of Antennapedia class vertebrate homeobox genes. Proc Natl Acad Sci USA 86:5459–5463

Keller RE, Trinkaus JP (1987) Rearrangement of enveloping layer cells without disruption of the epithelial permeability barrier as a factor in *Fundulus* epiboly. Dev Biol 120:12–24

Kelly SJ (1977) Studies of the developmental potential of 4- and 8-cell stage mouse blastomeres. J Exp Zool 200:365–376

Kenyon C, Wang B (1991) A cluster of Antennapedia-class homeobox genes in a nonsegmented animal. Science 253:516–517

Kessel M, Gruss P (1990) Murine developmental control genes. Science 249:374–379

Kimmel CB (1989) Genetics and early development of zebrafish. Trends Genet 5:283–288

Kimmel CB, Law RD (1985a) Cell lineage of zebrafish blastomeres. I. Cleavage pattern and cytoplasmic bridges between cells. Dev Biol 108:78–85

Kimmel CB, Law RD (1985b) Cell lineage of zebrafish blastomeres. II. Formation of the yolk syncitial layer. Dev Biol 107:86–93

Kimmel CB, Law RD (1985c) Cell lineage of zebrafish blastomeres. III. Clonal analysis of the blastula and gastrula stages. Dev Biol 107:94–101

Kimmel CB, Warga RM (1986) Tissue-specific cell lineages originate in the gastrula of the zebrafish. Science 231:365–368

Kimmel CB, Warga RM (1987a) Indeterminate cell lineage of the zebrafish embryo. Dev Biol 124:269–280

Kimmel CB, Warga RM (1987b) Cell lineages generating axial muscle in the zebrafish embryo. Nature 327:234–237

Kimmel CB, Warga RM (1988) Cell lineage and developmental potential of cells in the zebrafish embryo. Trends Genet 4:68–74

Kimmel CB, Sepich DS, Trevarrow B (1988) Developmental segmentation in zebrafish. Development 104 (Suppl):197–207

Kimmel CB, Kane DA, Walker C, Warga RM, Rothman MB (1989) A mutation that changes cell movement and cell fate in the zebrafish embryo. Nature 337:358–362

Kimmel CB, Warga RM, Schilling TF (1990) Origin and organization of the zebrafish fate map. Development 108:581–594

Kimmel CB, Hatta K, Eisen JS (1991) Genetic control of primary neuronal development in zebrafish. Development Suppl (in press)

Krauss S, Johansen T, Korzh V, Fjose A (1991a) Expression of the zebrafish paired box gene pax[zf-b] during early neurogenesis. Development 113:1193–1206

Krauss S, Johansen T, Korzh V, Moens U, Ericson JU, Fjose A (1991b) Zebrafish pax[zf-a]: a paired box-containing gene expressed in the neural tube. EMBO J 10:3609–3619

Krumlauf R, Holland PWH, McVey JH, Hogan BLM (1987) Developmental and spatial patterns of expression of the mouse homeobox gene, Hox 2.1. Development 99:603–617

Langille RM, Hall BK (1988) Role of the neural crest in development of the trabeculae and branchial arches in the embryonic sea lamprey, Petromyzon marinus. Development 102:301–310

Lawson KA, Menses JJ, Pederson RA (1986) Cell fate and cell lineage in the endoderm of the presomitic mouse embryo, studied with an intracellular tracer. Dev Biol 115:325–339

Lewis WH, Roosen-Runge EC (1942) The formation of the blastodisc in the egg of the zebrafish, Brachydanio rerio (illustrated with motion pictures). Anat Rec 84:463–464

Lin S, Long W, Chen J, Hopkins N (1992) Production of germ-line chimeras in zebrafish by cell transplants from genetically pigmented to albino embryos. Proc Natl Acad Sci USA 89:4519–4523.

Luther W (1936) Potenzprüfungen an isolierten Teilstücken der Forellenkeimscheibe. Wilhelm Roux' Arch Entwicklungsmech Org 135:359–383

Mahon KA, Westphal H, Gruss P (1988) Expression of homeobox gene Hox 1.1 during mouse embryogenesis. Development 104 (Suppl):187–195

Marcey D, Nüsslein-Volhard C (1986) New and views: embryology goes fishing. Nature 321:380–381

McEvoy T, Stack M, Keane B, Barry T, Sreenan J, Gannon F (1988) The expression of a foreign gene in salmon embryos. Aquaculture 68:27–37

McGinnis W, Levine MS, Hafen E, Kuroiwa A, Gehring WJ (1984) A conserved DNA sequence in homeotic genes of the Drosophila Antennapedia and bithorax complexes. Nature 308:428–433

McLaren A (1976) Mammalian chimeras. Cambridge Univ Press, New York

McMahon AP, Bradley A (1990) The Wnt-1 (int-1) proto-oncogene is required for development of a large region of the mouse brain. Cell 62:1073–1085

McMahon AP, Moon RT (1989) int-1 – a proto-oncogene involved in cell signalling. Development (1989 Suppl):161–167

Mendelson B (1986a) Development of reticulospinal neurons of the zebrafish. I. Time of origin. J Comp Neurol 251:160–171

Mendelson B (1986a) Development of reticulospinal neurons of the zebrafish. II. Early axonal outgrowth and cell body position. J Comp Neurol 252:172–184

Mintz B (1967) Gene control of mammalian pigmentary differentiation. I. Clonal origin of melanocytes. Proc Natl Acad Sci USA 58:344–351

Mintz B (1970) Gene expression in allophenic mice. In: Padykula HA (ed) Control mechanisms in the expression of cellular phenotypes. Academic Press, New York, p 15

Molven A, Wright CVE, BreMiller R, De Robertis EM, Kimmel CB (1990) Expression of a homeobox gene product in normal and mutant zebrafish embryos: evolution of the tetrapod body plan. Development 109:279–288

Molven A, Njølstad PR, Fjose A (1991) Genomic structure and restricted neural expression of the zebrafish *wnt-1 (int-1)* gene. EMBO J 10:799–807

Moody SA (1987) Fates of the blastomeres of the 32-cell-stage *Xenopus* embryo. Dev Biol 122:300–319

Morgan TH (1895) The formation of the fish embryo. J Morphol 10:419–472

Myers PZ, Eisen JS, Westerfield M (1986) Development and axonal outgrowth of identified motoneurons in the zebrafish. J Neurosci 6:2278–2289

Newport J, Kirschner M (1982) A major developmental transition in early *Xenopus* embryos. I. Characterization and timing of cellular changes at the midblastula stage. Cell 30:675–686

Nishida H (1987) Cell lineage analysis in ascidian embryos by intracellular injection of a tracer enzyme. III. Up to the tissue restricted stage. Dev Biol 121:526–541

Njølstad PR, Fjose A (1988) In situ hybridization patterns of zebrafish homeobox genes homologous to *Hox-2.1* and *En-2* of mouse. Bioch Biophys Res Comm 157:426–432

Njølstad PR, Molven A, Eiken HG, Fjose A (1988a) Structure and neural expression of a zebrafish homeobox sequence. Gene 73:33–46

Njølstad PR, Molven A, Hordvik I, Apold J, Fjose A (1988b) Primary structure, developmentally regulated expression and potential duplication of the zebrafish homeobox gene ZF-21. Nucl Acids Res 16:9096–9111

Njølstad PR, Molven A, Fjose A (1988c) A zebrafish homologue of the murine *Hox-2.1* gene. FEBS Lett 230:25–30

Njølstad PR, Molven A, Apold J, Fjose A (1990) The zebrafish homeobox gene *hox-2.2*: transcription unit, potential regulatory regions and in situ localization of transcripts EMBO J 9:515–524

Nusse R (1988) The *int* genes in mammary tumorigenesis and in normal development. Trends Genet 2:244–247

Oliver G, Wright CVE, Hardwicke J, DeRobertis EMM (1988a) Differential antero-posterior expression of two proteins encoded by a homeobox gene in *Xenopus* and mouse embryos. EMBO J 9:515–524

Oliver G, Wright CVE, Hardwicke J, DeRobertis EMM (1988b) A gradient of homeodomain protein in developing forelimbs of *Xenopus* and mouse embryos. Cell 55:1017–1024

Oliver G, Sidell N, Fiske W, Heinzmann C, Mohandas T, Sparkes RS, De Robertis EM (1989) Complementary homeo protein gradients in developing limb buds. Genes Dev 3:641–650

Oppenheimer JM (1936) Processes of localization in developing *Fundulus*. J Exp Zool 73:405–444

Oppenheimer JM (1937) The normal stages of *Fundulus heteroclitus*. Anat Rec 68:1–15

Oppenheimer JM (1938) Potencies for differentiation in the teleostean germ ring. J Exp Zool 79:405–444

Ozato K, Kondoh H, Inohara H, Iwamatsu T, Wakamatsu Y, Okada TS (1986) Production of transgenic fish: introduction and expression of chicken δ-crystallin gene in medaka embryos. Cell Differ 19:237–244

Papkoff J, Schryver B (1990) Secreted *int-1* protein is associated with the cell surface. Mol Cell Biol 10:2723–2730

Papkoff J, Brown AMC, Varmus HE (1987) The *int-1* proto-oncogene products are glycoproteins that appear to enter the secretory pathway. Mol Cell Biol 7:3978–3984

Pasteels J (1936) Etudes sur la gastrulation des vertebres meroblastiques. I. Teleosteens. Arch Biol 47:206–308

148 Eric S. Weinberg

Pastnik A, Vreeken C, Nivard MJM, Searles KK, Vogel EW (1989) Sequence analysis of N-ethyl-N-nitrosourea-induced vermillion mutations in *Drosophila melanogaster*. Genetics 123:123–129

Patel NH, Martin-Blanco E, Coleman KG, Poole SJ, Ellis MC, Kornberg TB, Goodman CS (1989a) Expression of engrailed proteins in arthropods, annelids, and chordates. Cell 58:955–968

Patel NH, Schafer B, Goodman CS, Holmgren R (1989b) The role of segment polarity genes during *Drosophila* neurogenesis. Genes Dev 3:890–904

Perkins DD (1955) Tetrads and crossing over. J Cell Physiol 45 (Suppl 2):119–149

Poole SJ, Kauvar L, Drees B, Kornberg T (1985) The *engrailed* locus of *Drosophila*: structural analysis of an embryonic transcript. Cell 40:37–43

Püschel AW, Balling R, Gruss P (1990a) Position-specific activity of the *Hox 1.1* promoter in transgenic mice. Development 108:435–442

Püschel AW, Balling R, Gruss P (1990b) Separate elements cause lineage restriction and specify boundaries of *Hox-1.1* expression. Development 112:279–287

Püschel AW, Gruss P, Westerfield M (1992) Sequence and expression pattern of *pax-6* are highly conserved between zebrafish and mice. Development 114:643–651

Reid L (1990) Meeting review: from gradients to axes, from morphogenesis to differentiation. Cell 63:875–882

Reinhard E, Nedivi E, Wegner J, Skene JHP, Westerfield M (1991) Functional conservation of GAP-43 gene regulatory elements between mammals and fish. Soc Neurosci Abstr 16:1310

Richardson KK, Richardson FC, Crosby RM, Swenberg JA, Skopek TR (1987) DNA base changes and alkylation following in vivo exposure of *Escherichia coli* to N-methyl-N-nitrosourea or N-ethyl-N-nitrosourea. Proc Natl Acad Sci USA 84:344–348

Rokkones E, Alestrom P, Skjervold H, Gautvik KM (1985) Development of a technique for microinjection of DNA into salmonid eggs. Acta Physiol Scand 124: Suppl 542, 417

Rokkones E, Alestrom P, Skjervold H, Gautvik KM (1989) Micro-injection and expression of a mouse metallothionein human growth hormone fusion gene in fertilized salmonid eggs. J Comp Physiol B 158:751–758

Roosen-Runge E (1936) Furchung und Primitiventwicklung von *Brachydanio rerio*. Verh Anat Ges, Anat Anz 81:297–301

Roosen-Runge E (1938) On the early development – bipolar differentiation and cleavage – of the zebra fish, *Brachydanio rerio*. Biol Bull 75:119–133

Roosen-Runge E (1939) Karyokinesis during cleavage of the zebrafish, *Brachydanio rerio*. Biol Bull 77:79–91

Rossant J (1985) Interspecific cell markers and cell lineage in mammals. Philos Trans R Soc Lond Ser B 312:91–100

Ruis i Altaba A (1991) Vertebrate development: an emerging synthesis. Trends Genet 7:276–280

Runstadler JA, Kocher TD (1991) A new *antennapedia*-class gene from the zebrafish. Nucl Acids Res 19:5434

Russel WL, Kelley EM, Hunsicker PR, Bangham JW, Maddux SC, Phipps EL (1979) Specific-locus test shows ethylnitrosourea to be the most potent mutagen in the mouse. Proc Natl Acad Sci USA 76:5818–5819

Sanes JR, Rubenstein JLR, Nicolas J-F (1986) Use of a recombinant retrovirus to study post-implantation cell lineage in mouse embryos. EMBO J 5:3133–3142

Schirone RC, Gross L (1968) Effect of temperature on early embryological development of the zebra fish, *Brachydanio rerio*. J Exp Zool 169:43–52

Schneider JF, Hallerman EM, Yoon SJ, He L, Gross ML, Liu Z, Faras AJ, Hackett PB, Kapuscinski AR, Guise KS (1989) Transfer of the bovine growth hormone gene into northern pike, *Esox lucius*. J Cell Biochem 13B (Abstr)

Schughart K, Pravtcheva D, Newman MS, Hunihan L, Jiang Z, Ruddle FH (1989) Isolation and regional localization of the murine homeobox-containing gene *Hox-3.3* to mouse chromosome region 15E. Genomics 5:76–83

Schulte-Merker S, Ho RK, Hermann BG, Nüsslein-Volhard C (1992) The protein product of the zebrafish homologue of the mouse *T* gene is expressed in nuclei of the germ ring and the notochord of the early embryo. Development (In press)

Scott MP, Weiner AJ (1984) Structural relationships among genes that control development: Sequence homology between the Antennapedia, Ultrabithorax, and fushi tarazu loci in *Drosophila*. Proc Natl Acad Sci USA 81:4115–4119

Shackleford GM, Varmus HE (1987) Expression of the proto-oncogene *int-1* is restricted to postmeiotic male germ cells and the neural tube of midgestational embryos. Cell 50:89–95

Smith JC, Price BMJ, Green JBA, Weigel D, Herrmann BG (1991) Expression of a *Xenopus* homolog of *Brachyury (T)* is an immediate-early response to mesoderm induction. Cell 67:1–20

Soriano P, Jaenisch R (1986) Retroviruses as probes of mammalian development: allocation of cells to the somatic and germ cell lineages. Cell 46:19–29

Streisinger G (1984) Attainment of minimal biological viability and measurements of genotoxicity: production of homozygous diploid zebra fish. Natl Cancer Inst Monogr 65:53–58

Streisinger G, Walker C, Dower N, Knauber D, Singer F (1981) Production of clones of homozygous diploid zebra fish (*Brachydanio rerio*). Nature 291: 293–296

Streisinger G, Singer F, Walker C, Knauber D, Dower N (1986) Segregation analyses and gene-centromere distances in zebrafish. Genetics 112:311–319

Streisinger G, Coale F, Taggart C, Walker C, Grunwald DJ (1989) Clonal origins of cells in the pigmented retinal of the zebrafish eye. Dev Biol 131:60–69

Stuart GW, McMurray JV, Westerfield M (1988) Replication, integration and stable germ-line transmission of foreign sequences injected into early zebrafish embryos. Development 103:403–412

Stuart GW, Vielkind JR, McMurray JV, Westerfield M (1990) Stable lines of transgenic zebrafish exhibit reproducible patterns of transgene expression. Development 109:577–584

Thomas KR, Capecchi MR (1990) Targeted disruption of the murine *int-1* proto-oncogene resulting in severe abnormalities in midbrain and cerebellar development. Nature 346:847–850

Treisman J, Harris E, Despan C (1991) The paired box encodes a second DNA-binding domain in the paired homeo domain protein. Genes Dev 5:594–604

Trevarrow B, Marks DL, Kimmel CB (1990) Organization of hindbrain segments in the zebrafish embryo. Neuron 4:669–679

Trinkaus JP (1984a) Mechanism of *Fundulus* epiboly – a current view. Am Zool 24:673–688

Trinkaus JP (1984b) Cells into organs. The forces that shape the embryo. Prentice-Hall, Englewood Cliffs

Van Raamsdonk W, Pool CW, teKronnie G (1978) Differentiation of muscle fiber types in the teleost, *Brachydanio rerio*. Anat Embryol 153:137–155

Van Raamsdonk W, van't Veer L, Veeken K, Heyting D, Pool CW (1982) Differentiation of muscle types in the teleost, *Brachydanio rerio*, the zebrafish. Anat Embryol 164:51–62

Veini M, Bellairs R (1986) Heat shock effects in chick embryos. In: Bellairs R, Ede DA, Lash J (eds) Somites and developing embryos. Plenum Press, New York, p 135

Walker C, Streisinger G (1983) Induction of mutations by gamma-rays in pregonial germ cells of zebrafish embryos. Genetics 103:125–136

Walther G, Guenet J-L, Simon D, Deutsch U, Jostes B, Goulding MD, Plachov D, Balling R, Gruss P (1991) *Pax*: a murine multigene family of paired box containing genes. Genomics 11:424–434

Warga RM, Kimmel CB (1990) Cell movements during epiboly and gastrulation in zebrafish. Development 108:569–580

Weisblat DA, Sawyer RT, Stent GS (1978) Cell lineage analysis by intracellular injection of a tracer enzyme. Science 202:1295–1298

Weisblat DA, Zackson SL, Blair SS, Young JD (1980) Cell lineage analysis by intracellular injection of fluorescent tracers. Science 209:1538–1541

Westerfield M (1989) The zebrafish book. Institute of Neurosciences, University of Oregon, Eugene, Oregon

Westerfield M, McMurray JV, Eisen JS (1986) Identified motoneurons and their innervation of axial muscles in the zebrafish. J Neurosci 6:2267–2277

Westerfield M, Liu DW, Kimmel CB, Walker C (1990) Pathfinding and synapse formation in a zebrafish mutant lacking functional acetylcholine receptors. Neuron 4:867–874

Westerfield M, Wegner J, Jegalian BG, DeRobertis EM, Püschel AW (1992) Specific activation of mammalian *Hox* promoters in mosaic transgenic zebrafish. Genes Dev 6:591–598

Wieschaus E (1978) Cell lineage relationships in the *Drosophila* embryo. In Gehring WJ (ed) Results and problems in cell differentiation, vol 9. Springer, Berlin Heidelberg New York, p 97

Wilkinson DG, Bailes JA, McMahon AP (1987) Expression of the proto-oncogene *int-1* is restricted to specific neural cells in the developing mouse embryo. Cell 50:79–88

Wilkinson DG, Bhatt S, Herrmann BG (1990) Expression pattern of the mouse *T* gene and its role in mesoderm formation. Nature 343:657–659

Wilson HV (1891) The embryology of the sea bass. Bull US Fish Comm 9:209–277

Wood A, Timmermans LPM (1988) Teleost epiboly: reassessment of deep cell movement in the germ ring. Development 102:575–585

Yamamoto T (1969) Sex differentiation. In: Hoar WS, Randall DJ (eds) Fish physiology, vol 3. Academic press, New York, p 117

Yanagisawa KO, Fujimoto H, Urushihara H (1981) Effects of the *Brachyury (T)* mutation on morphogenetic movement in the mouse embryo. Dev Biol 87: 242–248

Yoon SJ, Liu Z, Kapuscinski AR, Hackett PB, Faras A, Guise KS (1990) Successful gene transfer in fish. In: Verma I, Mulligan R, Beauset A (eds) UCLA Symposium on molecular and cellular biology, vol 87. Alan Liss, New York, p 29

Zelenin AV, Alimov AA, Barmintzev VA, Beniumov AO, Zelenina I, Krasnov AM, Kolesnikov VA (1991) The delivery of foreign genes into fertilized fish eggs using high-velocity microprojectiles. FEBS Lett 287:118–120

Zhang P, Hayat M, Joyce C, Gonzalez-Villasenor LI, Lin CM, Dunham R, Chen TT, Powers DA (1990) Gene transfer, expression and inheritance of pRSV-rainbow trout-GHcDNA in the carp, *Cyprinus carpio* (Linnaeus). Molec Reprod Dev 25:3–13

4 Early Mouse Development

Achim Gossler

1 Introduction

The mouse is well established as *the* experimental mammal for developmental and genetic studies. The large interest in mouse development and genetics is founded in our curiosity about human embryonic development and in our interest in understanding genetic diseases perturbing normal embryogenesis. The use of mice (as a mammalian model) is favoured due to their easy maintenance and handling as well as their relatively short generation time.

Mouse embryonic development has been studied in detail and has been well described (e.g. Theiler 1989; Rugh 1990). Culture conditions and micromanipulation techniques for pre- and postimplantation embryos have been developed and considerable progress has been made in the analysis of cell fates and the developmental potential of various cell types in the early mouse embryo.

A tremendous amount of genetic information has accumulated. Hundreds of genes and loci have been identified and genetically mapped and numerous mutations affecting normal development have been described (Lyon and Searle 1989). Still, relatively little is known about the genetic elements governing developmental decisions in the mouse, although a number of candidate genes have been isolated in recent years based on molecular, biochemical or other biological criteria. The analysis of their role in mouse development is currently a major objective in many laboratories.

Over the last 15 years advances have been made which have led to a new phase in mouse developmental genetics. First, new genes can be introduced into the mouse genome by DNA microinjection or retroviral vectors (Jaenisch 1976), methods which have become almost standard techniques in many laboratories. Second, pluripotent embryonic stem (ES) cell lines have been established from early embryos (Evans and Kaufman 1981; Martin 1981). ES cells can be put back into the embryo by injection into the blastocyst after which they participate in embryonic development and colonize all tissues of the mouse including the germ line (Bradley et al. 1984;

Max-Delbrück-Laboratorium in der MPG, Carl-von Linné-Weg 10, 5000 Köln 30, FRG

Results and Problems in Cell Differentiation 18
W. Hennig (Ed.)
Early Embryonic Development of Animals
© Springer-Verlag Berlin Heidelberg 1992

Gossler et al. 1986). The pioneering work of O. Smithies (Doetschman et al. 1987) and M. Capecchi (Thomas and Capecchi 1987) established the technical means of altering cloned genes by targeted mutagenesis in the genome of ES cells. These mutations can be introduced into the mouse germ line by injecting the cells carrying targeted genes into blastocysts, which allows the functional analysis of mutated genes in vivo after germ line transmission of the mutated allele. Establishment of transgenic mice and targeted mutagenesis have paved the way for the analysis of gene regulation and function in the mouse in vivo.

The aim of this chapter is to provide a basic introduction to mouse development until the early postgastrulation period and into mouse developmental genetics. It is intended to give an overview rather than being comprehensive and should help newcomers working with mice to better understand more detailed descriptions and discussions on various aspects of mouse development.

2 Early Mouse Development

Embryonic development takes 18 to 20 days in the mouse depending on the strain. Embryogenesis starts with fertilization. After 4.5 days of pre-implantation development the first differentiated cell types are present and implantation takes place shortly thereafter. Development proceeds rapidly after implantation and the embryo undergoes dramatic morphological changes. The three embryonic germ layers are formed during gastrulation beginning around day 6.5 post coitum (p.c.) (the morning after copulation, which can be detected by the appearance of a vaginal plug, is counted as day 0.5 p.c.). During gastrulation the basic body plan of the mouse is essentially established, organogenesis starts and most organ primordia are present around day 10 p.c.

Detailed descriptions and illustrations of embryonic development are found in Theiler (1989), Rugh (1990) and Kaufman (1990), instructions for all kinds of manipulations of pre- and post-implantation embryos as well as embryonic stem cells are given in Hogan et al. (1986), Monk (1987), Robertson (1987) and Copp and Cockroft (1990).

2.1 Gametes and Fertilization

Fertilization, the fusion of the female and male gametes, activates the egg to enter the complex process of embryogenesis. The gametes, oocyte and sperm are products of a lengthy and complicated series of events during gametogenesis. During oogenesis and spermatogenesis, the genetic material of the parental genomes is redistributed by meiotic recombination leading to a unique genetic complement in each gamete. In sperm the genetic

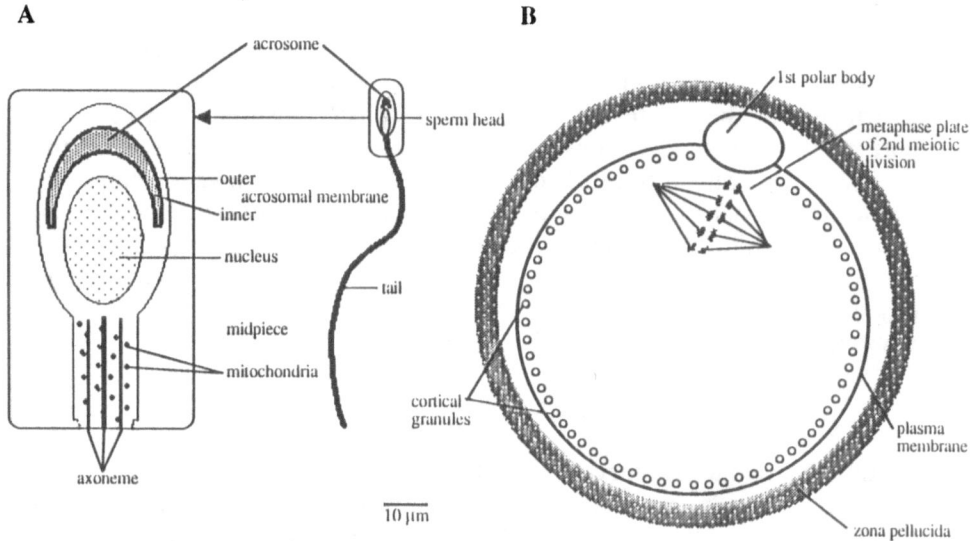

Fig. 1. Schematic drawings of sperm and oocyte before fertilization. **A** Sperm; **B** oocyte. For details, see text

material is reduced to a haploid set of chromosomes, while in the oocyte reduction to haploidy occurs only after fertilization (see below). In addition, during this differentiation and maturation process female and male germ cells adopt their highly specialized structures required for fertilization and the very early stages of development. As a detailed discussion of oogenesis and spermatogenesis is beyond the scope of this chapter, the development and structure of the mature mouse sperm and oocyte will be described only briefly, insofar as it is required for the understanding of fertilization and early development.

2.1.1 Structure of Gametes

2.1.1.1 Sperm

Mature sperm can be structurally subdivided into the head and the tail (Fig. 1A). The elongated mouse sperm head, about 8 µm long, contains the nucleus and the acrosome. The nucleus contains the tightly packed and reorganized chromatin in which histones have been replaced by protamines during late spermiogenesis. The acrosome is a membranous, lysosome-like organelle derived from the Golgi apparatus during spermiogenesis. It occupies the tip of the sperm head and covers and partly engulfs the nucleus. The acrosome contains various hydrolytic enzymes, which are released during the acrosomal reaction and digest a path for the sperm through the outer envelope of the egg (see below). The sperm tail is responsible for the sperm's motility and, based on structural differences, can be subdivided further into the neck (or connecting piece), the midpiece, the principal piece and the end piece. The connecting piece is the region immediately behind the sperm head. It is a complex structure which is responsible for the attachment of the tail to the head. The midpiece constitutes the proximal part of the tail and contains numerous mitochondria which provide the energy for sperm motility. The tail along its entire length contains the axoneme or axial filament complex. This microtubule structure consists of two central microtubules surrounded by evenly spaced doublet microtubules and provides the structural elements required for sperm motility. Detailed descriptions and discussions of sperm structure and function are given by Fawcett (1975) and Bellvé and O'Brien (1983).

2.1.1.2 Oocytes

A fully grown mouse oocyte ready to undergo the final maturation steps measures about 85 μm in diameter. It is surrounded by a 7 μm thick extra-cellular envelope, the zona pellucida, which is embedded in multiple layers of follicle or granulosa cells. The zona pellucida consists of three major sulphated glycoproteins (ZP1, ZP2 and ZP3) (Bleil and Wassarman 1980a). They are synthesized during oogenesis by the oocyte (Bleil and Wassarman 1980b) and secreted and assembled into filamentous fibres that form the porous structure of the zona pellucida (Greve and Wassarman 1985). Projections of the surrounding follicle cells reach through the zona pellucida and form junctional complexes with the oocyte. These connections allow metabolite transfer and probably mediate the nutritional role of the granulosa cells during egg growth (Heller et al. 1981). In addition they seem to be crucial for the regulation of meiotic control (Eppig 1985 and references therein). The large nucleus, called germinal vesicle, contains the chromosomes in the prophase of the first meiotic division. Approximately 4000 lysosome-like organelles, the cortical granules are located in the egg's cortex beneath the plasma membrane. They are derived from the Golgi apparatus during growth of the oocyte and contain various hydrolytic enzymes which are released from the egg after fertilization and are involved in the zona reaction (see below). The structure of the oocyte at fertilization is schematically shown in Fig. 1B.

2.1.2 Ovulation and Fertilization

During their growth, follicles and oocytes acquire the competence to respond to the appropriate hormonal stimuli for their final maturation and ovulation. In each cycle, which takes about 4–5 days in the mouse, only a few follicles respond to an increase in the level of FSH (follicle stimulating hormone). Follicle cells break contact with the oocyte and the follicle accumulates fluid. The swollen follicles, now called the Graafian follicles, move towards the periphery of the ovary. Shortly before ovulation the level of LH (luteinizing hormone) surges and the nuclear maturation of the egg commences with the disintegration of the nuclear membrane (germinal vesicle breakdown) and resumption of meiosis. One set of chromosomes is removed from the egg together with some cytoplasmic material forming the first polar body. In the metaphase stage of the second meiotic division nuclear maturation is arrested and only proceeds after fertilization.

Up to 20 oocytes ovulate during one cycle and the eggs are transported into the ampulla, the anteriormost part of the oviduct. This is achieved by the movements of the cilia of the epithelial cells lining the opening of the oviduct facing the ovary (infundibulum). Each egg is surrounded by its zona pellucida and accompanying follicle (cumulus) cells that are embedded in a matrix of proteins and hyaluronic acid. To achieve fertilization the sperm must first penetrate the viscous matrix of hyaluronic acid surrounding the cumulus cells and the egg. This might be aided by hyaluronidase, a hyaluronic acid hydrolyzing enzyme which is contained in the sperm fluid. Sperm associates with the surface of the zona pellucida and then binds more tightly in a species specific manner. This binding is brought about by interactions between egg binding proteins of the sperm head and the zona component ZP3, which is the sperm receptor in mice (Bleil and Wassarman 1980c; Shur and Hall 1982a,b). Binding to ZP3 elicits the acrosome reaction: the outer membrane of the acrosomal vesicle fuses with the overlying sperm plasma membrane at multiple sites. The fusion leads to the formation of numerous vesicles which are eventually shed, exposing the inner acrosomal membrane to the outside. Proteolytic and glycolytic enzymes are released allowing the sperm to penetrate the zona pellucida after limited proteolysis. One of these enzymes, acrosine, has a structure similar to trypsin. It exists as an inactive acrosomal form which is activated after the acrosomal reaction (Stambaugh and Buckley 1960; Wincek et al. 1979; Stambaugh and Mastroianni 1980). Acrosine possibly remains closely associated with the sperm head since only a narrow trail is digested through the zona pellucida. After penetrating the zona pellucida the posterior part of the sperm head fuses with the egg membrane and triggers a cascade of events which prevent polyspermy and ultimately lead to the formation of the diploid zygote. Stored Ca^{2+} is mobilized raising the free Ca^{2+} concentration within the egg. The transient wave of high Ca^{2+} stimulates the cortical granule reaction, a process similar to the acrosomal reaction. By fusion of the cortical granules with the overlying egg membrane hydrolytic enzymes are released from the egg into the perivitelline space between plasma membrane and zona pellucida. These enzymes alter the zona pellucida and ZP3, such that it becomes impermeable for sperm: bound sperm detach and new sperm can no longer bind to the zona pellucida. (For a more detailed discussion of the processes of fertilization, see Wassarman 1987 and references therein.)

Fertilization activates the egg and triggers the completion of meiosis. This results in the extrusion of the second polar body and leaves behind a haploid set of maternal chromosomes in the egg's female pronucleus. Meanwhile the nuclear membrane of the sperm nucleus breaks down, the chromatin decondenses, is reorganized and a new nuclear membrane is

formed around the male pronucleus. Then the two pronuclei move towards
each other. While they migrate DNA replication takes place. Upon meeting
the two pronuclei do not fuse but their nuclear membranes break down, the
chromosomes assemble on the metaphase plate and the first cell division
takes place.

2.2 Preimplantation Development

Preimplantation development takes about 4.5 days in the mouse (summarized in Fig. 2). During this time, the zygote migrates through the
oviduct towards the uterus which it reaches at around day 3 of embryogenesis. The embryo has completed cleavage and at day 3.5 has formed the
blastocyst. It contains 32 to 64 cells. The first differentiation step has already
occurred and the allocation of cells either to embryonic or extraembryonic
cell lineages has begun. (The diversification of embryonic and extraembryonic lineages is summarized in Fig. 3.) Two distinct cell populations, the
inner cell mass and trophectoderm, are discernable. The embryo is still surrounded by the zona pellucida which is shed shortly before implantation.
During this stage embryos are easily accessible and amenable to a variety of
manipulations. These include DNA transfer (by microinjection or retroviral
infection) or chimaera production (by aggregation of different embryos after
removal of the zona pellucida or injections of cells into the blastocoel).

2.2.1 Early Cleavage Development

Cleavage divisions in the mouse, as well as in other mammals, are slow. The
first cell division occurs about 20 h after fertilization. The next divisions
follow at approximately 12-h intervals but are not truly synchronous between the different cells, which are called blastomeres. Thus, embryos
having odd cell numbers are found frequently. Up to the eight cell stage
blastomeres represent totipotent, equivalent cells. Isolated single blastomeres
of two- and four-cell embryos can form blastocysts in vitro (Tarkowski
and Wróblewska 1967). Blastocysts formed in vitro from isolated two cell
embryos can develop into normal mice after transfer into the uteri of females
mated with vasectomized males (Tarkowski 1959). Individual blastomeres
from four- and eight-cell embryos cannot generate a mouse by themselves
(Rossant 1976a). This is probably due to the small number of cells in the
resulting experimental embryos which cannot form a proper blastocyst.
However, this does not reflect a lack of developmental potency, since when
single, isolated blastomeres are combined with (genetically) marked eight-
cell embryos, they can form normal chimaeric embryos and contribute to a
broad range of embryonic and extraembryonic tissues (Kelly 1977). The
position of the "single" blastomere during the aggregation seems to strongly
influence its developmental fate: labelled blastomeres which were placed on
the outside of aggregates of other blastomeres developed predominantly

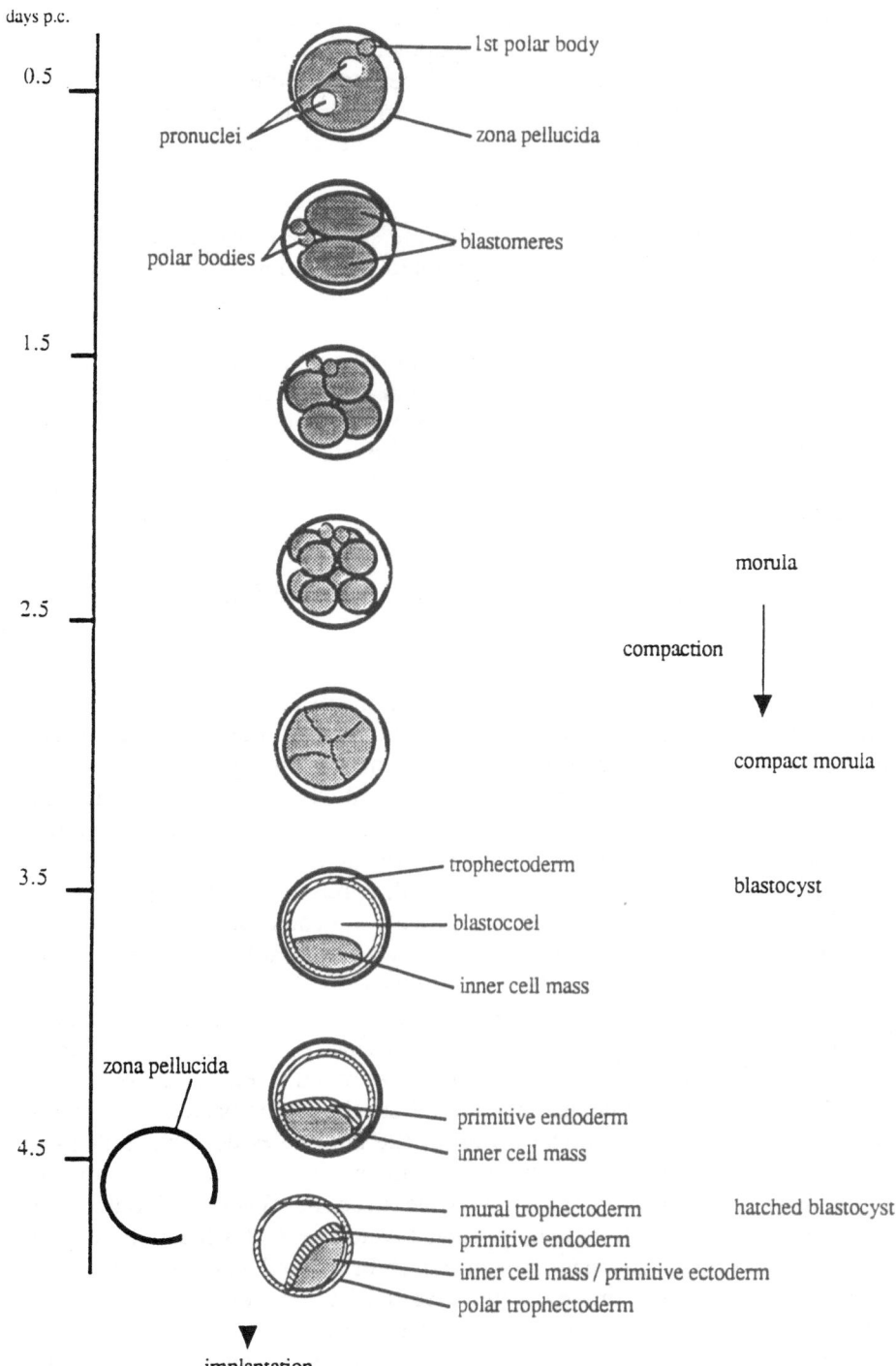

days p.c.

0.5 — 1st polar body

pronuclei — zona pellucida

polar bodies — blastomeres

1.5 —

2.5 — morula

compaction

compact morula

3.5 — trophectoderm — blastocyst

blastocoel

inner cell mass

zona pellucida

4.5 — primitive endoderm

inner cell mass

mural trophectoderm — hatched blastocyst

primitive endoderm

inner cell mass / primitive ectoderm

polar trophectoderm

implantation

Fig. 2. Schematic presentation of mouse preimplantation development

into trophectoderm cells in resulting blastocysts and were mainly found in trophoblast tissue at day 10 of development. When labelled blastomeres were surrounded by other blastomeres they contributed large numbers of daughter cells to the inner cell mass and formed parts of the embryo rather than extraembryonic tissues at later stages of development (Hillmann et al. 1972).

Isolated, single blastomeres of two-cell embryos are capable of forming a mouse and blastomeres of later stages are still able to contribute to probably all embryonic tissues when aggregated with other blastomeres. However their nuclei seem to have already lost the potential to support normal development from the one-cell stage onwards. When nuclei of zygotes were microsurgically replaced by nuclei from blastomeres of various preimplantation-stage embryos only a small percentage of embryos developed normally through cleavage development. This was not due to the manipulation technique, since when zygote nuclei were replaced by nuclei of other zygotes, embryos developed normally and gave rise to living mice (McGrath and Solter 1983, 1984, 1986)

2.2.2 Separation of Embryonic and Extraembryonic Lineages

2.2.2.1 Compaction and Polarization

During the first 2 days of development, blastomeres are spherical cells which are loosely attached at their sites of contact. At the eight cell stage blastomeres alter their adhesive behaviour and the embryo, now called the morula, undergoes a profound morphological change. The cells flatten towards each other, maximize cell-cell contacts and the former grape-like structure is transformed into a compact aggregate of cells. This phenomenon is called compaction and is dependent on the presence of calcium (Whitten 1971). One major component of this process is uvomorulin (UM), a transmembrane glycoprotein which is also known as E-cadherin (Hyafil et al. 1980; Vestweber and Kemler 1984; Yoshida-Noro et al. 1984). This protein was originally identified by polyclonal antibodies raised against F9 teratocarcinoma cells (Kemler et al. 1977). The presence of Fab fragments of anti-uvomorulin antibodies prevents compaction of cultured, early eight-cell morulae. Also, early compacted morulae can be decompacted by incubation with these antibodies. The action of the antibodies does not influence cell division or viability of blastomeres, but prevents the formation of the blastocyst. Decompaction is a reversible process. When antibodies are removed, compaction occurs and normal development proceeds: blastocysts are formed, which can generate normal embryos and mice after transfer into uteri of females mated with vasectomized males. Uvomorulin is involved in Ca^{2+} dependent cell-cell adhesion not only in preimplantation embryos, but also in many epithelial cells later during development and in adult tissues.

During compaction, tight junctions start to develop between blastomeres

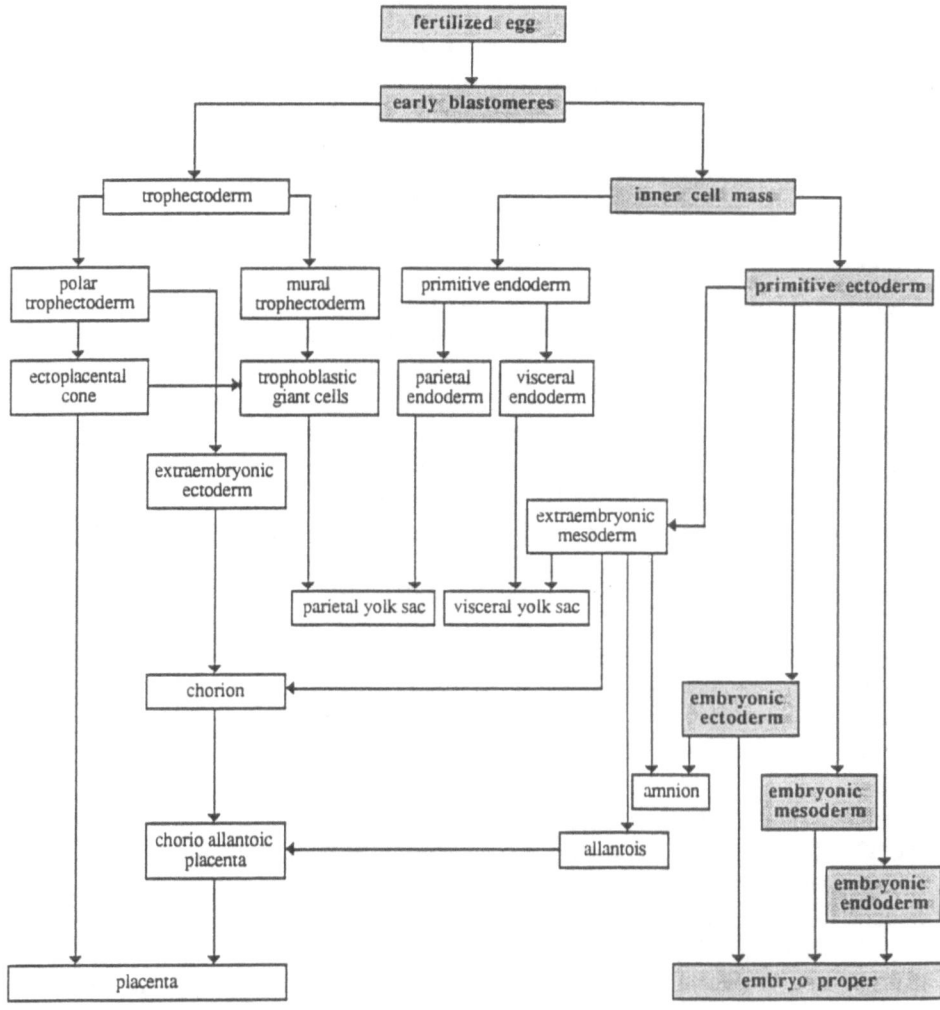

Fig. 3. Embryonic and extraembryonic lineages of the mouse embryo

(Ducibella and Anderson 1975). In the periphery of the compact morula junctional complexes eventually constitute the zona occludens, a tight permeability barrier which is required for the formation of the blastocoel (Ducibella et al. 1975).

Shortly before compaction blastomeres start to polarize and after compaction they show distinct apical and basolateral surfaces (Reeve and Ziomek 1981). The apical (outer) surface is covered by numerous microvilli but is essentially free of uvomorulin, which is redistributed on the cell surface of blastomeres prior to and during compaction. In the egg and in the early cleavage stage embryo uvomorulin is evenly distributed on the cell surface, but at the eight-cell stage it becomes restricted in outer cells to the basolateral surface and later is localized predominantly in the zona occludens of these cells (Vestweber et al. 1987). Concomitant with changes in adhesiveness, alterations in cytoskeletal architecture, lectin binding properties and distribution of membrane and cytoplasmic components occur during and after compaction (Johnson and Maro 1984; Fleming and Pickering 1985; Maro et al. 1985; Pratt 1985; and references therein). Once differences in the distribution of cellular components and structures (i.e. membrane proteins, cytoskeletal proteins and cell organelles) have been established within one cell, they can lead to differences in daughter cells after the next cell division. Divisions parallel to the basal-apical axis result in two daughter cells both of which have inner (basolateral) and outer (apical) surfaces, while divisions perpendicular to the this axis lead to "inner" and "outer" cells which inherit different molecules from their mother cell (Johnson and Ziomek 1981; Johnson 1986). There is good evidence that the "inner" cells generated during cell divisions in the late morula give rise to the inner cell mass (ICM) cells of the blastocyst while outer cells predominantly form trophectodermal cells (see below; Pedersen et al. 1986; Fleming 1987). This allocation of cells reflects the developmental fate of these cells rather than their developmental potential: when cells isolated from the "inside" of late morulae are reaggregated and cultured in vitro, they can form normal blastocysts with ICM and trophectoderm as is also true for "outer" cells (Rossant and Lis 1979; Rossant and Vijh 1980).

2.2.2.2 Formation of the Blastocyst

After compaction fluid accumulates between intercellular spaces and around the 32-cell stage the blastocoel becomes evident. Outer cells pump fluid into the nascent cavity which rapidly expands. The timing of cavitation seems to depend on the nucleo-cytoplasmic ratio or DNA or chromosomal replication, but does not depend on the absolute number of cells or cell divisions in the zygote. When the number of cells in the embryo was experimentally reduced or enlarged or cell divisions where suppressed with cytochalasin-B (which does not affect DNA replication) neither manipulation substantially affected the time of blastocyst formation (Smith and McLaren 1977).

Two distinct cell populations are present in the blastocyst: an outer layer of trophectodermal (TE) cells which represents a true epithelium surrounding the blastocoel, and the inner cell mass (ICM) cells, a group of cells which is attached to one side of the inner surface of the trophectoderm. As described above, allocation of cells to either the trophectoderm or inner cell mass occurs during the late morula stage. Trophectoderm and inner cell mass cells seem to remain totally distinct lineages from the onset of cavitation (Dyce et al. 1987). The trophectodermal cells give rise exclusively to extraembryonic tissue. Shortly before implantation some of the inner cell mass cells differentiate into a second epithelial cell type, the primitive endoderm, which arises on the free surface of the ICM facing the blastocoel. The remaining ICM cells will give rise to the embryo proper and to the extraembryonic mesoderm. The primitive endoderm will give rise to the embryonic membranes, i.e. the endodermal component of the visceral yolk sac and parietal yolk sac (see below).

2.2.3 Gene Activity During Cleavage Development

A variety of changes in the composition of RNA and proteins occur already early during cleavage development and are discussed in detail by Schultz (1986). The level of stored maternal RNA declines rapidly. During the first 24 h after fertilization about 30–40% of the total maternal RNA, about 70% of the polyadenylated RNA fraction and around 90% of the specific messages for histone H3 and actin are degraded (Levey et al. 1978; Bachvarova and DeLeon 1980; Piko and Clegg 1982; Giebelhaus et al. 1985; Graves et al. 1985). There seems to be only very little, if any, transcription from the embryonic genome during this time. In accordance with this, transcriptional inhibitors like α-amanitin do not block development of fertilized eggs to the two-cell stage (Braude et al. 1979).

The apparent lack of RNA polymerase II transcription in one-cell mouse embryos is not due to a general inability to transcribe DNA. One cell embryos can transcribe microinjected DNA and can translate newly synthesized mRNA into protein (Brinster et al. 1982).

At the two-cell stage transcription from the embryonic genome starts and is required for normal development to proceed: all classes of new RNA are synthesized (Clegg and Piko 1977, 1983) and further development is blocked by inhibition of transcription with α-amanitin (Warner and Versteegh 1974).

Newly synthesized proteins were found already upon fertilization by 2-D gel analysis of metabolically ^{35}S methionine labelled material from early embryos (Braude et al. 1979; Schultz et al. 1979; Cullen et al. 1980). Synthesis of these proteins does not require transcription since they were also produced when embryos were either incubated in the presence of the transcriptional inhibitor α-amanitin (Braude et al. 1979) or mechanically enucleated (Petzhold et al. 1980). This implies a post-transcriptional control of this class of genes. The mRNAs coding for at least some of these proteins accumulate during oogenesis and following fertilization, translation from these mRNAs is strongly increased (Cascio and Wassarman 1982).

Proteins translated from newly synthesized embryonic RNA were detected in two cell embryos using genetic variants of β-glucuronidase and β$_2$-microglobulin which were paternally encoded and could be discriminated from the maternally inherited form (Wudl and Chapman 1976; Sawicki et al. 1981). Since sperm does not contribute RNA to the zygote, the appearance of paternally encoded proteins confirms that newly transcribed embryonic RNA has been translated into protein.

In conclusion, the transition from maternal to embryonic gene expression occurs at the two-cell stage during mouse development and only the first cell division is exclusively under maternal control.

2.2.4 Implantation

Implantation occurs on the fifth day of development. Before the blastocyst can be implanted it has to escape from its zona pellucida. This process is called hatching and is brought about by localized proteolysis of the zona and contractions and expansion of the blastocyst. A small hole is lysed into the protein matrix of the zona by strypsin, a trypsin-like protease. Strypsin is associated with the mural trophectoderm, a subset of trophectodermal cells not overlaying the inner cells mass (see below). This enzyme seems to be membrane associated and is probably locally applied by cellular processes touching the zona (Perona and Wassarman 1986). The blastocyst hatches through this hole, and, once freed from its zona, attaches to the epithelium of one of the lateral uterine walls with the trophectoderm opposite (not facing) the ICM. The uterine wall attached to the blastocyst responds by bulging into the lumen which orients the ICM either to the anterior or posterior end of the uterine horn. This and the following reorganization results in an invariable orientation of the early embryo. The axis through the ICM towards the opposite pole of the blastocyst parallels the dorso-ventral axis of the mother, the ICM always facing the dorsal side. The future anterior-posterior axis of the embryo, which becomes evident around day 6.5 of development, is more or less perpendicular to the anterior-posterior (longitudinal) axis of the uterine horn (Smith 1980, 1985). How-ever, the significance of this invariant orientation of the embryo with respect to the uterus in the determination of the embryonic axes is not clear, since embryos can also develop normally in vitro from preimplantation stages up to the limb bud stage (Hsu 1982). A detailed analysis of the orientation of mouse embryos during implantation and a discussion of how this might be achieved and might be related to embryonic axis formation is given by Smith (1980, 1985).

The blastocysts are not implanted at preformed sites but rather at random after they have been spaced throughout the uterine horn by peristaltic movements of the uterine musculature (Hollander and Strong 1950). After attachment to the uterine wall the trophectodermal cells invade the degenerating uterine epithelium and penetrate into the stroma (also called endometrium) of the uterus. The mesenchymal stromal cells respond with increased proliferation and establishment of junctional complexes (Finn 1971). This results in the formation of a thick layer of mesenchymal tissue, the decidua, into which the embryo is embedded. The implantation sites are readily visible within 1 day after implantation by the decidual swellings of the uterus.

2.3 Postimplantation Development

Preimplantation development results in the formation of the blastocyst which contains around 200 cells shortly before implantation, the cell number

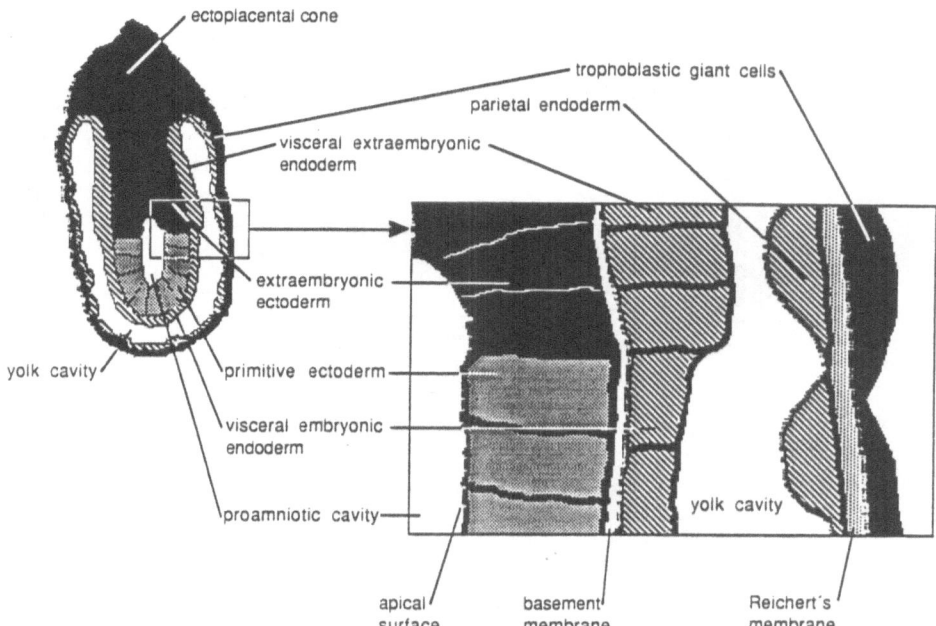

Fig. 4. Egg cylinder at around day 6 p.c. ■ Trophoblast tissue; ◨ primitive endoderm-derived tissue; ▨ primitive ectoderm

varying between different embryos of the same age. Three distinct cell types are present in the embryo, the trophectoderm, the inner cell mass, now called primitive (or embryonic) ectoderm, and the primitive endoderm. These cell types will rapidly diverge further after implantation and development progresses with profound morphological and cellular changes. Cells of the primitive ectoderm and the overlaying trophectoderm proliferate and form an elongated projection ventrally into the blastocoel now called yolk cavity (Fig. 4). This projection is called egg cylinder and shows a distinct junction between embryonic and extraembryonic ectoderm. The distal part (tip) of the egg cylinder consists of embryonic ectoderm cells. Only derivatives of these cells will form the embryo proper. The proximal part of the egg cylinder is composed of extraembryonic ectoderm derived from trophectodermal cells. Some trophectoderm cells stop proliferating, undergo endoreduplication and form trophoblastic giant cells. Other trophectoderm cells proliferate, extending the embryo dorsally, and form the ectoplacental cone. The extraembryonic ectoderm together with the ectoplacental cone will form most of the foetal part of the placenta. The primitive endoderm cells will give rise to the parietal and visceral yolk sac endoderm.

Gastrulation then leads to the formation and orderly distribution of the three definitive germ layers of the embryo and establishes the basic body plan of the mouse. Primitive ectoderm cells which migrate through the

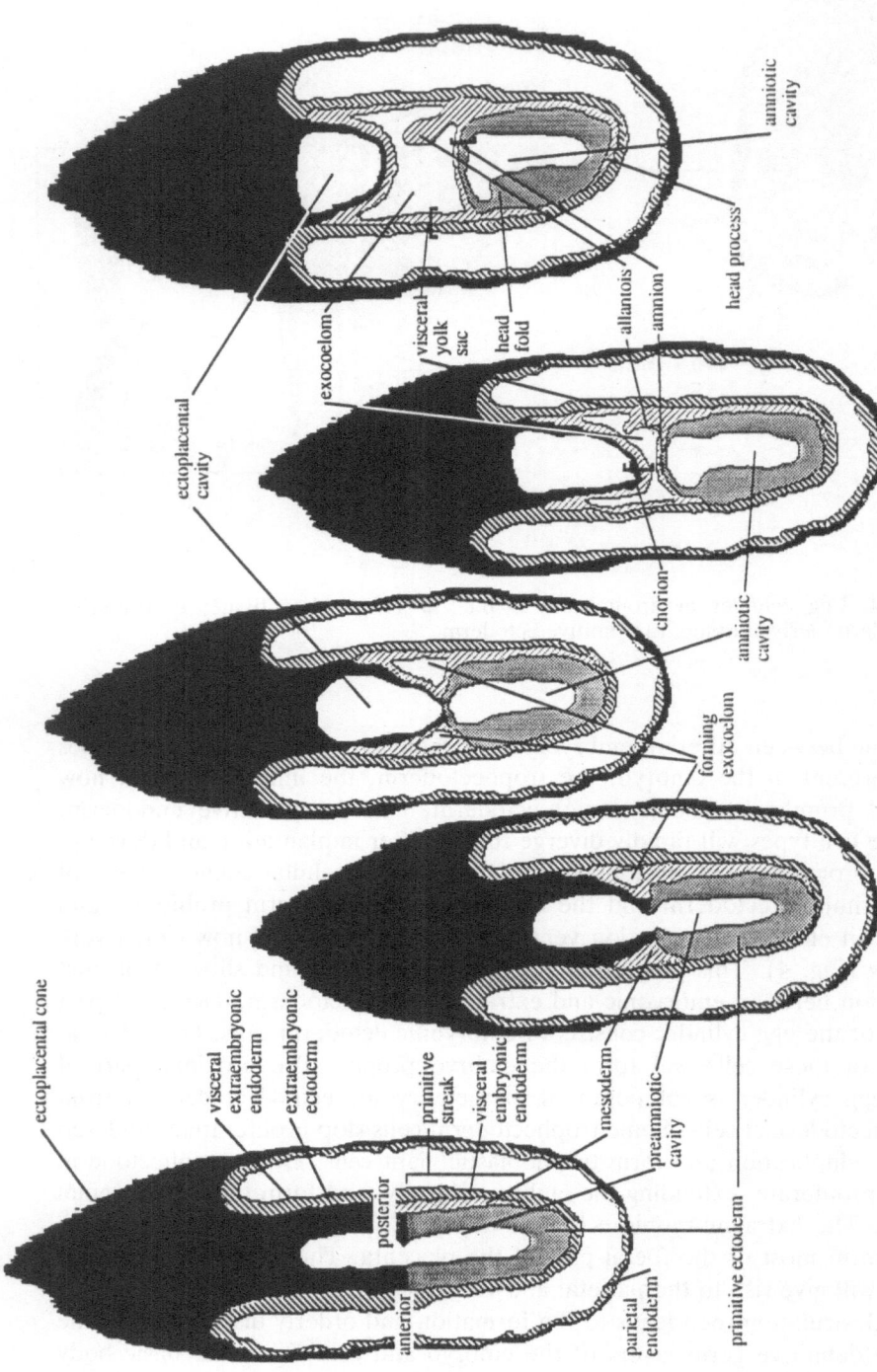

Fig. 5. Schematic drawings of primitive streak stage embryos between days 6.5 and 7.5 p.c., development advancing from *left to right*

164   Achim Gossler

primitive streak (see below) give rise to embryonic and extraembryonic mesoderm and the definitive (embryonic) endoderm. Primitive ectoderm not ingressed through the streak gives rise to neuroectoderm and surface ectoderm of the embryo and of the amnion. Extraembryonic mesoderm cells migrate out and form the allantois and the mesodermal components of the visceral yolk sac, the amnion and the chorion (Fig. 5). Embryonic mesoderm cells form the axial (notochord), paraxial (somites) and lateral mesoderm; embryonic endoderm cells form the foetal gut. Blood islands form in the (extraembryonic) mesoderm of the visceral yolk sac and haematopoiesis commences. Organogenesis starts during gastrulation in the anterior region of the embryo and progresses posteriorly as gastrulation proceeds. The primordial germ cells become discernable at the base of the allantois at the posterior end of the embryo. After that all the organ anlagen appear rapidly and the body plan is essentially established.

2.3.1 Trophectoderm and Its Derivatives

Already in the late preimplantation blastocyst the epithelial layer of trophectodermal cells does not consist of a homogeneous cell population. The cells in contact with and overlying the inner cell mass constitute the polar trophectoderm, the cells without contact to the ICM the mural trophectoderm. Mural trophectoderm cells stop proliferating and become large polyploid cells, the primary trophoblastic giant cells. In contrast, polar trophectoderm cells remain diploid, continue to proliferate and give rise to the ectoplacental cone and the extraembryonic ectoderm (Gardner and Johnson 1973; Gardner et al 1973; Rossant 1976b; Rossant et al. 1978; Papaioannou 1982). Polar trophectoderm cells that move away from the embryonic pole differentiate into mural trophectoderm. Up to day 3.5 mural trophectoderm cells can still revert to polar trophectoderm when placed in contact with the ICM (Gardner et al. 1973). However, around day 4, no descendants of mural trophectoderm cells could be found to colonize the polar trophectoderm while descendants of polar trophectoderm cells were still able to contribute to the mural trophectoderm (Cruz and Pedersen 1985). Thus, differences between the two subpopulations of trophectoderm seem to be established around day 4 of development and the mural cells seem to be committed to become giant cells. Contact or proximity to the ICM or its derivatives seem to control whether trophectoderm cells continue to proliferate or cease cell divisions and become polyploid giant cells. When trophoblast cells (these are the trophectodermal cells after implantation) were isolated from contact with ICM derivatives, they ceased proliferation and transformed into giant cells (Gardner 1972, 1975) and contact with ICM cells appears to prevent endoreduplication of trophectoderm cells (Rossant and Ofer 1977).

The extraembryonic ectoderm projects into the blastocystic cavity while the ectoplacental cone extends in the opposite direction. Cells from the periphery of the ectoplacental cone form additional (secondary) trophoblastic giant cells and can do so at least up to day 9.5 of development (Rossant 1977). Ectoplacental cone and extraembryonic ectoderm give rise to the majority of cells in the foetal part of the placenta. The extraembryonic ectoderm becomes epithelial, moves back towards the ectoplacental cone and together with extraembryonic mesoderm cells constitutes the chorion. The chorion together with the allantois, another mesodermal tissue which gives rise to the umbilical cord (see below), forms the chorioallantoic placenta, or labyrinthine region of the placenta where foetal blood vessels are bathed in maternal blood.

2.3.2 Primitive Endoderm and Its Derivatives

Shortly before implantation primitive endoderm cells differentiate on the surface of the inner cell mass facing the blastocystic cavity. The remaining cells of the inner cell mass constitute the primitive or embryonic ectoderm. Primitive endoderm cells can be distinguished from the embryonic ectoderm by various morphological features, notably their large and swollen endoplasmic reticulum (Nadijcka and Hillman 1974). The primitive endoderm does not contribute to the definitive endoderm of the embryo (Gardner and Rossant 1979). Soon after their appearance and as the egg cylinder forms, primitive endoderm cells undergo further differentiation into two morphologically and biochemically distinct cell types, the visceral and parietal endoderm. The cells remaining in contact with and covering the egg cylinder constitute the visceral endoderm. The parietal endoderm is formed by cells that grow out and migrate on to the inner surface of the mural trophectoderm (Snell and Stevens 1966). Visceral and parietal endoderm cells are part of the extraembryonic membranes, the visceral and parietal yolk sac respectively (see below).

Parietal endoderm cells start to grow and migrate onto the inner surface of the trophectoderm shortly after implantation and from day 6 onward they cover the inner surface of the trophectoderm as a lawn of evenly spaced individual cells (Snell and Stevens 1966; Enders et al. 1978). These cells are characterized by an enormous endoplasmic reticulum, which is related to the secretion of large amounts of extracellular matrix material as laminin, entactin, type IV collagen and heparan sulphate proteoglycan (Hogan et al. 1980, 1982; Carlin et al. 1981; Smith and Strickland 1981). This thick basement membrane, known as Reichert's membrane, is laid down between the parietal endoderm and the underlying trophoblastic giant cells. Parietal endoderm, Reichert's membrane and trophoblastic giant cells together constitute the parietal (outer) yolk sac of the embryo. Until it starts to break down around day 16 of gestation, Reichert's membrane is thought to act as a major barrier and coarse filter between maternal and foetal environments. Nutrients from the mother can pass through this barrier, while penetration of maternal cells is prevented (Smith and Strickland 1981). Parietal endoderm cells show marked differences in their appearance depending on their localization. Cells proximal to the boundary between visceral and parietal endoderm show a more blebby morphology and an intense motile activity compared with more distal cells, which are round and smooth and show only very little motility (Cockroft 1986). This might reflect recruitment of parietal endoderm cells from the visceral endoderm and their subsequent migration distally (Hogan and Newman 1984).

The visceral endoderm cells become organized into a distinct epithelium. The apical surface (facing the former blastocoel which is now called the yolk cavity) is covered by microvilli and cells are connected at their apical side by desmosomal junctions (Hogan and Tilly 1981). On the basal side visceral endoderm cells are separated from the underlying embryonic and extraembryonic ectoderm cells by a thin basement membrane (Clark et al 1982). Like parietal endoderm cells, visceral endoderm cells show marked morphological differences depending on their localization. In the region of the embryonic ectoderm (visceral embryonic endoderm) cells tend to be flat. In contrast, visceral endoderm cells around the extraembryonic ectoderm (visceral extraembryonic endoderm) are collumnar and contain more microvilli on their surface. Together with extraembryonic mesoderm cells visceral endoderm constitutes the visceral (inner) yolk sac. Visceral endoderm cells have important absorptive and secretory functions. Visceral embryonic endoderm cells take up substances from the mother which have filtered through Reichert's membrane at their apical surface, process them in their lysosomes and deliver them basally to the primitive ectoderm. In addition, visceral embryonic endoderm (but not visceral extraembryonic endoderm) cells synthesize and secrete α-fetoprotein (AFP), a foetal protein similar to serum albumin. Later the visceral yolk sac endoderm synthesizes serum proteins and other substances such as AFP, apolipoproteins, α1-antitrypsin and transferrin (Dziadek and Adamson 1978; Adamson 1982; Meehan et al. 1984). In addition, transcripts for insulin and insulin like growth factor II have been detected in the rat visceral yolk sac.

2.3.3 Embryonic Ectoderm

The cell types described thus far give rise only to extraembryonic tissues which serve important supportive functions for the developing embryo. The embryo proper is formed exclusively by descendants of primitive ectoderm cells which in addition give rise to the germ line and extraembryonic mesoderm (Gardner and Rossant 1979; for review see Beddington 1983a). Around the time of implantation, primitive ectoderm cells have a spherical or irregular shape, are apparently apolar and form a core of about 30–40 cells surrounded by primitive endoderm cells ventrally, and juxtaposed to extraembryonic ectoderm cells on the dorsal side. Shortly after implantation primitive ectoderm cells start to proliferate rapidly (Snow 1977) and form the ventral part (tip) of the egg cylinder. Between day 5.5 and 6 post coitum (p.c) a small lumen called the proamniotic cavity forms in the center of the primitive ectoderm and the cells form an epithelial layer of columnar cells. The apical side faces the lumen of the cavity and cells are joined apically by junctional complexes. The basal surface is attached to the thin basal membrane separating primitive ectoderm from visceral embryonic endoderm. Around day 6 p.c. the central cavity extends more dorsally into the extraembryonic ectoderm resulting in a small lumen throughout the egg cylinder.

2.3.4 Gastrulation and Development of the Germ Layers

Gastrulation begins around day 6.5 p.c. and within the next 24 h rapid changes occur leading from the egg cylinder to the primitive streak stage embryo. At the future posterior end of the embryonic portion of the egg cylinder the epithelial continuity of the primitive ectoderm breaks down. This region is called the primitive streak and extends soon after its appearance distally towards the tip of the egg cylinder thereby defining the future anterior-posterior axis of the embryo. During gastrulation, ectodermal cells delaminate from the epithelial cell layer, migrate through the primitive streak and as a new tissue layer between the ectoderm and primitive endoderm differentiate into mesoderm cells.

Shortly after their first appearance around day 6.5 p.c. mesoderm cells start to migrate towards the extraembryonic ectoderm into the extraembryonic region. These cells give rise to extraembryonic mesoderm. At the margin of the embryonic part of the egg cylinder ectodermal cells bulge into the lumen of the egg cylinder together with the underlying extraembryonic mesoderm and form the amniotic folds. Formation of the amniotic folds starts at the posterior end and progresses laterally towards the anterior end which leads to a continuous constriction of the central cavity that is most advanced at the posterior end. The amniotic folds grow towards each other, meet finally and fuse. Concomitantly the mesoderm within the folds develops a central cavity, the exocoelom, pushes the extraembryonic ectoderm towards the ectoplacental cone and separates it from the embryonic ectoderm. On day 7.5 p.c., after the amniotic folds have fused, the proamniotic cavity of the egg cylinder has been divided into the amniotic, exocoelomic and ectoplacental cavities (Fig. 5). The ectoplacental cavity is surrounded by extraembryonic ectoderm cells. The adjacent exocoelomic cavity is surrounded by extraembryonic mesoderm which contacts dorsally the extraembryonic ectoderm, laterally the visceral endoderm and ventrally the ectoderm of the amnion. The latter is the dorsal lining of the amniotic cavity, which is laterally and ventrally surrounded by embryonic ectoderm. Extraembryonic ectoderm with the underlying extraembryonic mesoderm constitutes the chorion, visceral endoderm with the attached extraembryonic mesoderm forms the visceral yolk sac. The amnion consists of an ectodermal cell layer covered by extraembryonic mesoderm. At the posterior end of the embryo extraembryonic mesoderm cells give rise to a finger like structure which grows through the exocoelom towards the chorion. This tissue is called allantois and will later fuse with the chorion, linking the embryo with the ectoplacental cone. The allantois will form the umbilical cord and together with the chorion will give rise to the chorioallantoic placenta.

The primitive streak progresses anteriorly until about day 7.5 p.c. when it extends from the posterior end of the embryo proper to the distalmost part of the egg cylinder. During extention of the streak mesodermal cells are continuously formed and move laterally and anteriorly away from the primitive streak (Snell and Stevens 1966; Poelman 1981a). In addition, during this time the definitive endoderm of the embryo is recruited from cells originating from the anterior region of the primitive streak. This new population of endoderm cells displaces the visceral embryonic endoderm cells (derived from the primitive endoderm of the late blastocyst) into the yolk sac, colonizes the midline region of the embryo and forms the midgut (Lawson et al. 1986; Lawson and Pedersen 1987).

At the anterior end of the streak the ectoderm thickens around day 7 p.c. This thickening is equivalent to Hensen's node (as this structure is called in birds) or the primitive knot or node (as it is called in the human embryo) but it is not very prominent in the mouse. Cells migrating through this area move anteriorly and anteriolaterally and constitute the head process. The mesodermal cells moving anteriorly along the anterior-posterior axis (axial mesoderm) form the cranial portion of the notochord, a transient embryonic tissue (Snell and Stevens 1966; Jurand 1974; Beddington 1981; Lawson et al. 1986). Endodermal cells from this region contribute to the trunk endoderm (Snell and Stevens 1966; Beddington 1981; Poelman 1981b; Lawson et al. 1986). The embryonic ectoderm cells overlying the head process form the neural plate which in the midline starts to form the neural groove. From day 7.5 p.c. onwards extensive anterior growth and regression of the primitive streak extend the neural plate posteriorly. Concomitantly the primitive node moves back and cells migrating through the regressing node form more posterior parts of the notochord. Ingression of cells through the primitive streak persists up to and through day 10 p.c. (midgestation).

Cells invaginating at different positions along the primitive streak have distinct developmental fates and give rise to different prospective mesodermal and endodermal tissues. Cells emerging from the anterior part of the primitive streak contribute mainly to notochord and gut, cells emerging anteriolaterally to the streak form paraxial mesoderm (which will give rise to the somites, see below). Cells from the middle region of the streak give rise to lateral mesoderm (mesoderm located laterally to the paraxial mesoderm) and cells from the posterior part of the streak move mainly into the extraembryonic mesoderm (Tam and Beddington 1987). In addition to the position-dependent allocation of cells to different mesodermal tissues, there is also a developmental stage dependent potential of the streak to form different mesodermal cell types. While the early primitive streak (day 6.5 to 8) produces both embryonic and extraembryonic mesoderm, the older primitive streak (from day 8.5 onwards) continues to produce embryonic mesoderm but ceases to contribute significantly to extraembryonic mesoderm (Tam and Beddington 1987).

Also, the fate of embryonic ectoderm cells in the day 7.5 embryo seems to depend on the position of the cells along the anterior-posterior axis. Cells from the anterior regions give rise to neuroectoderm of the prosencephalon and mesencephalon, cells flanking the anterior end of the streak give rise to neuroectoderm of the rhombencephalon and cells flanking the anterior and middle region of the streak give rise to the spinal cord. In addition, the future dorsoventral orientation of neuroectodermal cells in the neural tube seems to be already established at this stage of gastrulation. Cells which are closer to the midline end up in more ventral positions than cells which are located more laterally. Ectoderm from the posteriormost regions give rise to surface ectoderm and cells from positions most lateral to the midline are the presumptive neural crest cell precursors (Beddington 1981, 1982; Tam 1989). These cell fates, however, do not imply that cells are committed to specific lineages prior to gastrulation, since there is little regional restriction in the developmental potency of embryonic ectoderm cells (Beddington 1983b).

The arrangement of ectoderm inside and endoderm on the outside of the embryo which is found prior to and during early gastrulation, is known as inversion of the germ layers and common to mouse, rat, rabbit, guinea-pig and other closely related rodents. In sagittal sections the embryo has the appearance of a "U", the midgut endoderm lines the outer curvature of the "U" and fore- and hindgut follow at either end (Fig. 6). When the first six to eight somites have formed (see below) the inversion of the germ layers is reverted by a process known as turning. During turning, the embryo rotates (normally anticlockwise) around its anterior posterior axis. As a consequence, the curvature of the "U" is reversed and the ectoderm comes to lie at the outer aspect of the embryo. The embryo becomes surrounded by the extraembryonic membranes, the amnion and the visceral yolk sac, because the extraembryonic membranes are attached to the embryo along the boundary of the body wall and the future site of attachment of the umbilical cord. For a more detailed and illustrated description of turning, see Kaufman (1990).

Various regions of mesoderm, which contribute to different tissues of the embryo are generated during gastrulation. Mesoderm along the midline of the embryo (axial mesoderm) forms the head process and notochord, which, in turn, induces the overlying ectoderm to form the neural tube. Paraxial mesoderm flanks the notochord and the neural tube laterally on both sides as thick bands of mesoderm. Beginning at around day 7.75 p.c. somites condense from paraxial mesoderm. The first somites form in the posterior head fold region of the embryo and somite condensation progresses posteriorly, while caudally new mesoderm cells are still being generated from the primitive streak. The region posterior to the first condensed somites is called presomitic or unsegmented mesoderm. New somites condense from presomitic mesoderm caudally to the first somites at regular intervals as the primitive streak regresses. The total number of about 65 somite pairs is formed at around day 13 of development. The metameric

A

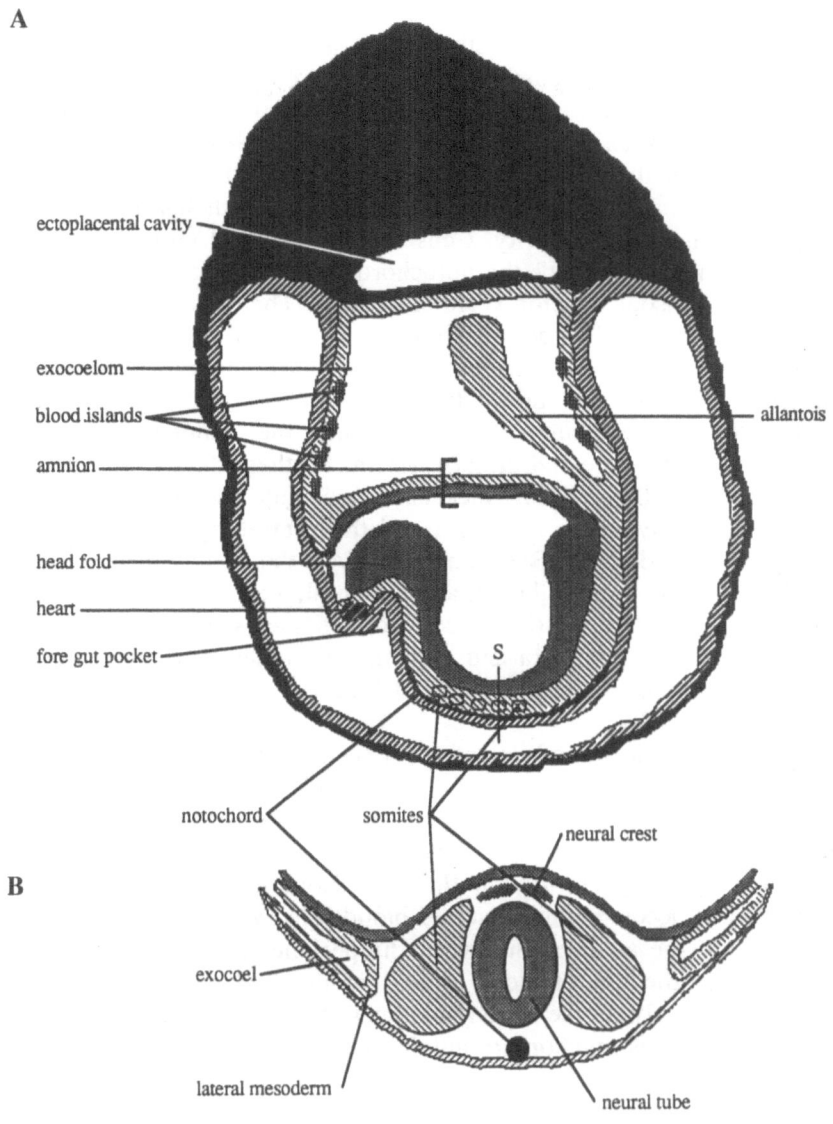

ectoplacental cavity

exocoelom

blood islands

amnion

allantois

head fold

heart

fore gut pocket

S

notochord   somites

neural crest

B

exocoel

lateral mesoderm

neural tube

Fig. 6A,B. Schematic drawing of early somite stage embryo before turning

somites can be considered as segmental subdivisions along the anterior-posterior axis (Hogan et al. 1985).

Preceeding somite formation discrete aggregates of mesoderm cells, the somitomeres, are discernable in the presomitic mesoderm (Meier and Tam 1982; Tam et al. 1982). When presomitic mesoderm explants are cultured in vitro, the number of developing somites corresponds to the number of somitomeres which were originally present in the explant (Tam 1986). This suggests that they represent the precursors of the metameric somites. It is less clear, however, whether cells present in a particular somitomere are already allocated to the somite which will develop from this somitomere, since there seems to be some cell mixing within the presomitic mesoderm during somite condensation (Tam and Beddington 1987).

Shortly after their condensation somites differentiate further. A central cavity forms within each somite and cells become epithelial the apical sides facing the lumen of the cavity. Cells from the ventral part of the somites detach and migrate towards the notochord, which they enclose. These cells give rise to the axial skeleton (vertebrae, ribs). The cells left behind form a bilayered structure called dermamyotome. The dorsal dermatome cells give rise to the connective tissue of the skin, the ventral myotome cells form the muscles of the body wall and the limbs.

Mesodermal cells immediately flanking the somites (intermediate mesoderm) form the genital urinary system. Further laterally lies the lateral plate mesoderm. Lateral plate mesoderm splits into the dorsal (or somatic) mesoderm underlying the ectoderm and the ventral (splanchic) mesoderm underlying the endoderm. Between these layers the coelom forms which will later be subdivided into the separate pleural, pericardial and peritoneal cavities. Lateral plate mesoderm cells form tissues such as the heart, connective tissues of the viscera and cartilage and bone of the limbs. Most of the mesenchyme of the head is derived from the cephalic neural crest cells (see below).

Extraembryonic mesoderm of the visceral yolk sac is the first site of haematopoiesis in the developing embryo (Snell and Stevens 1966). From the 7th day of gestation onwards blood islands appear on the inner side of the visceral yolk sac. These are condensations of mesenchymal cells which form an irregular girdle around the exocoelom. The inner cells of these condensations become embryonic red blood cells (which are nucleated cells in contrast to the adult erythrocyte), the peripheral cells differentiate and form the endothelium of blood vessels of the yolk sac. Yolk sac blood islands are the only source of red blood cells for the embryo until the foetal liver takes over this function around day 12 of development.

2.3.5 Neural Tube and Neural Crest

The nervous system develops from neural plate ectoderm which gives rise to the neural tube and the neural crest which in turn form all parts of the central and peripheral nervous system. Starting around day 7.5 the neural groove begins to form along the midline of the neural plate. While the primitive streak regresses and the neural plate is extended posteriorly, the neural groove deepens and the neural folds develop. In the cranial third of the embryo the head folds emerge rapidly and bulge deeply into the amniotic cavity, due to rapid growth and the indentation of the foregut pocket, which pushes the overlying neuroectoderm ahead of itself. As the folds become higher, the edges start to approach each other and finally meet and fuse to form the neural tube which underlies the surface ectoderm. Closure of the neural tube starts around day 8.25 at the position of the 4th to 5th somite and progresses anteriorly and posteriorly. The open ends of the neural tube are called the anterior and posterior neuropore. Development of the neural tube progresses more rapidly in the cranial region and the anterior neuropore is closed around day 9, while closure of the posterior neuropore is not complete until day 10 p.c. Transient periodic structures called the neuromeres are apparent in the anterior region of the neural tube. Cells from the edge of the neural folds between neuroectoderm and surface ectoderm give rise to the neural crest. The neural crest is a transient structure which is present only shortly after closure of the neural tube. The neural crest cells disperse rapidly and migrate through the embryo. They give rise to a variety of cell types depending from which part along the anterior-posterior axis they originated and where they finally settle (Le Douarin 1982). Among those are the neurons and glial cells of the spinal ganglia, the peripheral nervous system and the adrenal medulla, the melanocytes of the epidermis and most of the mesenchymal cells of the head (skeletal and connective tissue).

2.3.6 Development of the Germ Line

Primordial germ cells (PGCs), the ancestors of the gametes, originate in the mouse at least as early as on day 7 of development. They seem to separate from the embryonic ectoderm lineage via the caudal end of the primitive streak and are found first in the posterior part of the embryo at the base of the allantois (Copp et al. 1986; Ginsburg et al. 1990). They are large, round cells, which contain a high level of alkaline phosphatase activity. This enzymatic activity can be used to trace primordial germ cells in the early embryo (Chiquoine 1954). From day 8.5 onwards primordial germ cells migrate throught the hindgut and mesentery wall and colonize the genital ridges. The genital ridges, which give rise to the gonads, are a paired mesodermal tissue which lies beneath the dorsal mesentery of the body. By day 12.5 PGCs are largely confined to the developing gonads. In vitro studies suggest that colonization of the genital ridges is brought about by active, invasive movement of the PGCs and that PGCs lose their invasive motility after entering the gonad anlagen (Donovan et al. 1986). During their migration, the PGCs proliferate and the population of about 10–100 PGCs present around day 7–8 increases to more than 20000 in the colonized genital ridges around day 14 of development (Tam and Snow 1981; Wylie et al. 1985). Genital ridges seem to release intrinsic factors which stimulate the proliferation of PGCs and which act as chemoattractants for PGCs in vitro (Godin et al. 1990). At day 12.5 differences between male and female genital ridges become apparent and male and female germ cells embark on their specific developmental programs. Male primordial germ cells enter mitotic arrest around day 13 p.c. and continue development only after birth. In contrast, female mouse primordial germ cells enter meiosis from day 13 of development onwards and already about 3–5 days after birth all germ cells have undergone oogonial development and are in the diplotene stage of meiosis.

2.3.7 The Midgestation Embryo

At midgestation (11th day of development) the basic body plan is essentially established. Organogenesis has progressed and accelerated considerably. Separation of the four brain vesicles proceeds and the major divisions of the brain are clearly visible. The cellular differentiation of the nervous system, which begins around day 9 p.c., continues, proliferating neuroblasts are found in the walls of the entire central nervous system and the spinal ganglia are well formed. The major elements of the circulatory system have developed. Fore- and hindlimb buds are present, the anterior limbs (because they arise first), being more developed than the posterior ones. In the trunk region the development of the vertebrae commences and in this region somites start becoming less discernable. All major organ anlagen are present or emerge within the next few days.

The events described thus far are all subject to genetic control and considerable efforts are being made to identify and analyze mouse genes controling development. Although a number of genes have been isolated (see below) which are involved in the control of mouse embryogenesis, we are far from understanding the genetics of mouse development. Especially for very early developmental decisions (e.g. trophectoderm vs ICM, ICM vs primitive endoderm) no genes have been identified which could be implicated in these determinative processes. However, considerable progress has been made in mouse developmental genetics and exciting results can be expected in the years to come.

3 Mouse Developmental Genetics

More genetic information is available about the mouse and its development than about any other mammal except man. The history of mouse (developmental) genetics goes back to the beginning of this century when the applicability of Mendel's laws on inheritance in mice was confirmed using coat colour differences between different strains of mice (Castle 1903). The first linkage between different genetic markers [albino (c) and pink eye (p)] was demonstrated in 1915 (Haldane et al. 1915; for a brief history of mouse genetics see Russell 1985 and references therein). Since then, an increasing number of loci have been identified and have been placed on the mouse linkage map. Hundreds of mutants have been produced, collected and maintained, many of them affecting embryonic development. The phenotypes of a number of developmental mutations were thoroughly analyzed and traced back to their earliest appearance during embryogenesis.

Numerous chromosomal variants (structural changes in the mouse karyotype) have been found and have facilitated, for example, the assignment of linkage groups to the different chromosomes. Gene mapping techniques have been devised which allow the rapid chromosomal localization of cloned genes and more than 1800 genes have been placed on the mouse map. Almost 300 highly inbred mouse strains have been generated which provide virtually genetically identical individuals in an inbred line. A list of mouse loci, their map positions and compiled information about mouse strains and the mouse genome can be found in Lyon and Searle (1989). More recent information can be found in current issues of *Mouse Genome*.

Many mouse genes implicated in developmental processes have been isolated on the basis of analysis of mutants or based on molecular, biochemical or other biological criteria. It is not the aim of this chapter to give a comprehensive account of (putative) mouse developmental genes or to discuss the evidence supporting their importance for mouse embryogenesis. Rather some of the approaches which are in use to isolate and analyze these genes will be briefly described.

3.1 General Data Concerning the Mouse Genome

The diploid mouse genome contains 40 chromosomes, 19 pairs of autosomes and either two X-chromosomes in females or an X- and a Y-chromosomes in males. All chromosomes are acrocentrics. The diploid physical length is about 2.7×10^9 bp and its genetic length (which varies among species) is estimated to be about 1600 cM. Thus, one centiMorgan is equivalent to 1.7×10^6 bp in the mouse which is an average value for the whole genome. For a particular chromosomal region this can differ and 1 cM might represent a larger or smaller physical distance due to different crossing over frequencies of different chromosomal regions. Up to 65% of the genome contains repeated or highly repeated sequences (satellite DNA, B1, B2, L1 and others) (Hastie 1989). The number of single copy vital genes is estimated to range from 5000 to 10000 (Shedlovsky et al. 1986; Dove 1987).

3.2 Chromosomal Variants

Structural variants of chromosomes such as translocations, inversions, duplications and deletions have been described in the mouse and are maintained (Searle 1989). The interest in chromosomal variants is founded in their value for numerous genetic and cytogenetic studies. Structural rearrangements of mouse chromosomes were first recognized or suspected on the basis of genetic analyses (Painter 1927, Snell 1946). Cytogenetically, some can be identified by conspicuous alterations in the karyotype during mitosis (e.g. Robertsonian translocations by the appearance of metacentric chromosomes and reduced chromosome number or reciprocal translocations by abnormal size of chromosomes), while most become visible in germ cells due to the abnormal meiotic figures they produce. Banding techniques unambiguously identifying every chromosome have been developed (Caspersson et al. 1968; Schnedl 1971) and allow cytogenetic analysis also of less severe chromosomal rearrangements (Miller and Miller 1972; Davisson and Roderick 1976).

Reciprocal and Robertsonian translocations comprise by far the largest group of chromosomal variants in the mouse. Reciprocal translocations are symmetrical exchanges between non-homologous chromosomes. Robertsonian translocations, also called centric fusions, are fusions of two chromosomes at their centromeres. This results in a metacentric chromosome and provides an excellent cytogenetic marker for the two fused chromosomes (Gropp et al. 1972; reviewed by Gropp and Winking 1981). Translocations have frequently been induced by radiation, but many Robertsonian translocations have also been found in wild mouse populations. Translocations can be identified cytogenetically by the mitotic and meiotic figures they produce in heterozygous animals and, in addition, they can be detected genetically: when mice of either sex, heterozygous for a translocation, are mated to a wild type mouse, they produce abnormally small litters, a phenomenon called "semi-sterility". This is due to the formation of gametes with aneuploid chromosome complements (unbalanced gametes) which arise during meiosis and lead to embryonic death (Snell 1946; Carter et al. 1955, 1956).

Unbalanced gametes derived from animals heterozygous for translocations carry either duplications and deficiencies in part of the chromosomes (in the case of reciprocal translocations) or are disomic or nullisomic for particular chromosomes (in the case of Robertsonian translocations). This can be used to generate zygotes which are aneuploid for chromosomes or part of chromosomes (crosses between animals heterozygous for translocations and wild-type animals) and allows one to study the consequences of experimentally induced chromosomal imbalance such as monosomies or trisomies, for example (for review, see Epstein 1986). When two animals heterozygous for a particular translocation are mated, unbalanced gametes from both parents can combine and complement each other. Offspring can result which obtained from one parent both copies of the chromosomes or chromosomal parts affected in the translocation in an otherwise balanced situation. This has allowed one to study and define chromosomal regions that display functional differences when inherited either from the mother or from the father, a phenomenon called "imprinting" which is the subject of Reik (this Vol.).

Translocations involving the X chromosome have been used for analysis of X-inactivation and position effect variegation (Cattanach 1974, 1975; Rastan 1983; Russel 1983) and reciprocal and Robertsonian translocations have been extremely helpful in assigning linkage groups to chromosomes (Miller and Miller 1972, 1975).

Inversions can be induced by radiation and chemical mutagens (Roderick and Hawes 1970, 1974) but also occur naturally, and have been used to a limited extent for the isolation of recessive lethal genes (Lyon et al. 1982). Inversions are particularly useful for the development of tester stocks for mutagenesis experiments because crossing over between the inverted region of a chromosome and its structurally normal homologue is suppressed.

An extensively studied group of naturally occurring inversions, the so-called t-haplotypes on chromosome 17, has been isolated from wild mice and was originally identified by interaction with the dominant mouse mutation Brachyury (T; see also Sect. 3.5), which leads to tailless animals in a double heterozygous (T/t) condition. t-Chromosomes contain two inversions in the proximal region of chromosome 17 and display biological phenomena such as recessive embryonic lethality and transmission ratio distortion in heterozygous males (for reviews, see Bennett 1975; Frischauf 1985; Silver 1985).

3.3 The Genetic Map of the Mouse

The present genetic map of the mouse is a composite of phenotypic markers (mutations), biochemical variants, cloned genes and other DNA probes (Lyon and Searle 1989). A variety of techniques have been used to place genes and markers on this map. The assignment of mutations and biochemical variants requires specific test crosses to be set up to identify a possible linkage between a known marker and the gene to be tested. Special linkage testing stocks have been developed. They breed true for a series of marker genes which tag different chromosomes and which are easy to score and do not impair viability. These allow, in addition to chromosomal assignment, the new gene to be ordered on the chromosome with respect to the markers present in the testing stock.

3.3.1 Recombinant Inbred Strains

An important development for the mapping of polymorphic loci was the establishment of different recombinant inbred (RI) strains. These strains are produced by crossing two highly inbred progenitor strains and inbreeding any two individuals of the F2 generation (Bailey 1971). The two alleles for all the loci in which the progenitor strains differ segregate and assort independently (if they are not too close). Recombinations between homologous chromosomes derived from the two progenitor strains accumulate from generation to generation and become eventually fixed during the subsequent inbreeding. Every RI strain is homozygous for all its genes but represents a patchwork of chromosomal fragments (due to recombination during meiosis) on average derived in equal parts from the progenitor strains (Fig. 7). Loci that are polymorphic between the two progenitor strains can now be used to analyze which progenitor allele is present in the different RI strain derived from the original cross. Using markers with known map positions, the strain

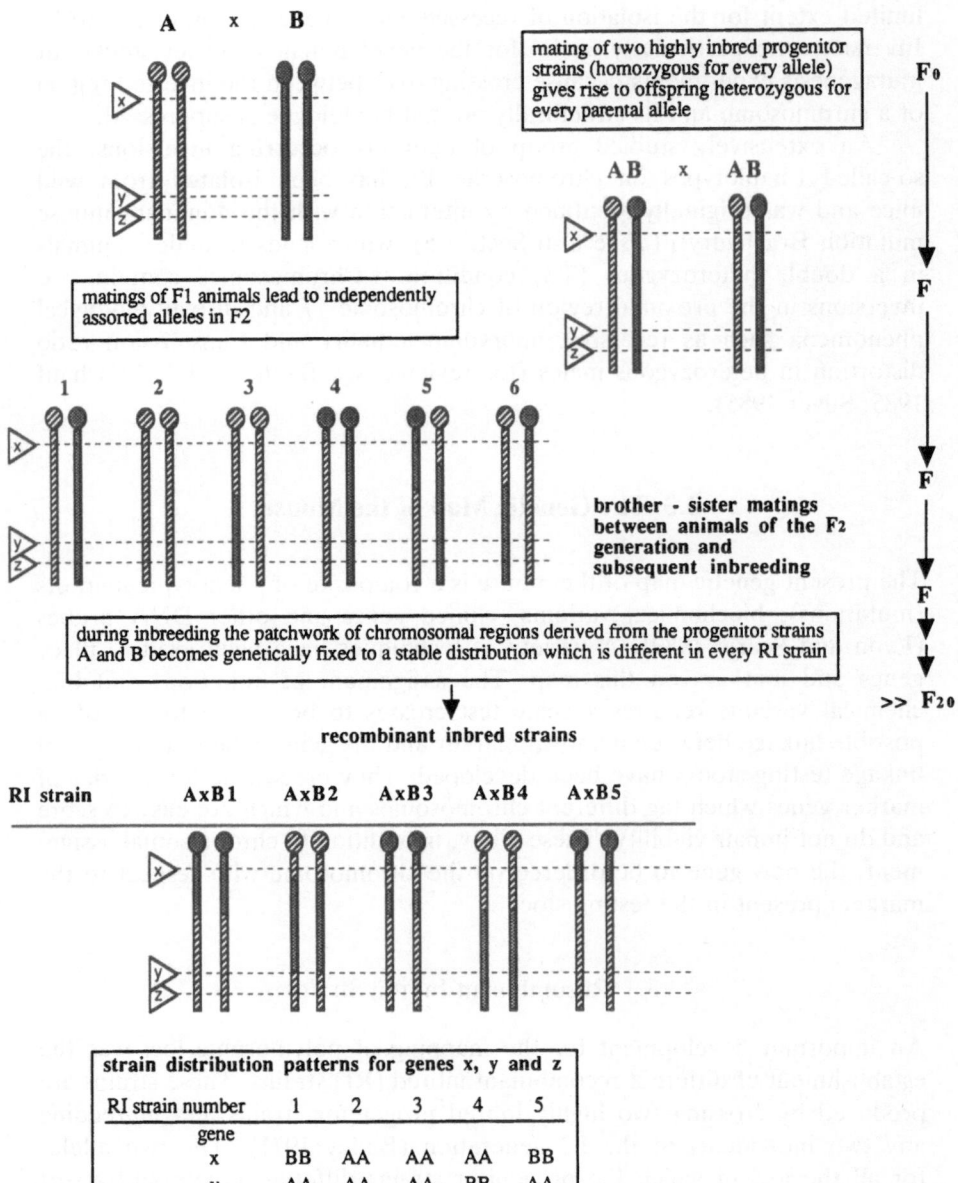

Fig. 7. Scheme of production of RI strains and the resulting strain distribution patterns for three arbitrary polymorphic genes (markers), x, y and z, in five arbitrary recombinant inbred strains. For simplicity, only one pair of chromosomes is depicted

distribution patterns (SDP) of the parts of the respective chromosomes can be established. Since the order of genes has not been changed during establishment of the RI strains, genes closely linked to a marker should show the same or a very similar SDP. Thus, any new locus that shows a polymorphism between the progenitor strains can now be localized with respect to known markers by comparing the SDPs of the new locus with the SDPs of already analyzed markers. DNA probes can be mapped using RI strains if the probe detects restriction fragment length polymorphisms (RFLPs) in the genomic DNAs of the progenitor strains. Since the inbred strains from which RI strains are available are derived from a small pool of mice (Bonhomme et al. 1987; Bonhomme and Guénet 1989), the main difficulty often encountered is finding an RFLP.

3.3.2 Interspecies Backcrosses

The problem of finding RFLPs between inbred strains has been overcome by an alternative mapping technique, which is based on backcrossing F1 animals obtained from breeding an inbred strain to the wild mouse species *Mus spretus* (Robert et al. 1985; Guénet et al. 1988; Copeland and Jenkins 1991). Since *M. spretus* and *M. musculus domesticus* (the species from which the laboratory strains are derived) are distantly related they show a high degree of polymorphism at the DNA level. Female inbred *M. musculus* mice are bred to *M. spretus* males (embryos obtained from the reciprocal cross die in utero) and resulting F1 females (F1 males are sterile) are backcrossed to inbred *M. musculus* males, resulting in animals which carry either one *M. spretus* and one *M. musculus* or two *M. musculus* alleles of any gene (Fig. 8). These animals are first typed with all available markers and "species distribution patterns" analogous to the strain distribution patterns (SDP) are established. DNA probes can be assigned to a particular chromosomal location by comparing their SDP with the panel of existing patterns. As with the RI strains the data are cumulative, every new probe increasing the resolution of the map, which in addition depends on the number of RI strains or backcross animals that have been established.

3.3.3 Synteny Relationships Between Mouse and Man

Establishment of linkage maps in mouse and man and comparison of map positions of homologous genes in both species have revealed that pairs of homologous genes which are closely linked in mouse are often (but not always!) also closely linked in man. Chromosomal regions in one species which contain two or more genes that are also found together in one chromosomal region of the genome of the other species are called syntenic. The order of homologous genes can be, but need not be conserved between species. Thus far, more than 60 homologous chromosome segments have been identified (Nadeau and Reiner 1989). Chromosomal mapping in one species therefore allows predictions as to where the homologous gene might

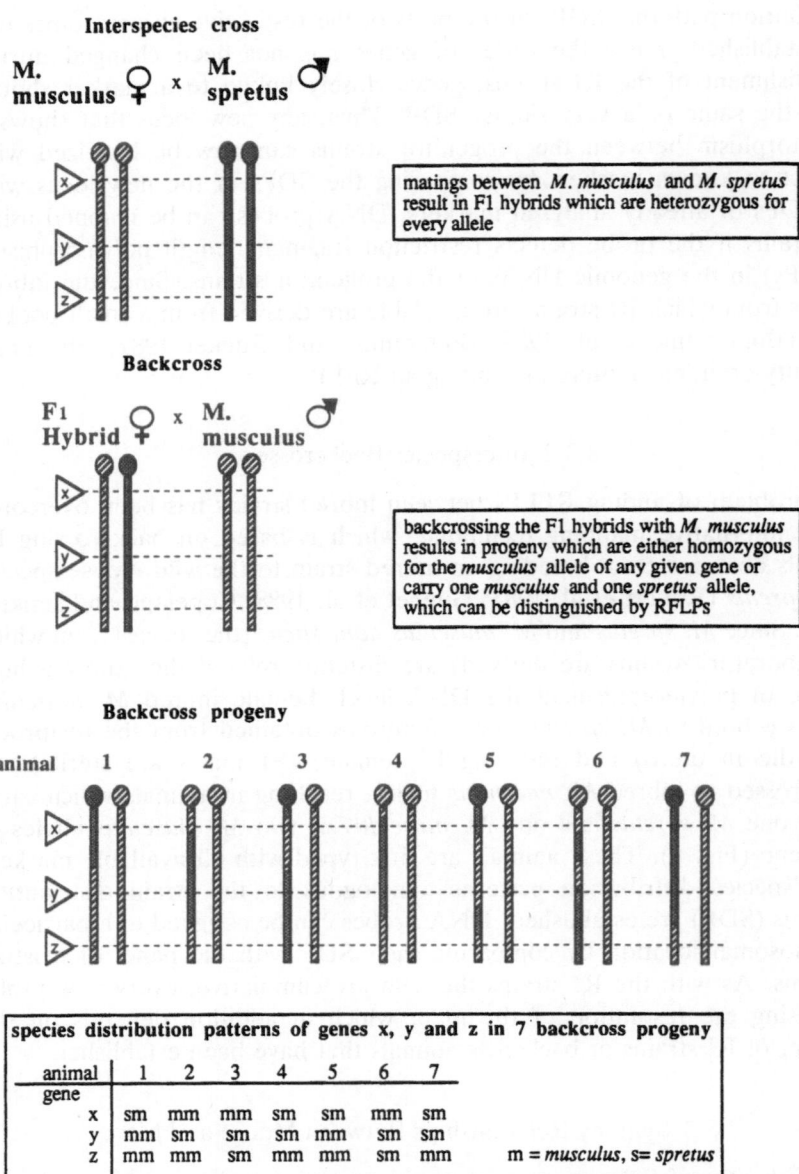

Interspecies cross

M. musculus ♀ x M. spretus ♂

> matings between *M. musculus* and *M. spretus* result in F1 hybrids which are heterozygous for every allele

Backcross

F1 Hybrid ♀ x M. musculus ♂

> backcrossing the F1 hybrids with *M. musculus* results in progeny which are either homozygous for the *musculus* allele of any given gene or carry one *musculus* and one *spretus* allele, which can be distinguished by RFLPs

Backcross progeny

animal   1   2   3   4   5   6   7

species distribution patterns of genes x, y and z in 7 backcross progeny

animal gene	1	2	3	4	5	6	7	
x	sm	mm	mm	sm	sm	mm	sm	
y	mm	sm	sm	sm	mm	sm	sm	
z	mm	mm	sm	mm	mm	sm	mm	m = *musculus*, s= *spretus*

Fig. 8. Scheme of interspecies crosses and the resulting species distribution patterns for three arbitrary polymorphic genes (markers), *x*, *y* and *z*, in seven arbitrary backcross animals. For simplicity, only one pair of chromosomes is depicted

map in the other species. Synteny relations have been helpful in unravelling the molecular basis of an interesting mouse mutation called W which affects stem cells of various lineages. W has a profound effect on haematopoiesis, pigmentation and germ cell development and the numerous alleles of this mutation produce various degrees of anaemia, white spotting and male sterility (Russell 1979). The chromosomal location of the human c-kit proto oncogene on human chromosome 4 and synteny relations between man and mouse suggested that the mouse homologue might map close to the W-locus and, consequently, the mouse c-kit gene could be mapped precisely to the W locus on mouse chromosome 5 (Chabot et al. 1988). The mouse c-kit gene, a member of the receptor tyrosine kinase family, was subsequently shown to be mutated in various W-alleles (Nocka et al. 1989; Reith et al. 1990).

The assignment of the c-kit gene to the W-mutation underlines the relevance of comparative gene mapping and of the analysis of synteny relationships between mouse and human genomes which will be of mutual benefit.

3.4 Mutational Analysis

A genetic analysis of mutants was the key towards identifying developmental control genes in *Drosophila* and *Caenorhabditis elegans* and in understanding their function and interaction. Although a genetic analysis of development is more difficult in mammals, even in mice, a large number of different dominant and recessive mutants have been collected to date. Mutant phenotypes include embryonic lethals at various stages of development, malformations of the axial skeleton and limbs, abnormal structure and function of the peripheral and central nervous system, defects in the immune and haematopoietic systems, abnormalities in gametogenesis and various other disturbances of normal development. Many of these mutations were described in detail, were phenotypically well characterized and have been placed on the mouse genetic map, but only for very few of these have the affected genes been identified and isolated.

Most of these mutants have been obtained either by physical (e.g. Russell 1962 and references therein) or chemical mutagenesis (e.g. Hitotsumachi et al. 1985; Rinchik et al. 1990a; Rinchik 1991) or are mutations which arose spontaneously in various mouse colonies. In addition to general mutagenesis experiments elaborate schemes for approaching saturation mutagenesis and fine-structure functional analysis of specific chromosomal regions have been worked out and performed (Shedlovsky et al. 1986, 1988; King et al. 1989; Rinchik et al. 1990b).

None of the mutagens mentioned above tag the mutated locus with a molecular marker which would allow more direct molecular access to the mutated genes. Thus, combinations of molecular and genetic techniques are required to narrow down the chromosomal location of the mutation and to identify small pieces of DNA containing the gene of interest. One of the most famous mouse mutations, T (Brachyury) has recently been cloned by this approach of "reversed genetics". This interesting dominant mutation maps to the proximal end of chromosome 17. Heterozygous animals can be easily recognized by their short tails. Homozygous embryos fail to develop structures posterior to the forelimb buds due to abnormal development of mesoderm, notochord and allantois and die around day 10 of development (for review, see Willison 1990 and references therein). The chromosomal region which carries T was narrowed down and finally a cosmid clone carrying a CpG island, which is often indicative of the presence of a gene, was used to identify a cDNA clone. The deduced amino acid sequence of the T gene shows minor sequence homology to the myoD protein sequences and might represent the first member of a new class of genes (Herrmann et al. 1990).

A new source of mouse mutants arose during the mid-1970s and early 1980s' when retroviral infection of preimplantation embryos and DNA microinjection into the pronucleus of fertilized eggs were implemented to introduce foreign DNA into the mouse germ line (see below). It soon became apparent that there were transgenic animals in which the transgenes had interrupted endogenous genes causing (developmental) mutations (e.g. Jaenisch et al. 1983; Woychik et al. 1985; Soriano et al. 1987; Kothary et al. 1989; Xin Xiang et al. 1990; for further references see Palmiter and Brinster 1986; Gridley et al. 1987; Babinet et al. 1989; Constantini et al. 1989). In insertional mutations the mutated locus is tagged by the transgene integration and can thus be cloned. Several of the transgene insertions obtained thus far have generated new alleles of already existing mouse mutations (Woychik et al. 1985; Kothary et al. 1989; Pohl et al. 1990; Xin Xiang et al. 1990). Cloning these genes will now permit a combination of molecular analysis allied to the previous genetic and embryological work. Until today, however, only a few of the loci which have been mutated in transgenic animals have been characterized. This is partly due to chromosomal rearrangements at the integration site or multiple copy integrations, which complicate the molecular analysis. The frequency of insertional mutagenesis in transgenic mice is around 10% and with increasing numbers of transgenic animals being produced many more interesting mutants can be expected.

Taken together, "classical" mouse mutations as well as insertional mutations provide a large pool of interesting developmental genes. Nevertheless, the contribition of these mutant stocks to the cloning of genes important for development is still small. However, mutant genes should become increasingly accessible in the future: with a steadily growing number of molecular markers becoming available a high density molecular genetic map of the mouse genome is being established which will allow rapid physical access to any chromosomal region and should greatly facilitate the cloning of mutant genes. Thus, the mutant stocks show promise of making major contributions towards our understanding of mammalian development.

3.5 Introduction of New Genetic Material into the Mouse Genome

The production of transgenic mice is a powerful means to address a variety of questions concerning gene regulation and function during development.

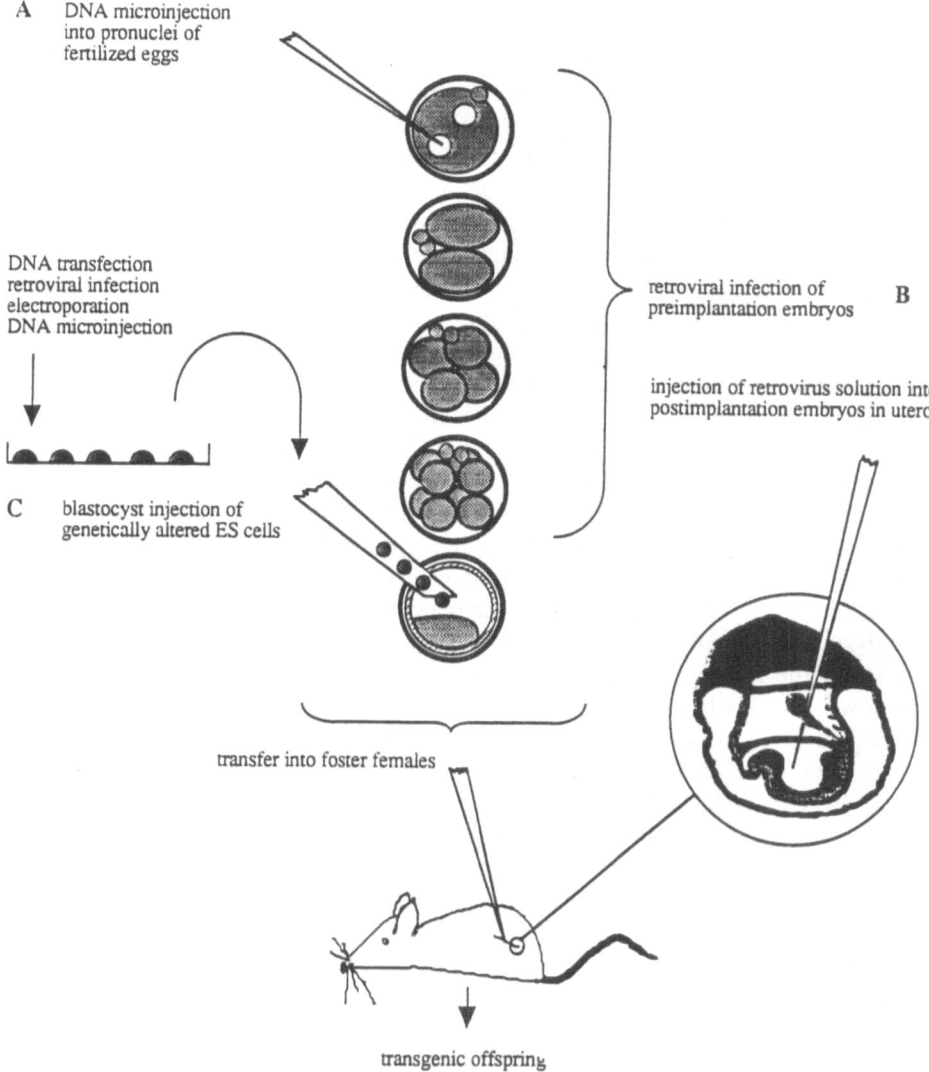

Fig. 9. Means to introduce new genetic information into mice. **A** DNA micro-injection into fertilized eggs; **B** retroviral infection of pre- or post-implantation embryos; **C** blastocyst injection of genetically altered ES cells

Three routes to introduce new genetic information into the mouse genome have been developed (Fig. 9). Besides technical differences, these methods vary with respect to their applicability to various biological questions. Of pivotal importance for all of them was the establishment of culture conditions allowing preimplantation embryos to be maintained in vitro and the development of techniques to transfer them back into foster mothers where normal embryonic development can proceed.

3.5.1 DNA Microinjection

The most extensively used and generally applied technique involves the
direct microinjection of DNA into the pronuclei of fertilized eggs (Gordon
et al. 1980; Brinster et al. 1981; Costantini and Lacy 1981; E.F. Wagner
et al. 1981; T.E. Wagner et al. 1981; Hogan et al. 1986). With DNA
microinjection any cloned piece of DNA (thus far, up to the size of cosmid
clones) can be used. The injected DNA seems to integrate at random.
Often, more than one copy integrates. In this case, usually all copies
integrate together into one chromosomal site in a tandem array. DNA
rearrangements are possible at the site of integration. Some of the parameters affecting the efficiency of introducing DNA into mice have been investigated showing that injection of a few hundred copies of linearized DNA
seems to be optimal (Brinster et al. 1985). Usually the transgenes are stably
integrated and transmitted to progeny, however, a few cases of extrachromosomal replication and episomal transmission have been reported
(Rassoulzadegan et al. 1986; Elbrecht et al. 1987). Very importantly, in
most cases the transgene is appropriately expressed although position effects
are frequently exerted on the expression of the transgene. Transgenic mice
obtained by DNA microinjection have been used in numerous studies to
analyze sequence elements required for tissue specific gene expression and
to study the function of genes by excess or ectopic expression of genes (for
reviews and further references, see Gordon 1983; Palmiter and Brinster
1986; Babinet et al. 1989). Two other interesting applications employing
transgenic mice have been developed: first, toxin genes (i.e. the diphteria
toxin A chain) have been fused to cell-type-specific promoters and have
been introduced into the genome of mice. Cell types expressing the toxin
gene were specifically killed by the toxin gene product ("genetic ablation"),
allowing one to study the consequences of depleting the embryo (or adult
animal) of particular cell types. This might help to analyze cellular interactions during development (i.e. the importance of a depleted cell type for
the morphogenesis of various tissues and organs usually containing these
cells) (Breitman et al. 1987, 1990; Palmiter et al. 1987). Second, LacZ
reporter constructs (enhancer traps) have been introduced into the mouse
genome to detect patterns of transcriptional activation during embryogenesis
in embryos derived from transgenic animals (Allen et al. 1988). LacZ
enhancer traps were originally devised in *Drosophila* and have been successfully used to detect and clone genes expressed in patterns during embryogenesis (O'Kane and Gehring 1987; Bellen et al. 1989; Wilson et al. 1989).
This strategy is based (1) on the introduction into the genome of (lacZ)
reporter gene constructs which require *cis*-acting elements close to or at the
site of integration in order to activate reporter gene expression and (2)
on the assumption that these elements regulate endogenous genes whose
expression patterns are visualized by the reporter gene. The enhancer trap
integration provides the physical link between the expression pattern and
the endogenous gene: the integration site and the corresponding wild type

locus can be cloned and transcribed sequences can be identified in the genomic DNA (for further details, see below).

3.5.2 Retroviral Infection

The second method to introduce new genetic material into mice involves retroviral infection. Retroviruses are RNA viruses, whose genome upon infection of a cell is transcribed into (so-called proviral) DNA which becomes integrated into the host genome and is stably maintained during DNA replication. Introduction of an exogenous retroviral proviral genome into the mouse germ line was first demonstrated by infection of preimplantation embryos with the replication-competent Moloney murine leukemia virus (Mo-MuLV) (Jaenisch 1976) and later by inoculation of postimplantation embryos (Jaenisch 1980). Also, inoculation of newborn SWR/J female mice (a strain which does not carry any endogenous murine leukemia virus proviral DNA) with replication-competent ecotropic leukemia virus can lead to transgenic offspring of the inoculated females (Panthier et al. 1988).

Analysis of the retroviral life cycle allowed the construction of replication-defective retroviral vectors capable of carrying exogenous DNA (Huszar et al. 1985; Jähner et al. 1985; Van der Putten et al. 1985). Retroviral vectors are replication-defective recombinant retroviruses which carry the sequences to be introduced into the embryo in their genome, but lack parts of the viral genome and do not code by themselves for the proteins required for production of infectious virus particles. Infectious retroviral particles containing the vector are produced in helper cell lines (packaging cell lines) which provide the viral proteins necessary for the virus particle to assemble. Introduction of foreign DNA into the genome by retroviral infection has the advantage that in most cases one copy of the retroviral vector is integrated with no or little rearrangement of the DNA at the site of insertion. However, the size of the exogenous DNA which can be carried by retroviral vectors is restricted due to the limited size of the retroviral genome that can be packaged. While retroviral vectors have been used only to a limited extent to express foreign genes in transgenic mice (e.g. Stewart et al. 1985, 1987; Soriano et al. 1986), they have proven to be powerful tools for cell lineage studies in various systems. For example, retroviral insertions have been used to estimate the number of cells constituting the embryo proper (Soriano and Jaenisch 1986) or to determine cell fates at later stages of development (Sanes et al. 1986; Price 1987; Price et al. 1987; Turner and Cepko 1987). In these studies retroviral vectors have been used which express the *E. coli lacZ* gene. Expression of *lacZ* can be easily detected by the enzymatic activity of its gene product, β-galactosidase. Viral infection can be performed at various stages of development with highly diluted virus to label individual cells and clones of cells derived from one infected cell can be identified and analyzed after some rounds of cell divisions and subsequent development. Expression of the *lacZ* gene in the target cell and all its descendants is required to detect all cells derived from one founder cell, and delayed integration of the virus into the genome or extensive cell mixing between cells derived from the target cell and neighbour cells can render the analysis more difficult.

3.5.3 Mouse Embryonic Stem Cells

The third technique for manipulating the mouse genome makes use of mouse embryonic stem (ES) cells, whose genome can be altered by various means (see below) to produce germ line chimaeric mice by blastocyst injection (Gossler et al. 1986; Robertson et al. 1986). Embryonic stem cells are pluripotent embryonic cells established from inner cell mass cells of mouse blastocysts (Evans and Kaufman 1981; Martin 1981). ES cells can spontaneously differentiate in vitro and give rise to a variety of differentiated cell types. In suspension cultures ES cells form complex cystic structures resembling visceral yolk sacs and some embryonic tissues (Doetschman et al. 1985). ES cells can be injected into blastocysts and then participate in embryonic development and efficiently colonize all tissues of the developing embryo including the germ line (Bradley et al. 1984; Gossler et al. 1986; Lallemand and Brulet 1990; Nagy et al. 1990). Foreign DNA can be intro-duced into the genome of ES cells in vitro by DNA transfection, elec-troporation, retroviral infection or DNA microinjection (Gossler et al. 1986, 1989; Robertson et al. 1986; Zimmer and Gruss 1989). Genetically altered ES cells can still be pluripotent, they can colonize the germ line in chimaeric mice and germ cells derived from genetically altered ES cells transmit the genetic alteration to subsequent generations (Robertson et al. 1986; Gossler et al. 1986). As described below, a particularly interesting aspect of using ES cells to introduce foreign DNA into the mouse germ line is the possibility of analyzing and selecting ES cells in vitro for a desired modification caused by the introduced DNA before introducing this modification into the germ line.

3.5.3.1 Targeted Mutagenesis

The analysis of mutant phenotypes is instrumental in understanding the role of a specific gene in the hierarchy of genes controling development. Unfor-tunately, for many cloned genes thought to be involved in the control of development, no natural mutants are available. Thus, it is desirable to produce mutations in these genes to allow analysis of the resulting phenotypes, which would help to understand gene function. This aim has been achieved by developing techniques to generate and detect homologous recombination events in ES cells which can occur between introduced and endogenous DNA during integration. The first gene to be mutated by homologous recombination in ES cells was the selectable HPRT gene (Doetschman et al. 1987; Thomas and Capecchi 1987). Since then various strategies have been devised to identify and isolate ES cells carrying targeted mutations in non-selectable genes. Mostly vectors are used in which coding sequences of the endogenous gene to be targeted are disrupted by the *neo* gene which also provides a dominant marker to select for cells that have taken up DNA (Thomas and Capecchi 1987). Homologous recombination events in neo^R ES cell clones are usually identified by a polymerase chain reaction (e.g. Joyner et al. 1989). Genes expressed in ES cells allow for enrichment of homologous integration versus random integration by using promoterless *neo* genes or in-frame fusions between the *neo* gene and translated exon sequences of the host gene (te Riele et al. 1990). Further enrichment of cells carrying homologous integrations can be obtained when a gene whose expression is lethal to the cell is placed at the end of the targeting vector (Mansour et al. 1988). During homologous recombination this part of the construct should be excluded, while it is retained in random integrations. This strategy is called positive-negative selection (PNS) and employs mostly the *HSV-tk* gene (Mansour et al. 1988) or more recently the diphteria toxin A chain (Yagi et al. 1990). Cells expressing the *HSV-tk* gene are selectively killed after addition to the culture medium of the nucleoside analogue gancyclovir. Expression of the diphteria toxin A chain is lethal to the cells by the direct action of the gene product. Thus, ES cells carrying random integrations are selectively killed, while cells having lost the negative selectable gene during homologous recombination survive. Homologous recombination has also been obtained after DNA microinjection into ES cells without any selection procedure at all (Zimmer and Gruss 1989).

Additional vectors have been designed that not only disrupt a gene but also concomitantly introduce the *lacZ* gene either as an in-frame fusion to endogenous exons (Mansour et al. 1990) or as a replacement for part of the endogenous gene (Le Mouellic et al. 1990). In both cases *lacZ* expression is directed by the regulatory elements of the endogenous gene which allows easy visualization of the expression pattern of the targeted locus by β-galactosidase staining in whole mount preparations.

A number of genes have been mutated by targeted mutation and in several cases the mutated allele was transmitted through the mouse germ line (e.g. Kuehn et al. 1987; Koller et al. 1989; Zijlstra et al. 1989; McMahon and Bradley 1990; Chisaka and Capecchi 1991; Joyner et al. 1991; see also references therein; for reviews, see Capecchi 1989a,b; Rossant and Joyner 1989). Thus, the ES-cell route for introducing new genetic information into the mouse germ line allows one to analyze in vivo the function of cloned genes mutated in vitro.

3.5.3.2 Enhancer and Gene Traps in ES Cells

As already mentioned, *lacZ* reporter gene constructs with weak promoters have been used in transgenic mice to detect patterns of transcriptional activation of endogenous genes during embryogenesis. Strategies have also been developed in ES cells to detect and isolate genes expressed in specific patterns during development, employing similar reporter constructs.

Reporter genes are introduced into the genome of ES cells and cell lines carrying reporter genes are used to produce chimaeric embryos which are subsequently stained for β-galactosidase activity. In addition to enhancer trap vectors, ES cells also allowed one to test and implement a different type of vector called "gene trap". Gene trap vectors contain a splice acceptor site in front of the *lacZ* gene but no promoter and require insertion into an intron to generate a fusion transcript between exon sequences of the endogenous host gene and the *lacZ* gene. In this vector functional β-galactosidase can only be expected when the reading frame of the *lacZ* gene is maintained after splicing, the fusion transcript providing direct access to endogenous exon sequences. ES cells can be preselected for gene trap integrations into genes (by analyzing *lacZ* expression in ES cells) and can then be further analyzed in chimaeras.

β-Galactosidase staining patterns were observed with a high frequency in chimaeric embryos with both types of vectors (Gossler et al. 1989) and corresponding endogenous genes have been cloned in several cases (Gossler et al. unpub.; A. Joyner and W. Skarnes pers. comm.). While some enhancer trap integration events might disrupt the endogenous gene, all gene trap integrations leading to *lacZ* expression must have occurred in the host gene and thereby very likely disrupted its expression. Thus, gene trap insertions should not only detect but also efficiently mutagenise endogenous genes. These mutations can then be analyzed in vivo after germ line transmission from chimaeric animals. In a preliminary study, two out of three gene trap insertions resulted in developmental defects in mice homozygous for the integrations (A. Joyner and W. Skarnes pers. comm.). Thus, the molecular analysis of novel genes can be combined with the analysis of the mutant phenotypes during embryogenesis.

In summary, ES cells promise to contribute not only to the analysis of known genes but also to the identification of novel genes which are expressed in specific patterns during embryogenesis.

3.6 Conserved Genes – Conserved Functions?

A large number of mouse putative developmental genes have been identified based on sequence similarities to developmental control genes in other species, notably *Drosophila*. When the first *Drosophila* genes affecting development, "homeotic" and "segmentation" mutations, were cloned, it soon became clear that they shared certain sequence motifs, i.e. the homeobox (Gehring and Hiromi 1986), paired-box (Bopp et al. 1986) or zinc-finger structure (Rosenberg et al. 1986), and that they could act as transcriptional regulators. These motifs were shown to be evolutionarily conserved from insects to mammals and probes containing these sequences were used to systematically identify and isolate a large number of vertebrate homeobox-, paired box- and zinc-finger-containing genes.

More than 40 mouse genes containing a homeobox (*Hox* genes) have been isolated. Many properties of the *Drosophila* homeobox-containing genes, e.g. clustered chromosomal arrangement or distinct anterior borders of expression depending on the location in the cluster are also found in their mouse counterparts. Expression of mouse *Hox* genes is restricted temporally and spatially to various domains along the anterior-posterior axis and consistent with a role of *Hox* genes interpreting positional information (for a recent review on mouse homeobox genes, see Shashikant et al. 1991 and references therein).

The best evidence for the importance of mouse homeobox-containing genes for development has been obtained by mutating homeobox genes in ES cells. The mouse *En-2* gene, the homologue of the *Drosophila* homeobox gene *engrailed*, and the *Hox 1.5* and *Hox 1.6* genes have been mutated and analyzed so far in mice homozygous for the mutated allele. En^-/En^- mice exhibit a distortion of the foliation pattern within the cerebellum, the tissue normally expressing *en-2* (Joyner et al. 1991). Hox-$1.5^-/Hox$-1.5^- and Hox-$1.6^-/Hox^-.6^-$ mice show complex lethal phenotypes, respectively, with multiple defects either predominantly in tissues derived from the four pharyngeal arches and/or pouches or in the hindbrain region (Chisaka and Capecchi 1991; Lufkin et al. 1991). These results clearly show that *Hox* genes are essential for normal embryonic development. Several other mouse *Hox* genes that have been mutated in ES cells await analysis in mutant mice.

To date, eight paired box (*Pax*)-containing genes of the mouse have been described (Walther et al. 1991). *Pax* genes are not clustered or confined to a certain chromosome. Unlike the *Hox* genes, most *Pax* genes are expressed along the entire length of the anterior-posterior axis (Deutsch et al. 1988; Dressler et al. 1990). In the neural tube, expression in specific domains along the dorsal-ventral axis is observed, whereas in mesoderm-derived tissues, expression appears to be confined to segmented structures. The most compelling evidence that *Pax* genes are developmental control genes comes from the assignment of one of the *Pax* genes, *Pax-1*, to the long known mouse mutation *undulated*, which affects the axial skeleton. In particular the intervertebral discs (Grüneberg 1954) are the primary sites of *Pax-1* expression during embryogenesis. *Pax-1* was mapped to chromosome 2, close to the mutation *undulated* (Dressler et al. 1988) and a subsequent molecular analysis of *undulated* revealed a point mutation in the paired box of *Pax-1* (Balling et al. 1988). The assignment of *Pax-1* to *undulated* underlines the importance of gene mapping in the mouse and exemplifies the successful combination of genetic and molecular analyses.

The developmental role of mouse "zinc-finger" genes identified by homology to *Drosophila* developmental genes (Chowdhury et al. 1987) is much less clear at present. No natural mutants of mouse zinc-finger genes have yet been identified, nor have zinc-finger genes been targeted and analyzed in mice homozygous for the mutated allele.

Many other "developmental" genes have been isolated in the mouse by a variety of approaches. These genes encode molecules such as transcription factors, growth factors and their receptors, retinoic acid receptors and binding proteins or proto-oncogenes. A number of them have been found to be conserved between mice and other mammals or vertebrates. In most cases their importance for embryonic development is inferred from expression patterns, relation to possible morphogenic molecules or activity of the gene products in various biological assays in different species. Mutations for these genes are not available in most species. With the possibility to produce targeted mutations in embryonic stem cells the mouse might serve as the genetic model to further study the function of these genes in vivo by mutating the mouse homologues.

4 Resumé and Outlook

With this chapter I hope to have provided a basic introduction and a summary of early mouse development and mouse developmental genetics which will be helpful to newcomers in the field and might serve to complement other more detailed presentations. I tried to combine a description of the early events during embryogenesis and of the mouse genome with a discussion of various aspects and properties of the early mouse embryo, of technical possibilities and of experimental approaches which have contributed to our current picture of mouse embryonic development or which promise to be helpful for future analysis. Many aspects, especially of the genes thought to control mouse embryogenesis and of their products have not been mentioned, since they are the subject of a variety of special articles and reviews. A variety of means to isolate and mutate "developmental" genes have become available. The regulatory elements conferring tissue- and cell-type-specific regulation on gene expression can be efficiently analyzed in transgenic mice, targeted mutations can be introduced into cloned genes and cell lineages can be traced by retroviral vectors. A high density molecular genetic map is being established and will allow rapid physical access to any chromosomal region.

In summary, the study of mouse development has entered a phase in which due to the collaborative interplay of embryologists, geneticists and molecular biologists the molecular analysis of genes and their products can be efficiently combined with the analysis of phenotypes resulting from mutations in these genes, allowing a comprehensive analysis of gene function during mouse embryogenesis.

Acknowledgements. I am thankful to Drs. Rosa Beddington, Jean-Louis Guénet and Rudi Balling for critically reading the manuscript and for their comments which helped to improve it.

References

Adamson ED (1982) The location and synthesis of transferrin in mouse embryos and teratocarcinoma cells. Dev Biol 91:227–234

Akam M (1989) Hox and HOM: homologous gene clusters in insects and vertebrates. Cell 57:347–349

Allen ND, Cran DG, Barton SC, Hettle S, Reik W, Surani A (1988) Transgenes as probes for active chromosomal domains in mouse development. Nature 333: 852–855

Babinet C, Morello D, Renard JP (1989) Transgenic mice. Genome 31:938–949

Bachvarova R, DeLeon V (1980) Polyadenylated RNA of mouse ova and loss of maternal RNA in early development. Dev Biol 74:1–8

Bailey DW (1971) Recombinant inbred strains, an aid to finding identity, linkage, and function of histocompatibility and other genes. Transplantation 11:325–327

Balling R, Deutsch U, Gruss P (1988) *undulated*, a mutation affecting the development of the mouse skeleton, has a point mutation in the paried box of *Pax 1*. Cell 55:531–535

Beddington RSP (1981) An autoradiographic analysis of the potency of embryonic ectoderm in the 8th day postimplantation mouse embryo. J Embryol Exp Morphol 64:87–104

Beddington RSP (1982) An autoradiographic analysis of tissue potency of different regions of embryonic ectoderm during gastrulation in the mouse embryo. J Embryol Exp Morphol 69:265–285

Beddington RSP (1983a) The origin of fetal tissues during gastrulation in the rodent. In: Johnson MH (ed) Development in mammals, vol 5. Elsevier, Amsterdam, pp 1–32

Beddington RSP (1983b) Histogenic and neoplastic potential of different regions of the embryonic egg cylinder. J Embryol Exp Morphol 75:189–204

Bellen HJ, O'Kane CJ, Wilson C, Grossniklaus U, Pearson RK, Gehring WJ (1989) P-element-mediated enhancer detection: a versatile method to study development in *Drosophila*. Genes Dev 3:1288–1300

Bellvé AR, O'Brien DA (1983) The mammalian spermatozoon: structure and temporal assembly. In: Hartmann JF (ed) Mechanism and control of animal fertilization. Academic Press, London, pp 55–137

Bennett D (1975) The t-locus of the mouse. Cell 6:441–454

Bleil JD, Wassarman PM (1980a) Structure and function of the zona pellucida: identification and characterization of the mouse oocyte zona pellucida. Dev Biol 76:185–202

Bleil JD, Wassarman PM (1980b) Synthesis of zona pellucida proteins by denuded and follicle-enclosed mouse oocytes during culture in vitro. Proc Natl Acad Sci USA 77:1029–1033

Bleil JD, Wassarman PM (1980c) Mammalian sperm and egg interaction: identification of a glycoprotein in mouse egg zonae pellucidae possessing receptor activity for sperm. Cell 20:773–882

Bonhomme F, Guénet J-L (1989) The wild house mouse and its relatives. In: Lyon MF, Searle AG (eds) Genetic variants and strains of the laboratory mouse. Oxford University Press, Oxford, pp 649–662

Bonhomme F, Guénet J-L, Dod B, Moriwaki K, Bulfield G (1987) The phylogenetic origin of laboratory inbred mice and their rate of evolution. Biol J Linn Soc Lond 30:51–58

Bopp D, Burri M, Baumgartner S, Frierio G, Noll M (1986) Conservation of a large protein domain in the segmentation gene paired and in functionally related genes of *Drosophila*. Cell 47:1033–1040

Bradley A, Evans M, Kaufman MH, Robertson E (1984) Formation of germ-line chimaeras from embryo-derived teratocarcinoma cell lines. Nature 309:255–256

Braude P, Pelham H, Flach G, Lobatto R (1979) Post-transcriptional control in the early mouse embryo. Nature 282:102–105

Breitman ML, Clapoff S, Rossant J, Tsui LC, Glode LM, Maxwell IH, Bernstein A (1987) Genetic ablation: targeted expression of a toxin gene causes microphthalmia in transgenic mice. Science 238:1563–1565

Breitman ML, Rombola H, Maxwell IH, Klintworth GK, Bernstein A (1990) Genetic ablation in transgenic mice with an attenuated diphteria toxin A gene. Mol Cell Biol 10:474–479

Brinster RL, Chen HY, Trumbauer ME, Senear AW, Warren R, Palmiter RD (1981) Somatic expression of herpes thymidine kinase in mice following injection of a fusion gene into eggs. Cell 27:223–231

Brinster RL, Chen HY, Trumbauer ME, Yagle MK, Palmiter RD (1985) Factors affecting the efficiency of introducing foreign DNA into mice by microinjecting eggs. Proc Natl Acad Sci USA 82:4438–4442

Brinster RL, Chen HY, Warren R, Sarthy A, Palmiter RD (1982) Regulation of metallotheionein-thymidine kinase fusion plasmids injected into mouse eggs. Nature 296:39–42

Capecchi M (1989a) The new mouse genetics: altering the genome by gene targeting. TIG 5:70–76

Capecchi M (1989b) Altering the genome by homologous recombination. Science 244:1288–1292

Carlin B, Jaffe R, Bender B, Chung AE (1981) Entactin, a novel basal lamina associated sulfated glycoprotein. J Biol Chem 256:5209–5214

Carter TC, Lyon MF, Phillips RJS (1955) Gene-tagged chromosome translocations in eleven stocks of mice. J Genet 53:154–166

Carter TC, Lyon MF, Phillips RJS (1956) Further genetic studies of eleven translocations in the mouse. J Genet 54:462–473

Cascio S, Wassarman PM (1982) Program of early development in the mammal: post-transcriptional control of a class of proteins synthesized by mouse oocytes and early embryos. Dev Biol 89:397–408

Caspersson T, Farber S, Foley GE, Kudynowski J, Modest EJ, Simonsson E, Wagh U, Zech L (1968) Chemical differentiation along metaphase chromosomes. Exp Cell Res 49:219–222

Castle WE (1903) Mendel's law of heredity. Proc Am Acad Arts Sci 38:535–548

Cattanach BM (1974) Position effect variegation in the mouse. Genet Res Camb 23:291–306

Cattanach BM (1975) Control of chromosome inactivation. Annu Rev Genet 9:1–18

Chabot B, Stevenson DA, Chapman VM, Besmer P, Bernstein A (1988) The proto-oncogene c-kit encoding a transmembrane tyrosine receptor maps to the mouse W locus. Nature 353:88–89

Chiquoine AD (1954) The identification, origin and migration of the primordial germ cells of the mouse embryo. Anat Rec 118:135–146

Chisaka O, Capecchi MR (1991) Regionally restricted developmental defects resulting from targeted disruption of the mouse homeobox gene Hox 1.5. Nature 350:473–479

Chowdhury K, Deutsch U, Gruss P (1987) A multigene family encoding several "finger" structures is present and differentially active in mammalian genomes. Cell 48:771–778

Clark CC, Crossland J, Kaplan G, Martinez-Hernandez A (1982) Location and identification of the collagen found in the 14.5 day rat embryo visceral yolk sac. J Cell Biol 93:251–260

Clegg KB, Piko L (1977) Size and specific activity of the UTP pool and overall rates of RNA synthesis in early mouse embryos. Dev Biol 58:76–95

Clegg KB, Piko L (1983) Quantitative aspects of RNA synthesis and polyadenylation in 1-cell and 2-cell mouse embryo. J Embryol Exp Morphol 74:169–182

Cockroft DL (1986) Regional and temporal differences in the parietal endoderm of the midgestation mouse embryo. J Anat 145:35–47

Constantini F, Lacy E (1981) Introduction of a rabbit globin gene into the mouse germ line. Nature 294:92–94

Constantini F, Radice G, Lee JJ, Chada KK, Perry W, Jin Son H (1989) Insertional mutations in transgenic mice. Prog Nucleic Acid Res 36:159–169

Copeland NG, Jenkins NA (1991) Development and applications of a molecular genetic linkage map of the mouse genome. TIG 7(4):113–118

Copp AJ, Roberts HM, Polani PE (1986) Chimaerism of primordial germ cells in the early postimplantation mouse embryo following microsurgical grafting of posterior primitive streak cells in vitro. J Embryol Exp Morphol 95:95–115

Copp AJ, Cockroft DL (eds) (1990) Postimplantation mammalian embryos: a practical approach. IRL Press, Oxford

Cruz YP, Pedersen RA (1985) Cell fate in the polar trophectoderm of mouse blastocysts as studied by microinjection of cell lineage tracers. Dev Biol 112:73–83

Cullen B, Emigholz K, Monohan J (1980) The transient appearance of specific proteins in one-cell embryos. Dev Biol 76:215–221

Davisson MT, Roderick TH (1976) Cytology of inversions in mice. Genetics 83:s19

Deutsch U, Dressler GR, Gruss P (1988) Pax 1, a member of a paired box homologous murine gene family is expressed in segmented structures during development. Cell 53:617–625

Doetschman TC, Eistetter H, Katz M, Schmidt W, Kemler R (1985) The in vitro development of blastocyst derived embryonic stem cell lines: formation of visceral yolk sac, blood islands and myocardium. J Embryol Exp Morphol 87:27–45

Doetschman T, Gregg RG, Maeda N, Hooper ML, Melton DW, Thompson S, Smithies O (1987) Targeted correction of a mutant HPRT gene in mouse embryonic stem cells. Nature 330:576–578

Donovan PJ, Stott D, Cairns LA, Heasman J, Wylie CC (1986) Migratory and postmigratory mouse primordial germ cells behave differently in culture. Cell 44:831–838

Dove WF (1987) Molecular genetics of Mus musculus: point mutagenesis and millimorgans. Genetics 116:5–8

Dressler GR, Deutsch U, Balling R, Simon D, Guenét JL, Gruss P (1988) Murine genes with homology to Drosophila segmentation genes. Development 104 (Suppl):181–186

Dressler GR, Deutsch U, Chowdhury K, Nornes HO, Gruss P (1990) Pax 2, a new murine paried-box-containing gene and its expression in the developing excretory system. Development 109:787–795

Ducibella T, Anderson E (1975) Cell shape and membrane changes in the eight-cell mouse embryo: prerequisites for morphogenesis of the blastocyst. Dev Biol 47:45–58

Ducibella T, Albertini DF, Anderson E, Biggers JD (1975) The preimplantation mammalian embryo: characterization of intercellular junctions and their appearance during development. Dev Biol 45:231–250

Dyce J, George M, Goddall H, Fleming T (1987) Do trophectoderm and inner cell mass cells in the mouse blastocyst maintain discrete lineages? Development 100:685–698

Dziadek M, Adamson ED (1978) The localization and synthesis of alpha-fetoprotein in postimplantation mouse embryos. J Embryol Exp Morphol 43:289–313

Elbrecht A, DeMayo FJ, Tsai MJ, O'Malley BW (1987) Episomal maintenance of a bovine papilloma virus vector in transgenic mice. Mol Cell Bio! 7:1276–1279

Enders AC, Given RL, Schlaffke S (1978) Differentiation and migration of endoderm in the rat and mouse at implantation. Anat Rec 190:65–78

Eppig JJ (1985) Oocyte-somatic cell interactions during oocyte growth and maturation in the mammal. In: Browder LW (ed) Developmental biology, vol 1. Oogenesis, Plenum Press, New York, pp 313–347

Epstein, CJ (1986) Generation and properties of mouse aneuploids. In: Epstein CJ (ed) The consequences of chromosome imbalance. Cambridge University Press, Cambridge, pp 207–232

Evans MJ, Kaufman MH (1981) Establishment in culture of pluripotential cells from mouse embryos. Nature 292:154–158

Fawcett DW (1975) The mammalian spermatozoon. Dev Biol 44:394–436

Finn CA (1971) The biology of decidual cells. Adv Reprod Physiol 5:1–26

Fleming TP (1987) A quantitative analysis of cell allocation to trophectoderm and inner cell mass in the mouse blastocyst. Dev Biol 119:520–531

Fleming TP, Pickering SJ (1985) Maturation and polarization of the endocytotic system in outside blastomeres during mouse preimplantation development. J Embryol Exp Morphol 89:175–208

Frischauf, AM (1985) The T/t complex of the mouse. TIG 1:100–103

Gardner RL (1972) An investigation of inner cell mass and trophoblast tissue following their isolation from the mouse blastocyst. J Embryol Exp Morphol 28:279–312

Gardner RL (1975) Analysis of determination and differentiation in the early mammalian embryo using intra- and inter-specific chimaeras. In: Markert CL (ed) The developmental biology of reproduction. 33rd Symp of the Society for Developmental Biology. Academic Press, New York, pp 207–238

Gardner RL, Johnson MH (1973) Investigation of early mammalian development using interspecific chimaeras between rat and mouse. Nature (New Biol) Lond 246:86–89

Gardner RL, Rossant J (1979) Investigation of the fate of 4.5 day post coitum mouse inner cell mass cells by blastocyst injection. J Embryol Exp Morphol 52: 141–152

Gardner RL, Papaiaonoou VE, Barton SC (1973) Origin of the ectoplacental cone and secondary giant cells in mouse blastocysts reconstituted from isolated trophoblast and inner cell mass. J Embryol Exp Morphol 30:561–572

Gehring, WJ, Hiromi Y (1986) Homeotic genes and the homeobox. Annu Rev Genet 20:147–173

Giebelhaus DH, Weitlauf HM, Schultz GA (1985) Actin mRNA content in normal and delayed implanting mouse embryos. Dev Biol 107:407–413

Ginsburg M, Snow MHL, McLaren A (1990) Primordial germ cells in the mouse embryo during gastrulation. Development 110:521–528

Godin I, Wylie C, Heasman J (1990) Genital ridges exert long range effects on mouse primordinal germ cell numbers and direction of migration in culture. Development 108:357–363

Gordon JW (1983) Transgenic mice: a new and powerful experimental tool in mammlian developmental genetics. Dev Genet 4:1–20

Gordon JW, Scangos GA, Plotkin DJ, Barbosa JA, Ruddle FH (1980) Genetic transformation of mouse embryos by microinjection of purified DNA. Proc Natl Acad Sci USA 77:7380–7384

Gossler A, Doetschman T, Korn R, Serfling E, Kemler R (1986) Transgenesis by means of blastocyst derived embryonic stem cell lines. Proc Natl Acad Sci USA 83:9065–9069

Gossler A, Joyner AL, Rossant J, Skarnes WC (1989) Mouse embryonic stem cells and reporter constructs to detect developmentally regulated genes. Science 244:463–465

Graham A, Papalopoulu N, Krumlauf R (1989) The murine and Drosophila homoebox gene complexes have common features of organization and expression. Cell 57:367–378

Graves RA, Marzluff DH, Giebelhaus DH, Schultz GA (1985) Quantitative and qualitative changes in histone gene expression during early mous development. Proc Natl Acad Sci USA 82:5685–5689

Greve JM, Wassarman PM (1985) Mouse egg extracellular coat is a matrix of interconnected filaments possessing a structural repeat. J Mol Biol 181:253–264

Gridley T, Soriano P, Jaenisch R (1987) Insertional mutagenesis in transgenic mice. TIG 3:162–166

Gropp A, Winking H (1981) Robertsonian translocations: cytology, meiosis, segregation patterns and biological consequences of heterozygosity. In: Berry RJ (ed) Biology of the house mouse. Academic Press, London, pp 141–181

Gropp A, Winking H, Zech U, Müller H (1972) Robertsonian chromosomal variation and identification of metacentric chromosomes in feral mice. Chromosoma 39:265–288

Grüneberg H (1954) Genetical studies on the development of the mouse. XII. The development of *undulated*. J Genet 52:441–455

Guénet J-L, Simon-Chazottes D, Avner PR (1988) The use of interspecific mouse crosses for gene localization: present status aand future perspectives. In: Mock B, Potter M (eds) Current topics of microbiology and immunology 137. Springer, Berlin Heidelberg New York, pp 13–17

Haldane JBS, Sprunt AD, Haldane NM (1915) Reduplication in mice. J Genet 5:133–135

Hastie ND (1989) Highly repeated DNA families in the genome of *Mus musculus*. In: Lyon M F, Searle AG (eds) Genetic variants and strains of the laboratory mouse. Oxford University Press, Oxford, pp 559–573

Heller DT, Cahill DM, Schultz RM (1981) Biochemical studies of mammalian oogenesis: metabolic cooperativity between granulosa cells and growing mouse oocytes. Dev Biol 84:455–464

Herrmann BG, Labeit S, Poustka A, King TR, Lehrach H (1990) Cloning of the T gene required in mesoderm formation in the mouse. Nature 343:617–622

Hillmann N, Sherman MI, Graham C (1972) The effect of spatial arrangement on cell determination during mouse development. J Embryol Exp Morphol 28:263–278

Hitotsumachi S, Carpenter DA, Russell WL (1985) Dose repetition increases the mutagenic effectiveness of N-ethyl-N-nitrosourea in mouse spermatogonia. Proc Natl Acad Sci USA 82:6619–6621

Hogan BLM, Newman R (1984) A scanning electron microscope study of the extraembryonic endoderm of the 8th-day mouse embryo. Differentiation 26:138–143

Hogan BLM, Tilly R (1981) Cell interactions and endoderm differentiation in culture mouse embryos. J Embryol Exp Morphol 62:379–394

Hogan BLM, Cooper AR, Kurkinen M (1980) Incorporation into Reichert's membrane of laminin-like extracellular proteins synthesized by parietal endoderm cells of the mouse embryo. Dev Biol 80:289–300

Hogan BLM, Taylor A, Cooper AR (1982) Murine parietal endoderm cells synthesize heparan sulphate and 170 k and 145 k sulphated glycoproteins as components of Reichert's membrane. Dev Biol 90:210–214

Hogan BLM, Holland P, Schofield P (1985) How is the mouse segmented? TIG 1:67–74

Hogan B, Constantini F, Lacy E (1986) Manipulating the mouse embryo. A laboratory manual. Cold Spring Harbor Laboratory, Cold Spring Harbor, New York

Hollander WF, Strong LC (1950) Intrauterine mortality and placental fusions in the mouse. J Exp Zool 115:131–150

Hsu YC (1982) Development of mouse embryos in vitro: preimplantation to limb bud stage. Science 218:66–68

Huszar D, Balling R, Kothary R, Magli MC, Hozumi N, Rossant J, Bernstein A (1985) Insertion of a bacterial gene into the mouse germ line using an infectious retrovirus vector. Proc Natl Acad Sci USA 82:8587–8591

Hyafil F, Morello D, Babinet C, Jacob F (1980) A cell surface glycoprotein involved in the compaction of embryonal carcinoma cells and cleavage stage embryos. Cell 21:927–934

Jaenisch R (1976) Germline integration and mendalian transmission of the exogenous Moloney leukemia virus. Proc Natl Acad Sci USA 73:1260–1264

Jaenisch R (1980) Retroviruses and embryogenesis: microinjection of Moloney leukemia virus into midgestation mouse embryos. Cell 19:181–188

Jaenisch R, Harbers K, Schnieke A, Lohler J, Chumakov I, Jähner D, Grotkopp D, Hoffmann E (1983) Germ line integration of Moloney murine leukemia virus at the Mov 13 locus leads to recessive lethal mutation and early embryonic death. Cell 32:209–216

Jähner D, Haase K, Mulligan R, Jaenisch R (1985) Insertion of the bacterial *gpt* gene into the germ line of mice by retroviral infection. Proc Natl Acad Sci USA 82:6927–6931

Johnson MH (1986) Manipulation of early mammalian development: what does it tell us about cell lineages? In: Gwatkin RBL (ed) Developmental biology: a comprehensive synthesis. Plenum Press, New York, pp 279–296

Johnson MH, Maro B (1984) The distribution of cytoplasmic actin in mouse eight cell blastomeres. J Embryol Exp Morphol 82:97–117

Johnson MH, Ziomek CA (1981) The foundation of two distinct cell lineages within the mouse morula. Cell 24:71–80

Joyner AL, Herrup K, Auerbach A, Davis CA, Rossant R (1991) Subtle cerebellar phenotype in mice homozygous for a targeted deletion of the *En-2* homeobox. Science 251:1239–1243

Joyner AL, Skarnes WC, Rossant R (1989) Production of a mutation in mouse *En-2* gene by homologous recombination in embryonic stem cells. Nature 338:153–156

Jurand A (1974) Some aspects of the development of the notochord in mouse embryos. J Embryol Exp Morphol 32:1–33

Kaufman MH (1990) Morphological stages of postimplantation development. In: Copp AJ, Cockroft DL (eds) Postimplantation mammalian embryos: a practical approach. IRL Press, Oxford, pp 81–91

Kelly SJ (1977) Studies on the developmental potential of 4- and 8-cell stage mouse blastomeres. J Exp Zool 200:365–376

Kemler R, Babinet C, Eisen H, Jacob F (1977) Surface antigen in early differentiation. Proc Natl Acad Sci USA 74:4449–4452

King TR, Dove WF, Herrmann B, Moser AR, Shedlovsky A (1989) Mapping to molecular resolution in the T to H-2 region of the mouse genome with a nested set of meiotic recombinants. Proc Natl Acad Sci USA 86:222–226

Koller BH, Hagemann LJ, Doetschman T, Hagaman JR, Huang S, Williams PJ, First NL, Maeda N, Smithies O (1989) Germ line transmission of a planned alteration made in a hypoxanthine phosphoribosyltransferase gene by homologous recombination in embryonic stem cells. Proc Natl Acad Sci USA 86:8927–8931

Kothary R, Clapoff S, Brown A, Campbell R, Peterson A, Rossant J (1989) A transgene containing a *lacZ* inserted into the dystonia locus is expressed in the neural tube. Nature 335:435–437

Kuehn MR, Bradley A, Robertson EJ, Evans MJ (1987) A potential animal model for Lesch-Nyhan syndrome through introduction of HPRT mutations into mice. Nature 326:295–298

Lallemand Y, Brulet P (1990) An in situ assessment of the routes and extents of colonization of the mouse embryo by embryonic stem cells and their descendants. Development 110:1241–1248

Lawson KA, Pedersen RA (1987) Cell fate, morphogenetic movement and population kinetics of embryonic endoderm at the time of germ layer formation in the mouse. Development 101:627–652

Lawson KA, Meneses JJ, Pedersen RA (1986) Cell fate and cell lineage in the endoderm of the presomite mouse embryo studies with an intracellular tracer. Dev Biol 115:325–339

Le Douarin N (1982) The neural crest. Cambridge University Press, Cambridge

Le Mouellic H, Lallemand Y, Brûlet P (1990) Targeted replacement of the homeobox gene *Hox 3.1* by the *Escherichia coli lacZ* in mouse chimeric embryos. Proc Natl Acad Sci USA 87:4712–4716

Levey IL, Stull GB, Brinster RL (1978) Poly(A) and synthesis of polyadenylated RNA in the preimplantation mouse embryo. Dev Biol 64:140–148

Lufkin T, Dierich A, LeMeur M, Mark M, Chambon P (1991) Disruption of the *Hox-1.6* homeobox gene results in defects in a region corresponding to its rostral domain of expression. Cell 66:1105–1119

Lyon MF, Searle AG eds (1989) Genetic variants and strains of the laboratory mouse. Oxford University Press, Oxford

Lyon MF, Phillips RJS, Fisher G (1982) Use of an inversion to test for induced X-linked lethals in mice. Mutat Res 92:217–228

Mansour SL, Thomas KR, Capecchi MR (1988) Disruption of the proto-oncogene int-2 in mouse embryo-derived stem cells: a general strategy for targeting mutations to non-selectable genes. Nature 336:348–352

Mansour SL, Thomas KR, Deng C, Capecchi MR (1990) Introduction of a *lacZ* reporter gene into the mouse int-2 locus by homologous recombination. Proc Natl Acad Sci USA 87:7688–7692

Maro B, Johnson MH, Pickering SJ, Louvard D (1985) Changes in the distribution of membranous organelles during mouse early development. J Embryol Exp Morphol 90:287–309

Martin GR (1981) Isolation of a pluripotent cell line from early mouse embryos cultured in medium conditioned by teratocarcinoma stem cells. Proc Nat Acad Sci USA: 78:7634–7638

McGrath J, Solter D (1983) Nuclear transplantation in the mouse embryo by micro-surgery and cell fusion. Science 220:1300–1302

McGrath J, Solter D (1984) Inability of mouse blastomere nuclei transferred to enucleated zygotes to support development in vitro. Science 226:1317–1319

McGrath J, Solter D (1986) Nucleocytoplasmic interactions in the mouse embryo. J Embryol Exp Morphol 97 (Suppl):277–289

McMahon AP, Bradley A (1990) The Wnt-1 (int-1) proto-oncogene is required for development of a large region of the mouse brain. Cell 62:1073–1085

Meehan RR, Barlow DP, Hill RE, Hogan BLM, Hastie ND (1984) Pattern of serum protein gene expression in mouse visceral yolk sac and fetal liver. EMBO J 3:1881–1885

Meier S, Tam PPL (1982) Metameric pattern development in the embryonic axis of the mouse. I. Differentiation of the cranial segments. Differentiation 21:95–108

Miller DA, Miller OJ (1972) Chromosome mapping in the mouse. Science 178:949–955

Miller OJ, Miller DA (1975) Cytogenetics of the mouse. Annu Rev Genet 9:285–303

Monk M (ed) (1987) Mammalian development: a practical approach. IRL Press, Oxford

Nadeau JH, Reiner AH (1989) Linkage and synteny relationships in mouse and man. In: Lyon MF, Searle AG (eds) Genetic variants and strains of the laboratory mouse. Oxford University Press, Oxford, pp 506–536

Nadijcka M, Hillman N (1974) Ultrastructural studies of the mouse blastocyst sub-stages. J Embryol Exp Morphol 32:675–695

Nagy A, Gócza E, Merentes Diaz E, Prideaux VR, Iványi E, Markkula M, Rossant R (1990) Embryonic stem cells alone are able to support embryonic development in the mouse. Development 110:815–821

Nocka K, Majumder S, Chabot B, Ray P, Cervone M, Bernstein A, Besmer P (1989) Expression of the c-kit gene products in known cellular targets of W mutations in normal and W mutant mice – evidence for an impaired c-kit kinase in mutant mice. Genes Dev 3:816–826

O'Kane C, Gehring W (1987) Detection in situ of genomic regulatory elements in *Drosophila*. Proc Natl Acad Sci USA 84:9123–9127

Painter TS (1927) The chromosome constitution of Gates' "non disjunction" (v-o) mice. Genetics 12:379–392

Palmiter RD, Brinster RL (1986) Germ line transformation of mice. Annu Rev Genet 20:465–499

Palmiter RD, Behringer RR, Quaife CJ, Maxwell IH, Brinster RL (1987) Cell lineage ablation in transgenic mice by cell-specific expression of a toxin gene. Cell 50:435–443

Panthier J-J, Condamine H, Jacob F (1988) Inoculation of newborn SWR/J females with an ecotropic murine leukemia virus can produce transgenic mice. Proc Natl Acad Sci USA 85:1156–1160

Papaioannou VE (1982) Lineage analysis of inner cell mass and trophectoderm using microsurgically reconstituted mouse blastocysts. J Embryol Exp Morphol 68:199–209

Pedersen RA, Wu K, Balakier H (1986) Origin of the inner cell mass in mouse embryos: cell lineage analysis by microinjection. Dev Biol 117:581–595

Perona RM, Wassarman PM (1986) Mouse blastocysts hatch in vitro by using a trypsin-like proteinase associated with cells of mural trophectoderm. Dev Biol 114:42–52

Petzhold U, Hoppe PC, Illmensee K (1980) Protein synthesis in enucleated fertilized and unfertilized mouse eggs. Wilhelm Roux's Arch Dev Biol 189:215–219

Piko L, Clegg KB (1982) Quantitative changes in total RNA, total poly(A) and ribosomes in early mouse embryos. Dev Biol 89:362–378

Poelman RE (1981a) The formation of embryonic mesoderm in the early post-implantation mouse embryo. Anat Embryol 162:29–40

Poelman RE (1981b) The head process and the formation of the diefinitive endoderm in the mouse. Anat Embryol 162:41–49

Pohl TM, Mattei MG, Rüther U (1990) Evidence for allelism of the recessive insertional mutation *add* and the dominant mouse mutation *extra-toes (Xt)*. Development 110:1153–1157

Pratt HPM (1985) Membrane organization in the preimplantation mouse embryo. J Embryol Exp Morphol 90:101–121

Price J (1987) Retroviruses and the study of cell lineages. Development 101:409–419

Price J, Turner DL, Cepko CL (1987) Lineage analysis in the vertebrate nervous system by retrovirus mediated gene transfer. Proc Natl Acad Sci USA 84:156–160

Rassoulzadegan M, Leopold P, Vailly J, Cuzin F (1986) Germ line transmission of autonomous genetic elements in transgenic mouse strains. Cell 46:513–519

Rastan S (1983) Non-random X inactivation in mouse X-autosome translocation embryos – location of the inactivation centre. J Embryol Exp Morphol 78:1–2

Reeve WJD, Ziomek CA (1981) Distribution of microvilli on dissociated blastomeres from mouse embryos: evidence for surface polarization at compaction. J Embryol Exp Morphol 62:339–350

Reith AD, Rottapel R, Giddens E, Brady C, Forrester L, Bernstein A (1990) W mutant mice with mild or severe developmental defects contain distinct point mutations in the kinase domain of the c-kit receptor. Genes Dev 4:390–400

Rinchik EM (1991) Cemical mutagenesis and fine-structure functional analysis of the mouse genome. TIG 7:15–21

Rinchik EM, Bangham JW, Hundsicker PR, Cacheiro NLA, Kwon BS, Jackson IJ, Russell LB (1990a) Genetic and molecular analysis of chlorambucil-induced germ-line mutations in the mouse. Proc Natl Acad Sci USA 87:1416–1420

Rinchik EM, Carpenter DA, Selby PB (1990b) A strategy for fine-structure functional analysis of a 6- to 11-centimorgan region of mouse chromosome 7 by high-efficiency mutagenesis. Proc Natl Acad Sci USA 87:896–900

Robert B, Barton P, Minty A, Daubas P, Weydert A, Bonhomme F, Catalan J, Chazottes D, Guénet J-L, Buckingham M (1985) Investigation of genetic linkage between myosin and actin genes using an interspecific mouse backcross. Nature 314:181–183

Robertson E, Bradley A, Kuehn M, Evans M (1986) Germ line transmission of genes introduced into cultured pluripotential cells by retroviral vector. Nature 323:445–448

Robertson EJ (ed) (1987) Teratocarcinomas and embryonic stem cells: a practical approach. IRL Press, Oxford

Roderick TH, Hawes NL (1970), Two radiation induced chromosomal inversions in mice (*Mus musculus*). Proc Natl Acad Sci USA 67:961–967

Roderick TH, Hawes NL (1974) Nineteen paracentric chromosomal inversion in mice. Genetics 76:109–117

Rosenberg UB, Schröder C, Preiss A, Kienlin A, Cote S, Riede I, Jäckle H (1986) Structural homology of the *Drosophila* Krüppel gene with *Xenopus* transcription factor IIIA. Nature 319:336–339

Rossant J (1976a) Postimplantation development of blastomeres isolated from 4- and 8-cell mouse eggs. J Embryol Exp Morphol 36:283–290

Rossant J (1976b) Investigation of inner cell mass determination by aggregation of isolated rat inner cell masses with mouse morulae. J Embryol Exp Morphol 36:163–174

Rossant J (1977) Cell commitment in early rodent development. In: Johnson MH (ed) Development in mammals, vol 2. North Holland, Amsterdam, pp 119–150

Rossant J, Joyner AL (1989) Towards a molecular genetic analysis of mammalian development. TIG 5:277–283

Rossant J, Lis WT (1979) The possible dual origin of the ectoderm of the chorion in the mouse embryo. Dev Biol 70:249–254

Rossant J, Ofer L (1977) Properties of extraembryonic ectoderm isolated from postimplantation mouse embryos. J Embryol Exp Morphol 39:183–194

Rossant J, Vijh KM (1980) Ability of outside cells from preimplantation mouse embryos to form inner cell mass derivatives. Dev Biol 76:475–482

Rossant J, Gardner RL, Alexandre HL (1978) Investigation of the potency of cells from the mouse postimplantation embryo by blastocystinjection: a preliminary report. J Embryol Exp Morphol 48:239–247

Rugh R (1990) The mouse. Its reproduction and development. Oxford University Press, London

Russell ES (1979) Hereditary anemias of the mouse: a review for geneticists. Adv Genet 20:357–459

Russell ES (1985) A history of mouse genetics. Annu Rev Genet 19:1–28

Russel LB (1983) X-autosome translocations in the mouse: their characterization and use as tools to investigate gene inactivation and gene action. In: Sandberg AA (ed) Cytogenetics of the mammalian X chromosome, Part A. Alan Liss, New York, pp 205–250

Russell WL (1962) An augmenting effect of dose fractionation on radiation induced mutation rate in mice. Proc Natl Acad Sci USA 48:1724–1727

Sanes JR, Rubenstein JLR, Nicolas JF (1986) Use of a recombinant retrovirus to study post-implantation cell lineage in mouse embryos. EMBO J 5:3133–3142

Sawicki JA, Magnuson T, Epstein CJ (1981) Evidence for expression of the paternal genome in the two-cell mouse embryo. Nature 295:450–451

Schnedl W (1971) The karyotype of the mouse. Chromosoma 35:111–116

Schultz GA (1986) Molecular biology of the early mouse embryo. Biol Bull 171:291–309

Sawicki JA, Magnuson T, Epstein CJ (1982) Evidence for expression of the paternal genome in the two-cell mouse embryo. Nature 294:450–451

Schultz RM, Letourmeau GE, Wassarman PM (1979) Program of early development in the mammal: changes in patterns and absolute rates of synthesis of tubulin and total protein synthesis during oogenesis and early embryogenesis in the mouse. Dev Biol 68:341–359

Searle AG (1989) Chromosomal variants. In: Lyon MF, Searle AG (eds) Genetic variants and strains of the laboratory mouse. Oxford University Press, Oxford, pp 582–616

Shashikant CS, Utset MF, Violette SM, Wise TL, Einat M, Einat J, Pendleton JW, Schughart K, Ruddle FH (1991) Homeobox genes in mouse development. Crit Rev Eucaryotic Gene Expression 1(3):207–245

Shedlovsky A, Guénet J-L, Johnson LL, Dove WF (1986) Introduction of induced recessive lethal mutations in the T/t-H2 region of the mouse genome via a point mutagen. Genet Res 47:135–142

Shedlovsky A, King TR, Dove WF (1988) Saturation germ line mutagenesis of the murine t region including a lethal allele at the quaking locus. Proc Natl Acad Sci USA 85:180–184

Shur BD, Hall NG (1982a) Sperm surface galactosyltransferase activities during in vitro capacitation. J Cell Biol 95:567–573

Shur BD, Hall NG (1982b) A role for mouse sperm surface galactosyltransferase in sperm binding to the egg zona pellucida. J Cell Biol 95:574–579

Silver, L (1985) Mouse t-haplotypes. Annu Rev Genet 19:179–208

Smith KK, Strickland S (1981) Structural components ond characteristics of Reichert's membrane, an extra-embryonic basement membrane. J Biol Chem 256:4654–4661

Smith LJ (1980) Embryonic axis orientation in the mouse and its correlation with blastocyst relationships to the uterus, Part I. Relationships between 82 hours and 4¼ days. J Embryol Exp Morphol 55:257–272

Smith LJ (1985) Embryonic axis orientation in the mouse and its correlation with blastocyst relationships to the uterus, Part II. Relationships from 4 ¼ to 9½ days. J Embryol Exp Morphol 89:15–35

Smith R, McLaren A (1977) Factors affecting the time of formation of the blastocoele. J Embryol Exp Morphol 41:79–92

Snell GD (1946) An analysis of translocations in the mouse. Genetics 31:157–180

Snell GD, Stevens LC (1966) Early embryology. In: Green EL (ed) Biology of the laboratory mouse, 2nd edn. McGraw-Hill, New York, pp 205–245

Snow MHL (1977) Gastrulation in the mouse: growth and regionalization of the epiblast. J Embryol Exp Morphol 42:293–303

Soriano P, Jaenisch R (1986) Retroviruses as probes for mammalian development: Allocation of cells to the somatic and germ cell lineage. Cell 46:19–29

Soriano P, Cone RD, Mulligan RC, Jaenisch R (1986) Tissue specific and ectopic expression of genes introduced into transgenic mice by retroviruses. Science 234:1409–1413

Soriano P, Gridley T, Jaenisch R (1987) Retroviruses and insertional mutagenesis in mice: proviral integration at the Mov 34 locus leads to early embryonic death. Genes Dev 1:366–375

Stambaugh R, Buckley U (1960) Identification and subcellular localization of the enzymes affecting the penetration of the zona pellucida by rabbit spermatozoa. J Reprod Fertil 19:423–432

Stambaugh R, Mastroianni L (1980) Stimulation of rhesus monkey (Macaca mulatta) proacrosin activation by oviduct fluid. J Reprod Fertil 59:479–484

Stewart CL, Schuetze S, Vanek M, Wagner EF (1987) Expression of retroviral vectors in transgenic mice obtained by embryo infection. EMBO J 6:383–388

Stewart CL, Vanek M, Wagner EF (1985) Expression of foreign genes from retroviral vectors in mouse teratocarcinoma chimaeras. EMBO J 4:3701–3709

Tam PPL (1986) A study on the pattern of prospective somites in the presomitic mesoderm of mouse embryos. J Embryol Exp Morphol 92:269–285

Tam PPL (1989) Regionalisation of the mouse embryonic ectoderm: allocation of prospective ectodermal tissues during gastrulation. Development 107:55–67

Tam PPL, Beddington RSP (1987) The formation of mesodermal tissues in the mouse embryo during gastrulation and early organogenesis. Development 99: 109–126

Tam PPL, Snow MHL (1981) Proliferation and migration of primordial germ cells during compensatory growth in mouse embryos. J Embryol Exp Morphol 64: 133–147

Tam PPL, Meier S, Jacobson AG (1982) Differentiation of the metameric pattern in the embryonic axis of the mouse. II. Somitomeric organization in the presomitic mesoderm. Differentiation 21:109–122

Tarkowski AK (1959) Experiments on the development of isolated blastomeres of mouse eggs. Nature 184:1286–1287

Tarkowski AK, Wróblewska J (1967) Development of blastomeres of mouse eggs isolated at the 4- and 8- cell stage. J Embryol Exp Morphol 18:155–180

te Riele H, Maandag ER, Clarke A, Hooper M, Berns A (1990) Consecutive inactivation of both alleles of the pim-1 proto-oncogene by homologous recombination in embryonic stem cells. Nature 348:649–651

Theiler K (1989) The house mouse. Atlas of embryonic development. Springer, Berlin Heidelberg New York

Thomas KR, Capecchi M (1987) Site-directed mutagenesi by gene targeting in mouse embryo-derived stem cells. Cell 51:503–512

Turner DL, Cepko CL (1987) A common progenitor for neurons and glia persists in rat retina late development. Nature 328:131–136

Van der Putten H, Botteri FM, Miller AD, Rosenfeld MG, Fan H, Evans RM, Verma IM (1985) Efficient insertion of genes into the mouse germ line via retroviral vectors. Proc Natl Acad Sci USA 82:6148–6152

Vestweber D, Kemler R (1984) Rabbit antiserum against a purified surface glycoprotein decompacts mouse preimplantation embryos and reacts with adult tissues. Exp Cell Res 152:169–178

Vestweber D, Gossler A, Boller K, Kemler R (1987) Expression and distribution of cell adhesion molecule uvomorulin in mouse preimplantation embryos. Dev Biol 124:451–456

Wagner EF, Stewart TA, Mintz B (1981) The human β-globin gene and a functional thymidine kinase gene in developing mice. Proc Natl Acad Sci USA 78:5016–5020

Wagner TE, Hoppe PC, Jollik JD, Scholl DR, Hodinka RL, Gault JB (1981) Microinjection of a rabbit β-globin gene into zygotes and its subsequent expression in adult mice and their offspring. Proc Natl Acad Sci USA 78:6376–6380

Walther C, Guenét JL, Simon D, Deutsch U, Jostes B, Goulding MD, Plachov D, Balling R, Gruss P (1991) Pax: a murine multigene family of paired box containing genes. Genomics 11:424–434

Warner CM, Versteegh LR (1974) In vivo and in vitro effect of α-amanitin on preimplantation mouse embryo RNA polymerase. Nature 248:678–680

Wassarman PM (1987) The biology and chemistry of fertilization. Science 235:553–560

Whitten W (1971) Nutrient requirements for the culture of preimplantation embryos in vitro. Adv Biosci 6:129–140

Wilson C, Pearson RK, Bellen HJ, O'Kane CJ, Grossniklaus U, Gehring WJ (1989) P-element-mediated enhancer detection: an efficient method for isolating and characterizing developmentally regulated genes in Drosophila. Genes Dev 3:1301–1313

Willison K (1990) The mouse brachyury gene and mesoderm formation. TIG 6:104–105

Wincek TJ, Parrish RF, Polakoski KL (1979) Fertilization: a uterine glycosaminoglycan stimulates the conversion of sperm proacrosin to acrosin. Science 203:553–554

Woychik RW, Stewart TA, Davis LG, D'Eustachio P, Leder P (1985) An inherited limb deformity created by insertional mutagenesis in a transgenic mouse. Nature 318:36–40

Wudl L, Chapman V (1976) The expression of β-glucuronidase during preimplantation development of mouse embryos. Dev Biol 48:104–109

Wylie CC, Stott D, Donovan PJ (1985) Primordial germ cells migration. In: Browder LW (ed) Developmental biology, vol 2. Plenum Press, New York, pp 433–448

Xin Xiang, KF Benson, Chada K (1990) Mini mouse: disruption of the Pygmy locus in a transgenic insertional mutant. Science 247:967–969

Yagi T, Ikawa Y, Yoshida K, Shigetani Y, Takeda N, Mabuchi I, Yamamoto T, Aizawa S (1990) Homologous recombination at c-fyn locus of mouse embryonic stem cells with use of diphteria toxin A fragment gene in negative selection. Proc Natl Acad Sci USA 87:9918–9922

Yoshida-Noro C, Suzuki N, Takeichi M (1984) Molecular nature of calcium-dependent cell adhesion system in mouse teratocarcinoma and embryonic cells studied with a monoclonal antibody. Dev Biol 101:19–27

Zijlstra M, Li E, Sajjadi F, Subramani S, Jaenisch R (1989) Germ line transmission of a disrupted β2-microglobulin gene produced by homologous recombination in embryonic stem cells. Nature 342:435–428

Zimmer A, Gruss P (1989) Production of chimaeric mice containing ES cells carrying a homeobox Hox 1.1 allele mutated by homologous recombination. Nature 338:150–153

5 Genomic Imprinting in Mammals

Wolf Reik

1 Introduction

This is a good time to write a general article on imprinting. In the last 5 years or so imprinting has progressed from an intriguing embryological observation to a firmly established biological principle with far-reaching consequences for mammalian development, genetics, and human disease. We now know why parthenogenesis in some mammals is not possible. We know of some imprinted genes and we even know what they do. We know that disomy can cause disease in humans. We know of genetic phenomena that extend classical Mendelian concepts. And we are gaining knowledge of the molecular mechanism of imprinting, knowledge that makes us believe that imprinting-like epigenetic controls guide differentiation in all multi-cellular organisms.

In this chapter I shall focus almost exclusively on autosomal imprinting in mammals. I do this because I believe that while epigenetic mechanisms may be shared by a wide variety of organisms, the biological purposes of employing these mechanisms can be very different. For example, imprinting of paternal chromosomes is a sex-determining device in mealy bugs (*Pseudococcidae*). Sex determination in mammals, however, does not involve any imprinting mechanisms as far as we know. I shall also largely ignore the imprinting of the mammalian X-chromosome, simply because that imprinting is a whole chromosome phenomenon, whereas autosomal imprinting is now thought to work on a gene by gene basis. Some of these omissions are entirely arbitrary; their only purpose is to make this chapter more concise (and shorter). I shall also try to present general principles rather than details whenever this is possible without distorting the truth.

2 What Is Imprinting?

One may get as many answers to this question as people that one asks. The most general definition of imprinting is that of a chromosomal memory.

Department of Molecular Embryology, Institute of Animal Physiology and Genetics Research, Babraham, Cambridge CB2 4AT, UK

Results and Problems in Cell Differentiation 18
W. Hennig (Ed.)
Early Embryonic Development of Animals
© Springer-Verlag Berlin Heidelberg 1992

Epigenetic information is introduced into chromosomes and is stably replicated together with the chromosome as cells divide. Many generations later this information can be retrieved from all of the progenitor chromosomes and can be used, for example, to switch genes on or off. A central feature of imprinting by epigenetic modification is therefore its clonal stability. For example, patterns of DNA methylation at CpG residues, but also other features of chromatin such as DNase I hypersensitivity, can be clonally stable. The second important property is that these modifications can affect gene expression. Hence, imprinting can be seen as a remote control. A switch is turned and many cell generations later gene expression, and thus phenotype, depend on this initial switch. And lastly, imprinting is reversible. Unlike other clonally stable switches that convey information, such as for example the somatic rearrangement of immunoglobulin genes, imprints can normally be taken off the chromosomes again without any permanent alteration of the genetic material.

This definition, however, is too wide for our purpose. We consider a specific class of imprints, namely those that mark the parental origin of genomes, chromosomes and genes in mammals (Monk and Surani 1990). Some genes in mammals are epigenetically marked in the germline or very early in embryogenesis, such that the maternal and the paternal copies of these genes behave differently even when they are in the same diploid cell (Cattanach 1986; Solter 1988; Reik et al. 1990; Sapienza 1990; Surani et al. 1990a). In the mouse, the paternal copy of the gene for insulin-like growth factor 2 (Igf2) is expressed in the embryo, whereas the maternal copy of the same gene is largely repressed (DeChiara et al. 1991; Ferguson-Smith et al. 1991). The imprinting that we consider here is therefore the parent-specific imprinting of genes and chromosomes. With this concept in mind, we shall now look at the manifestations and the consequences of imprinting.

3 Both Parental Genomes Are Needed for Normal Development

Despite some early claims to the contrary it is now clear that the development of monoparental diploid mouse embryos is restricted and cannot proceed to term (Barton et al. 1984; Mann and Lovell-Badge 1984; McGrath and Solter 1984; Surani et al. 1984). Hence, androgenetic embryos with two paternal genomes constructed from fertilized eggs by pronuclear transplantation (Fig. 1) when transferred to foster mothers will invariably die at early postimplantation stages at the latest. Embryos with a diploid maternal genome can be made either by pronuclear transplantation (gynogenones), or by parthenogenetic activation of unfertilized eggs followed by diploidization (parthenogenones); these embryos also achieve postimplantation development at the most. The development of gynogenetic and parthenogenetic embryos is indistinguishable and these embryos will be

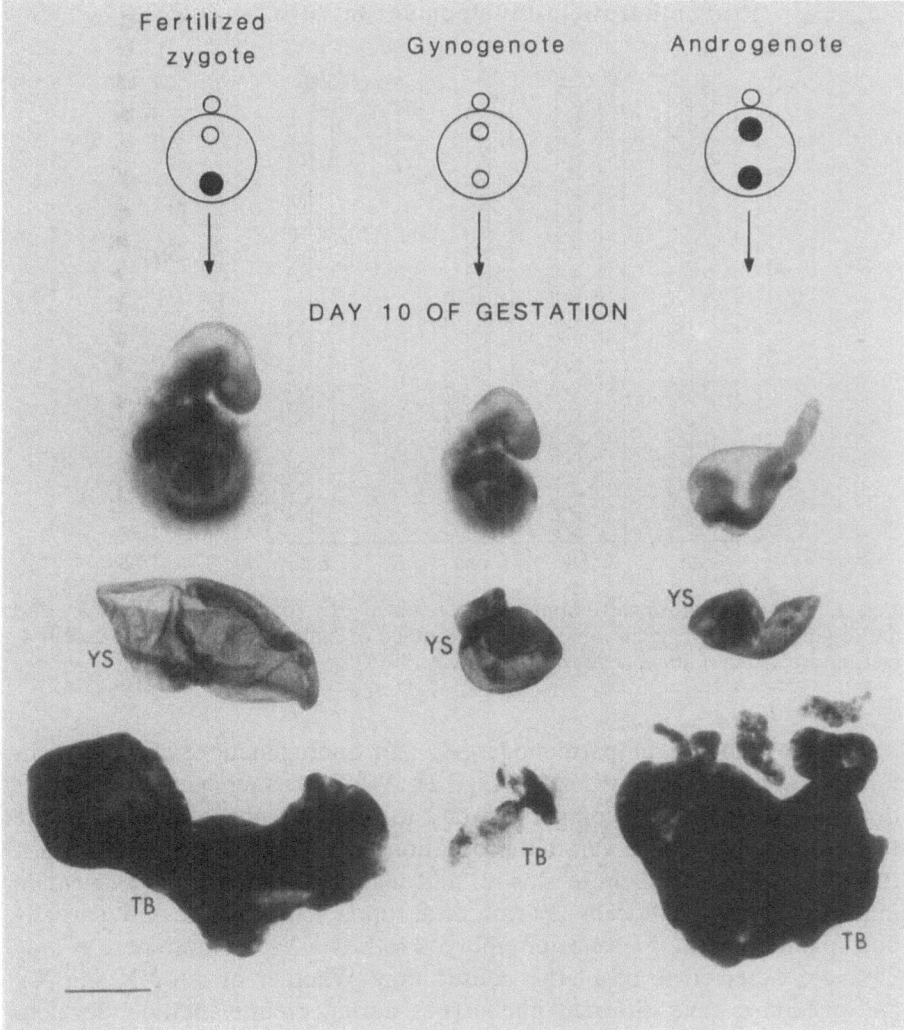

Fig. 1. Androgenetic, gynogenetic and normal embryos and their developmental potency. The fertilized zygote has a male pronucleus (*filled circles*) and a female pronucleus (*open circles*) containing paternal and maternal chromosomes respectively. Experimental zygotes can be constructed at this stage by pronuclear transplantation to result in diploid maternal or diploid paternal genotypes. A typical example of embryos developing from these zygotes is shown. *YS* Yolk sac; *TB* trophoblast; *scale bar* = 1 mm. (Howlett 1991)

considered as synonymous. It follows from this observation that the deficit in developmental potency resides exclusively in the genomes, at least from the pronuclear stage onwards, and that the unfertilized cytoplasm is fully capable of supporting early development without any need for extragenomic contributions from spermatozoa.

Preimplantation development in vitro

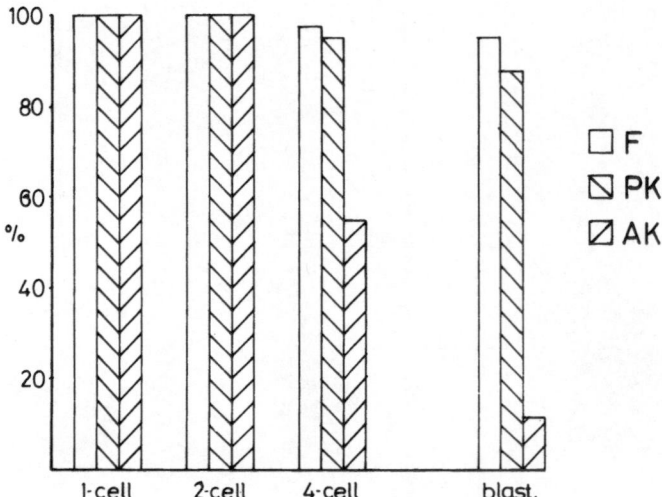

Fig. 2. Preimplantation development of fertilized (*F*), parthenogenetic (*PK*) and androgenetic (*AK*) eggs in vitro. The percentage of embryos reaching the 2-cell, 4-cell, and blastocyst stage is shown. (Howlett 1991)

The phenotype of parthenogenetic and androgenetic embryos is different from the earliest stages (Fig. 2). While parthenogenetic embryos develop in culture at a comparable rate to normal embryos at least to the blastocyst stage, only a variable proportion of androgenetic embryos reach that stage, in general more slowly, and many blastocysts are deficient in inner cell mass (ICM) cells (Barton et al. unpubl.). This can only partially be explained by the presence of embryos with two Y-chromosomes, as only 25% are expected to be of this constitution. Whether or not XX and XY androgenones have different phenotypes during preimplantation development is not known at present, but it appears that only XY embryos can achieve any significant postimplantation development (Kaufman et al. 1989).

The developmental potential of parthenogenetic embryos is largely independent of their genotype, that is parthenogenones made from different inbred strains behave in much the same way (Fundele et al. 1991). With androgenones this may not be the case. Androgenones made from male pronuclei formed in a DBA/2 strain egg cytoplasm develop particularly poorly, irrespective of the genotype of the recipient egg cytoplasm to which they are transferred (Latham and Solter 1991). This may suggest that some of the paternal imprints are modified by interactions between egg cytoplasm and the genome as the male pronucleus is formed. An extreme, but possible, view is that all the paternal imprinting is achieved following fertilization and repackaging of the paternal chromosomes using components from the egg cytoplasm. Some of the genotype-specific modifier genes may work in this way (Reik et al. 1990, see below), and this may contribute to postzygotic isolation mechanisms in speciation (Klose and Reik 1992; Reik 1992).

Following implantation, parthenogenetic and androgenetic embryos have very different phenotypes indeed (Barton et al. 1984; McGrath and Solter 1984, 1986; Surani et al. 1986a,b). While parthenogenetic conceptuses can progress to the 25-somite stage on day 10 of gestation with a relatively well-developed embryo proper, their extra-embryonic tissues are severely deficient, especially the trophoblast which is crucially important for placentation and hence nutrient transfer from the mother (for details, see Gossler, this Vol.). Androgenetic embryos, by contrast, show a very much opposite phenotype: extra-embryonic structures are well developed but the embryo itself is severely retarded and does not progress beyond the 4–6 somite stage. For parthenogenetic embryos the defect cannot exclusively be in the extraembryonic lineages, however, as elaborate lineage reconstitution experiments fail to rescue parthenogenetic development (Barton et al. 1985; Gardner et al. 1990). This point is also borne out by looking at the potency of parthenogenetic cells in detail (see below). Both types of embryos do not usually progress beyond halfway through development, that is day 10 of gestation.

In the human, androgenetic conceptuses develop into complete hydatidiform moles, characterized by hyperplasia of the trophoblastic tissues and little or disorganized foetal tissue (Kajii and Ohama 1977). At this level, therefore, imprinting in mouse and human appears to involve similar genetic traits. Parthenogenetic eggs can show some preimplantation development in the human, but foetuses with a diploid maternal constitution have so far not been detected.

It can be concluded that in the mouse, and probably in other mammals including the human (especially in placental mammals, see below), both parental genomes are necessary for development and only with the two of them a correct balance of gene dosage at imprinted loci is achieved. The opposite phenotypes observed with the two types of experimental embryos are indicative of the cumulative actions of maternally or paternally imprinted genes (Barton et al. 1991). (For the purpose of this chapter, I will establish the convention of calling genes repressed on the maternal chromosome 'maternally imprinted', because I believe that the repressed state is the state achieved by imprinting. However, this is not a generally accepted convention.)

We shall see below that much more can be learned from a detailed analysis of parthenogenetic and androgenetic cells, however, we shall first turn to genetics and to imprinted genes, as it is easier then to put these observations into perspective.

4 Testing the Chromosomes for Imprinting

In a classical study published in 1978, Searle and Beechey suggested that there could be functional haploidy for some autosomal genes in the mouse.

Note that this was before evidence for imprinting was obtained by embryological studies in 1983 and 1984. How was this conclusion reached?

When mice carrying chromosome translocations, such as Robertsonian translocations or reciprocal translocations, are bred, the presence of the translocation will invariably lead to a certain degree of nondisjunction in their gametes (when the translocation is heterozygous). They will produce some unbalanced gametes, that is gametes that do not contain the translocation chromosome and its normal partner, or that contain both of them. Mating can therefore produce unbalanced, and hence normally lethal, genotypes. Importantly, however, when a translocation heterozygote is crossed with another heterozygote carrying the same translocation, as a chance event one gamete with a disomy, or duplication of a chromosome or part of a chromosome, may meet with another gamete that is nullisomic or deficient for that chromosome. In this case the normal diploid constitution is restored, but one chromosome, or a region thereof, is present in two copies from one parent, and absent from the other parent (Lyon and Glenister 1977; Searle and Beechey 1978, 1990; Cattanach and Kirk 1985; Cattanach 1986; Cattanach and Beechey 1990). Such uniparental disomic embryos have been created for most chromosomes, and their development has been analyzed (see Fig. 3 as an example).

As one would expect from the development of androgenetic and parthenogenetic embryos, in which the whole genome is disomic, specific phenotypes are seen in disomic embryos that are caused by a subset of imprinted genes, those that are located on the disomic chromosome segment. For example, mice paternally disomic for proximal chromosome 11 are larger than their normal littermates, whereas mice maternally disomic for the same chromosome are smaller (Fig. 3; Cattanach and Kirk 1985). Clearly, this suggests that an imprinted gene or genes is present on chromosome 11 whose difference in dosage causes growth differences. Most chromosomes have now been tested in this way, and those regions that cause specific phenotypes when disomic indicate the presence of imprinted genes (Fig. 4).

A number of points can be made concerning these "imprinted regions" of the genome (Cattanach and Beechey 1990). First, there are segments that do not show overt phenotypes, suggesting that they either do not contain imprinted genes, or alternatively that imprinted genes in these regions have no major effect on phenotype. Indeed, several recent experiments involving targetted mutagenesis suggest that there is a surprising extent of "gene redundancy" in the mouse. Second, all imprinted regions investigated so far cause phenotypes associated with growth and viability of the embryo, and possibly with behaviour of neonates. Third, any of the imprinted regions may only contain a small number of imprinted genes. This is suggested by the observation that different translocation breakpoints in the same chromosome will cause the same phenotype, or that the phenotype will be lost completely (although there are one or two exceptions to this). A minimum estimate of imprinted genes (with phenotypic effects) is therefore simply

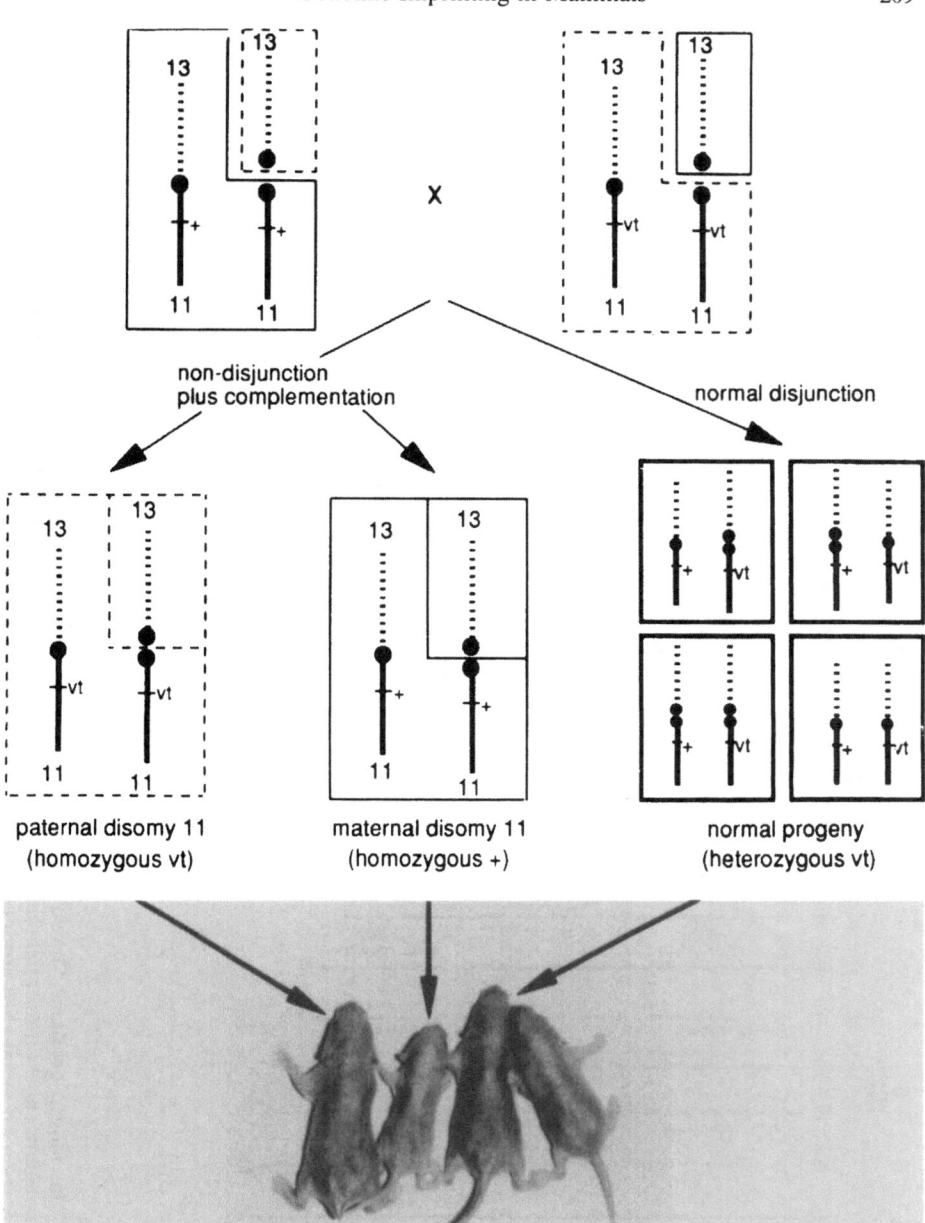

Fig. 3. Uniparental disomy of chromosome 11. When heterozygotes carrying a Robertsonian translocation of chromosome 11 and 13 are intercrossed, normal, balanced offspring are produced with normal disjunction in both parents. With non-disjunction and complementation, however, offspring with maternal or paternal disomy of chromosome 11 are produced. These can be identified by appropriate markers on chromosome 11, in this case *vt* (vestigial tail). Maternally disomic offspring are smaller than normal littermates, whereas paternally disomic ones are larger. (Cattanach and Beechey 1990)

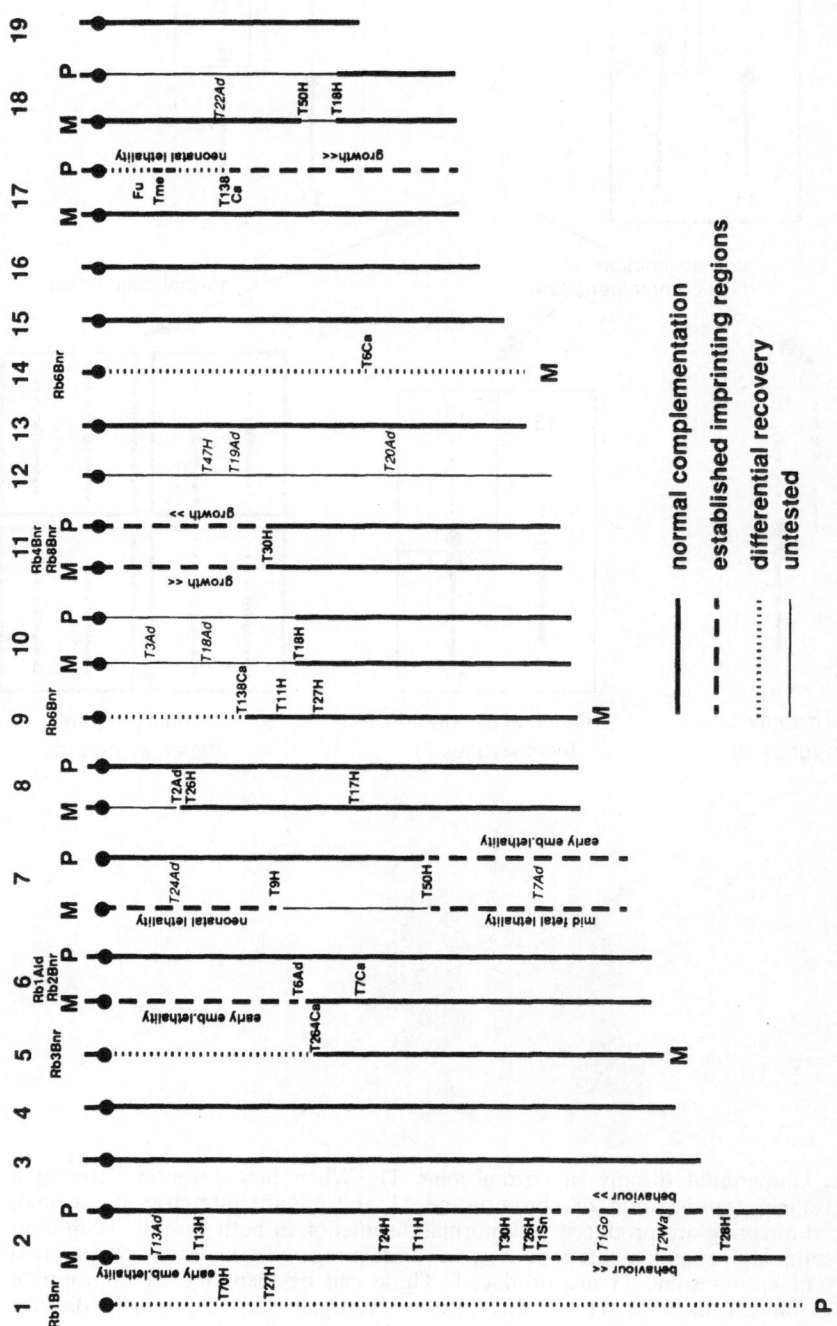

Fig. 4. Imprinting map of the mouse genome. Shown are imprinted regions as defined by recognizable phenotype or absence when uniparentally disomic. Some regions show differential recovery. The imprinted genes Igf2 and H19 are on distal chromosome 7, whereas the imprinted gene Igf2 receptor is on proximal chromosome 17 (*Tme*). (Cattanach and Beechey 1990)

Table 1. Uniparental disomy in the human (see text for references)

Chromosome	Phenotype
Maternal 7	Growth deficiency
Maternal 15q 11–13	Prader-Willi syndrome
Paternal 15q 11–13	Angelman syndrome
Paternal 11p 15.5	Beckwith-Wiedemann syndrome
Maternal 14	Short stature, hydrocephalus, small testes
Paternal 14	Mental retardation, multiple congenital abnormalities
Maternal 22	No phenotype

given by the number of imprinted chromosome segments. Fourth, most of the phenotypes seen are relatively independent of genetic background, suggesting that the major imprinted genes are not appreciably modified by background genes. Three imprinted genes have been identified on proximal chromosome 17 and distal chromosome 7 and we will turn our attention to these genes shortly.

Some of the strongest evidence for imprinting in the human genome comes from the realization that uniparental disomy can cause specific aberrant phenotypes, and hence disease (Table 1; Engel 1980; Reik 1989; Hall 1990). Disomy in the human conceptus can arise by different mechanisms, for example by initial trisomy followed by loss of the chromosome of which there is only one parental copy. Prader Willi syndrome (PWS) and Angelman syndrome (AS) can thus be the result of maternal or paternal disomy of chromosome 15q 11–13 respectively (Nicholls et al. 1989; Malcolm et al. 1991). PWS is characterized by hypotonia, hypogonadism and short stature, whereas AS patients suffer from severe mental retardation, inappropriate laughter and ataxic movements. Several candidate sequences for the disease genes have now been isolated, one of which (DN34) shows differences in DNA methylation when normal individuals and PWS and AS patients are compared (Nicholls et al. 1992). The homologous locus is on chromosome 7 in the mouse and it will be of great interest to see whether or not the methylation imprints are conserved.

Beckwith Wiedemann syndrome (BWS), a foetal overgrowth syndrome in the human, can also be caused by disomy. It was noted before that some sporadic BWS cases have a partial trisomy of chromosome 11p 15.5, which is most often paternally derived (Reik 1989; Brown et al. 1990). Not surprisingly, therefore, a number of BWS cases have now been detected with a paternal disomy of chromosome 11p 15.5 (Henry et al. 1991). Strikingly, the gene for insulin-like growth factor 2 (Igf2), is contained in this segment, and is known to be maternally imprinted in the mouse; thus, an excess of Igf2, resulting from paternal disomy of the chromosome, could conceivably contribute to the phenotype of foetal overgrowth in the human (see below).

Now that the association of uniparental disomy and genetic disease is firmly established, any recurrent but nonfamilial syndromes that do not show obvious cytogenetic abnormalities should be suspected to arise by this mechanism, and appropriate tests should be performed. In addition to the chromosomes mentioned above, maternal disomy of human chromosome 7 has been found in association with growth retardation (Spence et al. 1988; Voss et al. 1989), whereas a maternal disomy of chromosome 14 has recently been discovered to be associated with short stature, hydrocephalus and small testes (Temple et al. 1991). Maternal disomy for chromosome 22 seems to be without any obvious phenotypic effect (Kirkels et al. 1980; Palmer et al. 1980). We suggest that particular attention should be paid to the possibility that uniparental disomy could be responsible for a proportion of all spontaneous abortions in human pregnancy.

5 Imprinted Genes

Some workers in the field, including myself, have spent a considerable amount of time and sweat thinking of suitable strategies to identify imprinted genes, whose number, as pointed out before, may be small. Three genes have now been identified as undergoing imprinting, one by serendipity and two by being smart (Surani 1991). Fortunately, two of these have known functions so that one can now begin to interpret the observed phenotypes in the context of expression or repression of these genes. In addition, the identity of these two genes has given rise to the first coherent and consistent evolutionary interpretation of imprinting (see below).

When a null mutation was made by homologous recombination in the gene for the insulin-like growth factor 2 (Igf2), it was observed that mice inheriting the mutant allele from their father were considerably smaller than their littermates. But when the mutant allele was maternally inherited, offspring showed no growth deficiency (DeChiara et al. 1991; Fig. 5). Consistent with this phenotype, Igf2 transcripts were found to be produced from the paternal allele, but were almost completely repressed on the maternal allele of the gene. The only exception to this imprinting was found to occur in the leptomeninges and choroid plexus in brain, where both alleles are expressed (DeChiara et al. 1991). (Thus, not only the germline can reverse the imprint, but this can also occur in a highly tissue-specific fashion in somatic lineages. Alternatively, the existing imprint can in certain circumstances be "ignored".)

Additional evidence for imprinting of the Igf2 gene was obtained by the observation that RNA levels of this gene are extremely low in embryos maternally disomic for the distal region of chromosome 7, where the Igf2 gene resides (Ferguson-Smith et al. 1991). These embryos are smaller than their littermates, but in addition die *in utero* during the last third of the gestation period (Searle and Beechey 1990). Because of this additional

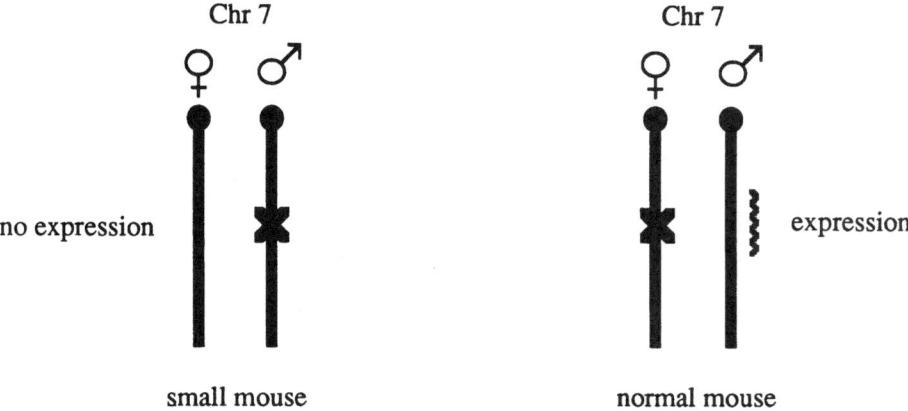

Fig. 5. Imprinting of the insulin-like growth factor 2 gene. A null mutation in the gene for Igf2 results in smaller mice when this mutation is paternally inherited. When the mutation is maternally inherited, the offspring are of normal size. Igf2 RNA is detected from the paternal chromosome, but is much reduced from the maternal chromosome. See DeChiara et al. (1991) and Ferguson-Smith et al. (1991) for details

phenotype, distal chromosome 7 must contain at least one other imprinted gene (see below). Hence, the Igf2 gene is maternally imprinted, and is repressed on the maternal, but expressed from the paternal chromosome. Igf2 is expressed widely throughout the embryo, predominantly but not exclusively in extraembryonic tissues and in the mesodermal lineage, and is known to be an embryonic mitogen (see Barton et al. 1991 for details). As we shall see later, this tissue distribution can explain some of the phenotypes of androgenetic and parthenogenetic cells.

Near the Igf2 gene, inseparable by recombination and perhaps very close to it, is another imprinted gene, H19 (Bartolomei et al. 1991). Interestingly, this gene, whose protein product is at present unknown, is imprinted in the opposite way. Hence, it is expressed from the maternal chromosome but repressed on the paternal one. Its overexpression (two-fold) in maternally disomic embryos for distal chromosome 7 could conceivably contribute to the lethality of these embryos, as embryos transgenic for additional copies of the H19 gene die at around the same stage (Brunkow and Tilghman 1991). The expression pattern of H19 during embryogenesis very closely parallels that of Igf2, however, after birth H19 expression is restricted to skeletal muscle (Bartolomei et al. 1991).

The third imprinted gene identified to date was found through the analysis of the only clear-cut parental origin-dependent mutation in the mouse, Tme (T maternal effect). Deletions of the Tme locus such as T^{hp} or T^{Wlub2} produce foetal lethality, again at late foetal stages, when maternally inherited but not when paternally derived. Strikingly, the gene for the insulin-like growth factor 2 receptor (Igf2r) has been mapped to this region of chromosome 17 and has been shown to be imprinted with the maternal

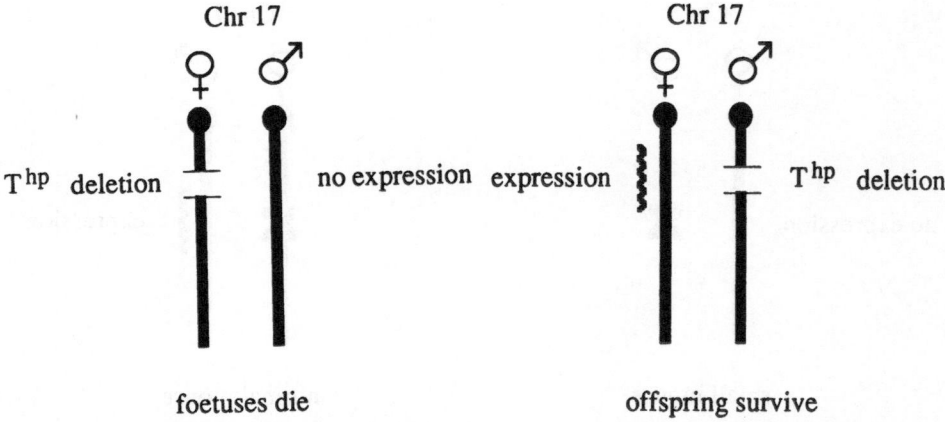

Fig. 6. Imprinting of the insulin-like growth factor 2 receptor gene. A deletion of a proximal region of chromosome 17(T^{hp} or T^{wlub2}) results in foetal lethality when maternally derived, but offspring survive when the same deletion comes from the father. The gene for Igf2r maps to this deletion and is expressed predominantly from the maternal chromosome. See Barlow et al. (1991) for details

copy being the active one (Fig. 6; Barlow et al. 1991) (as was expected from the behaviour of the mutation). However, it is not known at present whether the absence of Igf2r is the sole factor responsible for the lethality of the foetuses.

The molecular mechanism of imprinting of these three genes is at present unknown. Indeed, it is not even known yet at precisely which level of gene regulation repression occurs. Their regulation as well as their possible epigenetic modification, for example by CpG methylation, are presently under intense scrutiny and exciting results are expected in the near future.

The imprinting of Igf2 ligand and receptor in opposite directions, presumably not a coincidence, has led to intense speculations on the role of imprinted genes in development and evolution, especially since the receptor has little if any role in signal transduction but may act negatively by binding, thus sequestering the potentially active ligand. We shall address these ideas later but must first turn our attention to a more precise understanding of the phenotypes seen with parthenogenetic and androgenetic cells.

6 Potency and Phenotype of Androgenetic and Parthenogenetic Cells

Parthenogenetic and androgenetic embryos invariably die at relatively early postimplantation stages. However, their cells have been tested for potency

Table 2. Phenotype of parthenogenetic (PG) and androgenetic (AG) cells in chimaeras

	Tissue contribution	Tissue selection	Effects on phenotype of chimaera	Expected imprinting status[a]
PG	Brain Epidermis	Muscle Extra-embryonic	Decreased size Lethality with large contribution	Igf2 −/− Igf2r +/+ H19 +/+
AG	Muscle Skeleton Extra-embryonic	Brain Epidermis	Increased size Elongated AP axis in foetuses Skeletal abnormalities (lethal with large contribution)	Igf2 +/+ Igf2r −/− H19 −/−
AG-ES[b]	No apparent selection		Skeletal abnormalities	Igf2 +/+ Igf2r −/− H19 −/−
PG-ES[b]	Possibly no selection		Not known (possibly none)	Igf2 −/− Igf2r +/+ H19 +/+
MAT DIS dist 7	No selection		None	Igf2 −/− Igf2r +/− H19 +/+
PAT DIS dist 7	No selection		Increased size	Igf2 +/+ Igf2r +/− H19 −/−

[a] + indicates expression, − repression.
[b] Note that both PG-ES cells and AG-ES cells may have an altered imprinting status. It is at present impossible to predict what these altered states would look like. It is possible, for example, that repression (−) is lost for some alleles in culture. Igf2 might therefore be expected to go from −/− to +/+ (in PG-ES). However, it is also possible that epigenetic modification is necessary to achieve expression, loss of epigenetic modification in ES cells would then result in repression. See text for references.

and phenotype by incorporating them into chimaeric embryos that also contain normal cells and will thus rescue the early lethality. The two components can be distinguished later on in development through the use of various genetic markers. The routes by which such chimaeras have been made include blastomere aggregation, blastocyst injection and the use of androgenetic embryonic stem cells (ES cells, see Gossler, this Vol., for details) injected into host blastocysts. The following pattern emerges (Table 2). In all cases, the initial allocation to different lineages does not seem to be affected, rather selection against cells occurs once lineages have been set

apart (Thomson and Solter 1987, 1988). Parthenogenetic cells (PG) are
consistently lost from the extra-embryonic tissues (when analyzing aggregation chimaeras, Thomson and Solter 1987; Surani et al. 1988a), and in the
foetus from mesodermal tissues (Fundele et al 1989, 1990; Nagy et al. 1989).
However, they are usually well represented in neuroectodermal tissues. By
contrast, androgenetic cells (AG) are selected against in neuroectodermal
tissues but are contributing substantially, perhaps even to a greater extent
than normal cells, to mesodermal tissues such as skeletal muscle, heart
muscle and the skeleton (Barton et al. 1991). The basis of cell selection is at
present unknown, but there could be a greater or reduced propensity to cell
proliferation within a lineage due to overexpression or lack of autocrine or
short-range paracrine growth factors (such as Igf2). It is certainly suggestive
that parthenogenetic cells are notoriously absent in the very tissues in which
Igf2, Igf2r and H19 are predominantly expressed (Barton et al. 1991).

    In addition to cell selection, there are also recognizable phenotypes of
the embryos themselves caused by the presence of the experimental cells.
PG chimaeras with a substantial contribution of PG cells are considerably
smaller than normal littermates (Fundele et al. 1989), whereas AG chimaeras
can be up to 50% larger than controls (Barton et al. 1991). In the AG
chimaeras, there is also a characteristic alteration in embryonic shape with
an elongated anterior-posterior axis (Barton et al. 1991). AG chimaeras that
survive to term (because they have a sufficiently small contribution of AG
cells) have severe abnormalities of the skeleton, partly owing to hyperplasia
of the cartilage (Barton et al. 1991).

    There are some interesting differences in phenotype when chimaeras
made with AG embryonic cells (from the inner cell mass) are compared
to those made with AG embryonic stem cells (ES cells; Table 2). While
AG-ES chimaeras show skeletal abnormalities similar to those of AG-ICM
chimaeras, they do not seem to have an effect on growth and viability of the
embryo (Mann et al. 1990). This tells us that while some imprints are stably
maintained in ES cells over a prolonged period of culture, others are
probably lost in these cells. Support for this notion has recently also been
obtained from experiments with normal ES cells, which when injected into
tetraploid recipient blastocysts could give rise to normal foetuses, but the
offspring produced all succumbed to unexplained neonatal death. This
phenotypic abnormality has been ascribed to possible changes in epigenetic
modification in these cells (Nagy et al. 1990). This may reflect the more
fragile nature of some imprints at the blastocyst stage when ES cells are
derived or, alternatively, a mosaic distribution of imprints where different
cells in the ICM carry different epigenetic modifications. Since any ES cell
line is likely to be clonal, different ES lines could have different imprinting
patterns. This should quickly become apparent when a number of lines are
tested.

    However, we favour the notion that some imprints are more fragile than
others at these stages or over prolonged periods of culture. This may reflect
differences in the precise epigenetic modifications at that stage. Of par-

ticular interest is the possible susceptibility of some imprints to environmental influences (temperature?), as this may give us some clues as to the
molecular nature of the epigenetic modifications involved (see below).

The possible lability of some imprints at early stages of development
from the fertilized egg to the blastocyst (possibly Igf2) contrasts sharply with
the enormous stability of the Igf2 imprint at later stages. Thus, cell cultures
obtained at foetal stages from normal and from embryos maternally disomic
for distal chromosome 7 maintain their Igf2 expression characteristics ($+/-$
versus $-/-$) for at least 25 passages (P. Jones, A. Ferguson-Smith, H.
Sasaki, A. Surani, unpubl.). This key observation may suggest that imprints
evolve through different molecular phases. Indeed, a more easily reversible
phase may be necessary early in development, as the imprints have to be
reset in gametogenesis, and, at least for Igf2, also in some somatic tissues.

It is of interest to compare the phenotypic properties of PG and AG
chimaeras to those of chimaeras made with maternal or paternal disomic
embryos of distal chromosome 7 (Table 2; Ferguson-Smith et al. 1991), as
these reflect the altered activity of a subset of imprinted genes (Igf2, H19
and possibly others). For example, a double dose of Igf2 expression seems
to be sufficient to cause a size increase, whether or not there is a normal
dose ($+/-$) or absence ($-/-$) of the Igf2 receptor (Table 2). This is quite
comparable to the situation in the human where paternal disomy of chromosome 11p 15.5 (containing the Igf2 gene) leads to foetal overgrowth in the
Beckwith-Wiedemann syndrome (Henry et al. 1991). By contrast, absence
of Igf2 ($-/-$) leads to a much more severe phenotype when the Igf2
receptor is overexpressed ($+/+$). Although no potential role of H19 is
addressed, these studies underscore the importance of appropriate combinatorial action of growth factors and their receptors, some of which are
imprinted, in regulating normal development. They also show that the use
of disomic cells is a particularly simple yet powerful system to create dosage
differences at imprinted loci and to examine the phenotypic effects of these
dosage differences.

7 Why Imprinting?

There are a number of tentative answers to this question, none of which
is wholly satisfactory in my view. However, evolutionary biologists now
believe that they have the solution and it must be admitted that their
explanation is both elegant and consistent with most of the observations that
we have (Haig and Graham 1991; Moore and Haig 1991). Consider, in a
placental mammal a gene that acts in the foetus to acquire nutritional
resources from the mother. Any growth factor that acts during intra-uterine
life, such as for example Igf2, would qualify. Paternally derived alleles at
this locus will be interested in increasing growth of the foetus, as this would

maximize their chances of spreading through the population. By contrast, maternal alleles, while in principle having the same desire, must also consider the burden for the mother of increased resource transfer, as this may compromise reproductive success for *all* the offspring produced by one mother. Over evolutionary time there will therefore be a trade-off, a parental "tug-of-war", over the most successful combination of expression levels of maternally and paternally derived alleles. The story is really more complicated than this, but the interested reader may be referred to the primary literature for details (Haig and Graham 1991; Moore and Haig 1991; Haig 1992).

The appeal of this theory lies in the fact that it predicts the behaviour of Igf2 and its receptor very accurately. Dosage of Igf2, as we mentioned before, can increase or decrease the size of the foetus. It should therefore be imprinted to give higher expression when paternally inherited. The receptor, by contrast, presumably acts to bind Igf2 and thereby to decrease the availability of ligand for binding to the type 1 receptor, which is thought to be its main route of action (Moore and Haig 1991). Igf2 receptor expression thus acts negatively on the Igf2 axis, and is therefore predicted to be imprinted in the opposite direction. The Igf2 receptor, whose primary function is in mannose binding and transport (mannose-6-phosphate receptor), may have acquired its Igf2 binding capacity during the evolution of mammals, as for instance in *Xenopus laevis* it does not seem to bind Igf2 (Moore and Haig 1991). Appropriately, *Xenopus* is an oviparous species with little resource transfer from the mother postfertilization and hence, according to the theory, no need to imprint foetal growth factors.

It is also implicit in this theory that when alleles arise in a population that modify the imprinting mechanism (for example that lead to the expression of the maternal Igf2 allele), there will be very strong selection of such alleles, so that they will either be eliminated rapidly, or spread rapidly throughout the population. This may explain why, on the whole, the developmental potential of parthenogenetic embryos seems to be independent of the strain genotype in which they are made (Fundele et al. 1991).

There are a number of tests to which this theory should be submitted. First, it will be important to see what kind of molecules other imprinted genes encode, and whether or not they conform to the predictions (although it is possible that other categories of genes could be imprinted for different reasons). Second, it will be interesting to examine in which mammalian species genes like Igf2 are imprinted; a comparison of placental mammals with marsupials might be particularly informative. It is also worth checking that these genes are not imprinted in nonmammalian species. And third, it will be well worth examining phenotypes of parthenogenones and androgenones in nonmammalian species (e.g. Haack and Hodgkin 1991). Unequal resource transfer from parents to offspring is not limited to foetal development; any phenotypes observed with monoparental offspring could thus reveal the action of imprinted genes involved in such resource acquisition.

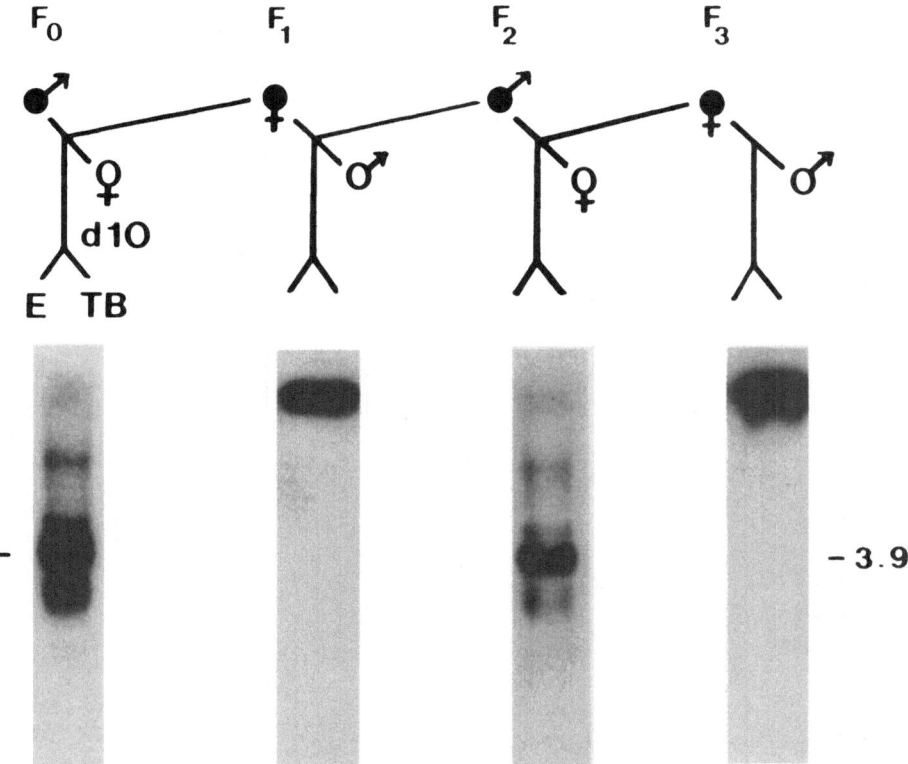

Fig. 7. Reversibility of transgene imprinting. Maternal and paternal transmission of the CAT17 transgene over several generations results in alternating patterns of high methylation and low methylation respectively. DNA was digested with HpaII and probed with a transgene probe. See Reik et al. (1987) for details

8 Imprinting of Transgenes and the Molecular Nature of Epigenetic Inheritance

The analysis of epigenetic modifications of transgenes in the mouse has resulted in the identification of a possible molecular mechanism for imprinting (Reik et al. 1990; Sapienza 1990). While it is at present unclear whether or not this molecular mechanism is involved in the imprinting of endogenous loci such as Igf2, these experiments tell us much about principles of epigenetic inheritance, the genetic control of epigenetic modifications and the general importance of epigenetic mechanisms in gene control. Many of these experiments have been recently discussed in detail (Surani et al. 1988b; Reik et al. 1990; Sapienza 1990; Surani et al. 1990a,b; Reik 1992); here, we only give a general summary insofar as it is relevant to parental imprinting. A typical example of the behaviour of an imprinted transgene is shown in Fig. 7, and Table 3 summarizes some of the properties

Table 3. Properties of imprinted transgenes

Designation of locus	Location	Imprinting[a]	Influence of modifier loci	Methylation differences in germline[b]	Mosaicism	References
CAT17		m > p	+			Reik et al. (1987, 1990)
TROPONIN I 379		m > p	+			Sapienza et al. (1987, 1989)
TROPONIN I		m > p				
TROPONIN I		p > m				
RSV-Myc		m > p	−	oo > sp		Swain et al. (1987) Chaillet et al. (1991)
HBsAg	Chr 13	m > p	+			Hadchouel et al. (1987)
TKZ-751		m(B/c) > p(B/c)	+		−	Allen et al. (1990)
OX1–5	Chr 7	m > p	+			Reik et al. (1990)
MPA434	Chr 11, A5	m > p	−	oo > sp		Sasaki et al. (1991)
Tg4	Chr 1	m > p	+		+	McGowan et al. (1989)
CMZ12[c]		m(B/c) > p(B/c)	+		+	Surani et al. (1990b)
Adp[c]		m > p				DeLoia and Solter (1990)
pHRD		m > p	+ (Ssm-1)		+	Engler et al. (1991)

[a] m > p: Maternal transmission of the transgene results in higher methylation than does paternal transmission. m(B/c) > p(B/c): Maternal transmission of Balb/c modifier results in higher methylation of the transgene than does paternal transmission. [b] oo: Oocyte; sp: sperm. [c] Note that in the case of CMZ12 and Adp no differences in methylation have been detected (see text for details).

of imprinted transgenes analyzed to date. The epigenetic modification involved in transgene imprinting is in all, but possibly two, cases DNA methylation at CpG residues, a known heritable modification that is associated with gene expression (Bird 1986). Typically, the methylation pattern on the transgene is characteristic of its mode of transmission, and in most cases paternal transmission leads to hypomethylation, whereas maternal transmission to hypermethylation. These methylation differences can be associated with expression or repression of the transgene in the offspring. For some transgenes the imprinting is reversible in the germline (Fig. 7; Reik et al. 1987), whereas with others epigenetic modification can persist, and indeed be cumulative, over several generations (Hadchouel et al. 1987; Allen et al. 1990). With some transgene constructs, imprinting seems to be position-dependent, whereas others show some degree of autonomy from the position of insertion (Chaillet et al. 1991). Using these constructs, it may be possible to begin to define some of the sequence requirements for an imprinted locus (similar experiments are of course possible using the endogenous imprinted genes, but only if they become imprinted in a position-independent fashion when reintroduced into the genome). The nature of the position dependence of other imprinted transgenes is unclear: however, allelic differences in methylation in the region of integration (in the absence of the transgene) have so far not been detected (Sasaki et al. 1991). Also, for a transgene to become imprinted, it does not appear to be necessary to be inserted into one of the imprinted regions of the genome (Reik et al. 1990).

One of the most important distinctions amongst imprinted transgenes is whether imprinting is retained on a homozygous genetic background (Table 3; Reik et al. 1990; Reik 1992). Two transgenes have been identified where this seems to be the case, and these may therefore constitute appropriate molecular models for some endogenous imprinted genes (Chaillet et al. 1991; Sasaki et al. 1991). Is DNA methylation the primary imprinting signal for these transgenes? That is, do methylation differences already exist in the gametes prior to fertilization, and do they persist during development? In the two cases studied, this seems to be the case. Thus, Chaillet et al. (1991) and Sasaki et al. (pers. comm.) have found that the imprinted transgenes RSV-Myc and MPA 434 respectively are unmethylated in foetal germ cells of both sexes irrespective of parental descent; in the female the transgenes then become methylated during the late stages of oogenesis (growth and maturation). Interestingly, some endogenous sequences also become methylated during late oogenesis, and the oocyte contains an extraordinarily high level of methylase enzyme (Howlett and Reik 1991; Monk et al. 1991). The fully methylated maternal copy of these transgenes is then kept methylated during preimplantation and postimplantation stages.

Methylation changes on the paternal copy are more complex. From the completely unmethylated form in day 14 male germ cells an increase in methylation is first observed during foetal development, and an additional increase during postnatal development of the testis (Chaillet et al. 1991).

Mature spermatozoa thus show a partially methylated pattern. Following fertilization, methylation is lost again from the paternal copy during pre-implantation development, the unmethylated state is reached in blastocysts. It is only after implantation and just prior to gastrulation that methylation increases again and the adult paternal pattern is established at that stage (Chaillet et al. 1991). The paternal transgene thus follows a pattern of methylation and demethylation which is very similar to that of some endogenous sequences (Howlett and Reik 1991). The maternal transgene, however, reaches a fully methylated state and remains locked into this methylated state until the formation of primordial germ cells in the embryo. Since some differences in methylation may now have been observed in maternal and paternal copies of endogenous imprinted genes (Chaillet 1992, and our unpubl. results), it will be crucial to find out whether or not these differences already exist in germ cells just prior to fertilization.

A number of imprinted transgenes do not retain imprinting on a homozygous genetic background (Allen et al. 1990; Reik et al. 1990; Engler et al. 1991). We believe that their imprinting can be achieved in one of the following two ways, both involving the action of genotype-specific modifier loci. The transgene pHRD becomes fully methylated in a single generation when exposed to the C57BL/6 (B6) allele of an unlinked modifier locus, Ssm1 (Engler et al. 1991). When the B6 allele of Ssm1 is segregated away from the methylated HRD transgene (and is replaced by DBA/2 alleles) methylation is lost from the paternally inherited transgene more rapidly than from the maternally derived one. How precisely this parental effect is brought about is at present unclear, but it tells us that there might be a number of loci that show parental effects in populations in which different modifier alleles segregate, but not in populations in which modifier alleles are homozygous and are the same throughout the population (such as in inbred strains of mice; Reik et al. 1990; Sapienza 1990; Reik 1992). For example, there are many genetic disorders in the human where variable penetrance and expressivity are observed together with parental trans-mission effects, suggesting that modifier genes of this type are involved (Reik 1988, 1989, 1992; Sapienza 1989, 1990; Hall 1990; Laird 1990).

A different situation is observed with the TKZ751 transgene (Allen et al. 1990). Here, the transgene becomes methylated on the BALB/c back-ground, but remains unmethylated on the DBA/2 background. Intriguingly, methylation upon exposure to the BALB/c modifier allele occurs as a dominant maternal effect (Fig. 8). Hence, the transgene becomes methylated when the BALB/c modifier allele is maternally derived, but not when it is paternally derived. The function of the BALB/c modifier allele is thus dependent upon its own parental origin, the parental origin of the transgene does not appear to matter. This imprinting of modifier alleles will lead to differences in phenotype in reciprocal crosses between inbred strains of mice. It is not known at present whether the maternal effect is caused by a nuclear gene, or whether the egg cytoplasm also plays a role (Surani et al. 1990b). Similar nonreciprocal expression phenotypes have also been observed with

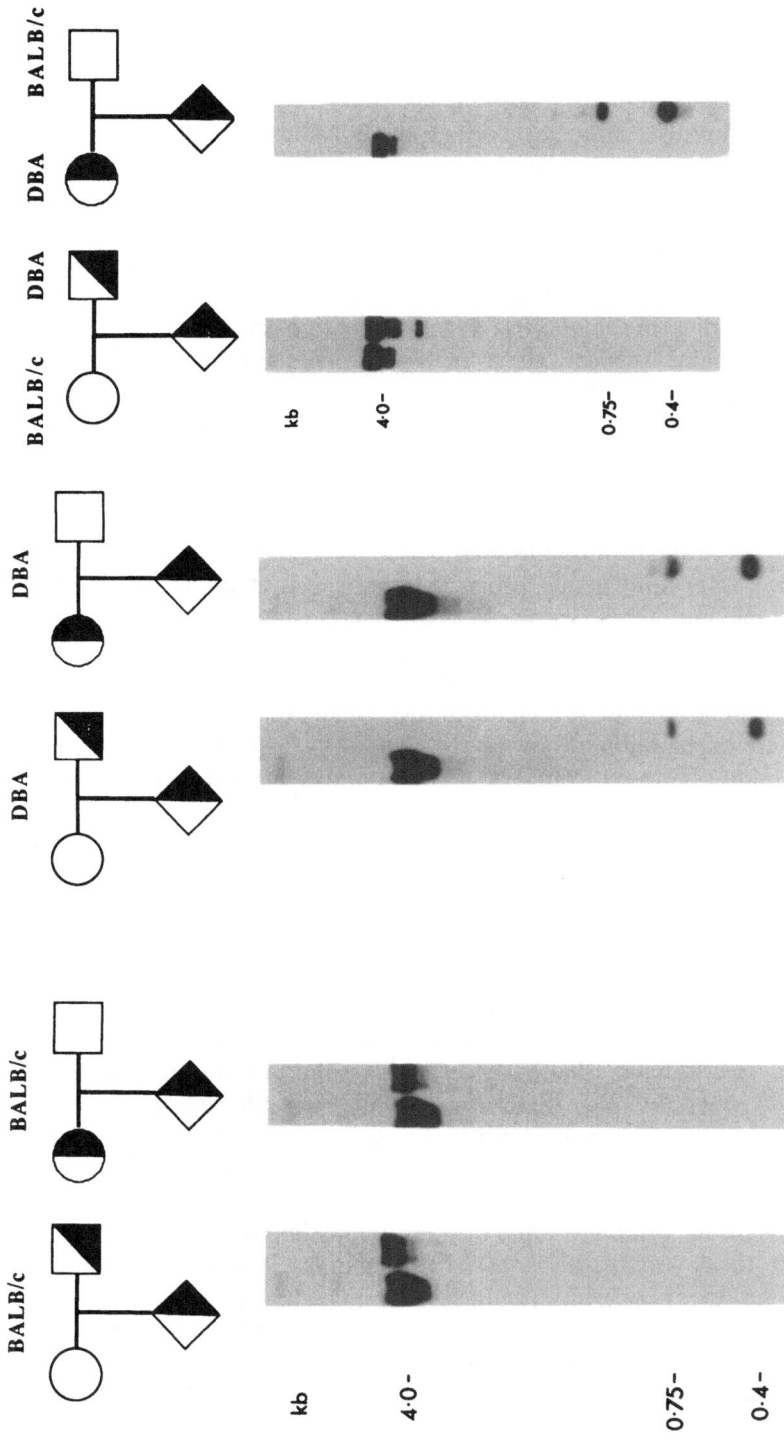

Fig. 8. Imprinting of the TKZ 751 transgene by modifier genes. In the BALB/c strain, the transgene is highly methylated irrespective of its own parental origin. In the DBA/2 strain, the transgene is hypomethylated regardless of parental origin. However, methylation increases in a reciprocal cross when BALB/c is the maternal genotype, but remains low when BALB/c is the paternal genotype. □ Males; ○ females; *half-filled symbols* transgene heterozygotes. Digestion was with SacI, and SacI + HpaII in each case and probed with a transgene probe. (After Allen et al. 1990)

endogenous loci (Klose and Reik 1992), and it has been suggested that some of them are caused by nucleo-cytoplasmic interactions between components of the egg cytoplasm and the parental chromosomes (Surani et al. 1990b; Reik 1992).

The nature of the modifier genes that can affect epigenetic programming is at present unknown, although one of them has been chromosomally mapped (Engler et al. 1991). However, because of overt similarities between transgene modification in the mouse and position effect variegation in *Drosophila melanogaster*, it is possible that they work in a similar way to the enhancers and suppressors of position effect variegation. In the fly, a number of these genes have now been cloned, and some of them seem to encode proteins that are involved in the formation of heterochromatic chromosome domains (which can form stable epigenetic switches; Paro 1990). Interestingly, a family of genes with homology to the *Drosophila* heterochromatin protein gene HP1 has recently been identified in the mouse (Singh et al. 1991). It will be important to see what kind of phenotype variant alleles of these genes can elicit.

Not surprisingly, the genetic identification of modifiers of epigenetic processes has led to the proposal that all parental imprinting may be caused by such genes ("imprinting genes"), particularly by those that are differently expressed in male and female gametogenesis. The most explicit proposal, by Sapienza (1990), contests that there are dosage-sensitive modifiers on the X-chromosome. This would lead to a double dose of their gene products in oogenesis as compared with spermatogenesis. This is unlikely, however, to be the complete explanation of how imprinting works, as it is possible to get perfectly normal offspring from female mice with an XO chromosome constitution.

9 Outlook

In this chapter I have tried to give a fairly comprehensive account of what is known about imprinting of autosomes in mammalian organisms. In the course of this description we have come to realize that astonishing progress has been made over the last couple of years in our understanding of the developmental significance of imprinting. We have also come to realize that there are a number of rather pressing questions that need to be addressed and that I believe will guide work over the next few years. As an outlook I shall therefore try to formulate some of these points and questions.

We need to devise methods to isolate additional imprinted genes, mainly in order to know what products they encode and what function they have in mammalian development. We need to see whether imprinting of some genes is conserved between mammalian species, in particular in placental mammals. Is there any imprinting of these or other genes in nonmammalian species?

Whenever new imprinted genes become identified, their influence on development can be tested by looking at the appropriate disomic mice, and phenotype and potency can be further analyzed in chimaeras. Precisely what type of developmental processes do imprinted genes influence?

The molecular mechanism of imprinting is and remains of the greatest concern. Are *cis*-acting sequences near imprinted genes responsible and can they be defined by reintroduction of parts of imprinted genes into the mouse genome? With some transgenes it now seems possible to begin to define some of the target sequence requirements. Is the opposite imprinting of the H19 and the neighbouring Igf2 gene a coincidence? Or can they only be imprinted in this arrangement, and thus would one only be able to recreate the imprinting in transgenic mice by transferring both genes on the same piece of DNA? Are there similar, oppositely imprinted genes elsewhere, for example near the Igf2 receptor gene?

What is the exact molecular level of repression of imprinted genes? Is it likely to be transcriptional or could post-transcriptional mechanisms also operate? Do different imprinted genes employ different levels of regulation?

Can epigenetic modifications, such as CpG methylation or DNAse I hypersensitivity, be detected that distinguish maternal and paternal alleles of imprinted genes? This question is most efficiently addressed using disomic embryos or even cell lines derived from such embryos. If methylation differences can be detected, can they be traced back to the gametes? Or are primary imprints of a different molecular nature and do imprints progress through different molecular phases? Could some of these phases be more stable than others, and hence can imprints be lost under specific circumstances?

Is all the imprinting essentially complete before fertilization or do imprints become modified, or even established, postfertilization and what is the role of nucleocytoplasmic interactions in this modification?

What is the molecular nature of modifier genes and how can we isolate the ones that have been genetically identified? Conversely, what is the function of the chromobox genes in mouse and man which have been isolated on the basis of homology with *Drosophila* heterochromatin proteins?

Is it possible to experimentally manipulate the imprinting process? Is it possible to change some of the phenotypes of parthenogenetic and androgenetic embryos or even to achieve parthenogenetic development to term?

If I were to venture a guess I would think that most of these questions, except perhaps the last one, will be answered in the next 5 to 10 years. Exciting times lie ahead.

Acknowledgements. First and foremost I would like to thank Azim Surani for the great influence he has had on my work and thinking over the years. Many thanks to Anne Ferguson-Smith, Nick Allen, Sarah Howlett, Robert Feil, Achim Gossler, Hiro Sasaki and Azim Surani for discussions and helpful comments on this manuscript. Even from Prim Singh I may have learned one or two useful things. I would also like to take this opportunity and acknowledge Davor Solter, Carmen Sapienza and Reinald Fundele for their continuing influence on my work. I am grateful to my colleagues Sarah Howlett, Jessica Penbeth, Lucy Bowden, Robert Feil, Irmgard Gurtmann and Joachim Klose for their contributions to this work and for allowing me to spend time away from the lab. Thanks to Bruce Cattanach, Nick Allen and Sarah Howlett for allowing me to use some of their illustrations. Finally, I gratefully acknowledge Dianne Styles and Linda Notton for their heroic struggles with my handwriting. The author is a Fellow of the Lister Institute of Preventive Medicine.

References

Allen ND, Norris ML, Surani MAH (1990) Epigenetic control of transgene expression and imprinting by genotype-specific modifiers. Cell 61:853–861

Barlow DP, Stoger R, Herrmann BG, Saito K, Schweifer N (1991) The mouse Igf-II receptor maps to the *maternal-effect* locus on chromosome 17 and is expressed only from the maternally inherited chromosome. Nature 349:84–87

Bartolomei MS, Zemel S, Tilghman SM (1991) Parental imprinting of the mouse H19 gene. Nature 351:153–155

Barton SC, Surani MA, Norris ML (1984) Role of paternal and maternal genomes in mouse development. Nature 311:374–376

Barton SC, Adams CA, Norris ML, Surani MAH (1985) Development of gynogenetic and parthenogenetic inner cell mass and trophectoderm tissues in reconstituted blastocysts in the mouse. J Embryol Exp Morphol 90:267–285

Barton SC, Ferguson-Smith AC, Fundele R, Surani MA (1991) Influence of paternally imprinted genes on development. Development 113:679–688

Bird AP (1986) CpG-rich islands and the function of DNA methylation. Nature 321:209–213

Brown KW, Williams JC, Maitland NJ, Mott MG (1990) Genomic imprinting and the Beckwith-Wiedemann syndrome. Am J Hum Genet 46:1000–1001

Brunkow ME, Tilghman SM (1991) Ectopic expression of the H19 gene in mice causes prenatal lethality. Genes Dev 5:1092–1101

Cattanach BM (1986) Parental origin effects in mice. J Embryol Exp Morphol (Suppl): 137–150

Cattanach BM, Beechey CV (1990) Autosomal and X-chromosome imprinting. Development (Suppl):63–72

Cattanach BM, Kirk M (1985) Differential activity of maternally and paternally derived chromosome regions in mice. Nature 315:496–498

Chaillet JR (1992) DNA methylation and genomic imprinting in the mouse. Sem Dev Biol 3:99–105

Chaillet JR, Vogt.TF, Beier DR, Leder P (1991) Parental-specific methylation of an imprinted transgene is established during gametogenesis and progressively changes during embryogenesis. Cell 66:77–84

DeChiara TM, Robertson EJ, Efstratiadis A (1991) Parental imprinting of the mouse insulin-like growth factor II gene. Cell 64:849–859

DeLoia JA, Solter D (1990) A transgene insertional mutation at an imprinted locus in the mouse genome. Development (Suppl):73–80

Engel E (1980) A new genetic concept: uniparental disomy and its potential effect. Am J Med Genet 6:137–143

Engler P, Haasch D, Pinkert CA, Doglio L, Glymour M, Brinster R, Storb U (1991) A strain-specific modifier on mouse chromosome 4 controls the methylation of independent transgene loci. Cell 65:939–948

Ferguson-Smith AC, Cattanach BM, Barton SC, Beechey CV, Surani MA (1991) Embryological and molecular investigations of parental imprinting on mouse chromosome 7. Nature 351:667–670

Fundele R, Norris ML, Barton SC, Reik W, Surani MA (1989) Systematic elimination of parthenogenetic cells in mouse chimeras. Development 106:29–35

Fundele R, Norris ML, Barton SC, Fehlau M, Howlett SK, Mills SK, Surani MA (1990) Temporal and spatial selection against parthenogenetic cells during development of fetal chimeras. Development 108:203–211

Fundele R, Howlett SK, Kothary R, Norris ML, Mills WE, Surani MA (1991) Developmental potential of parthenogenetic cells: role of genotype-specific modifiers. Development 113:941–946

Gardner, RL, Barton SC, Surani MA (1990) Use of triple tissue blatocyst reconstitution to study the development of diploid parthenogenetic primitive ectoderm in combination with fertilization-derived trophectoderm and primitive endoderm. Genet Res 56:209–222

Haack H, Hodgkin J (1991) Test for parental imprinting in the nematode C. elegans. Mol Gen Genet 228:482–485

Hadchouel M, Farza H, Simon D, Tiollais P, Pourcel C (1987) Maternal inhibition of hepatitis B surface antigen gene expression in transgenic mice correlates with de novo methylation. Nature 329:454–456

Haig D (1992) Genomic imprinting and the theory of parent-offspring conflict. Sem Dev Biol 3:153–160

Haig D, Graham C (1991) Genomic imprinting and the strange case of the insulin-like growth factor-II receptor. Cell 64:1045–1046

Hall JG (1990) Genomic imprinting: review and relevance to human diseases. Am J Hum Genet 46:857–873

Henry I, Bonaiti-Pellie C, Chehensse V, Beldjord C, Schwartz C, Utermann G, Junien C (1991) Uniparental paternal disomy in a genetic cancer-predisposing syndrome. Nature 351:665–666

Howlett SK (1991) Genomic imprinting and nuclear totipotency during embryonic development. Int Rev Cytol 127:175–192

Howlett SK, Reik W (1991) Methylation levels of maternal and paternal genomes during preimplantation development. Development 113:119–127

Kajii T, Ohama K (1977) Androgenetic origin of hydatidiform mole. Nature 268:633–634

Kaufman MH, Lee KKH, Spiers S (1989) Postimplantation development and cytogenetic analysis of diandric heterozygous diploid mouse embryos. Cytogenet Cell Genet 52:15–18

Kirkels VGHJ, Hustinx TWJ, Scheres JMJC (1980) Habitual abortion and translocation (22q; 22q): unexpected transmission from a mother to her phenotypically normal daughter. Clin Genet 18:456–461

Klose J, Reik W (1992) Expression of maternal and paternal phenotypes at the protein level. Sem Dev Biol 3:119–126

Laird CD (1990) Proposed genetic basis of Huntington's disease. Trends Genet 6:242–247

Latham KE, Solter D (1991) Effect of egg composition on the developmental capacity of androgenetic mouse embryos. Development 113:561–568

Lyon MF, Glenister P (1977) Factors affecting the observed number of young resulting from adjacent-2 disjunction in mice carrying a translocation. Genet Res 29:83–92

Malcolm S, Clayton-Smith J, Nicholas M, Robb S, Webb T, Armour JAL, Jeffreys AJ, Pembrey ME (1991) Uniparental disomy in Angelman's syndrome. Lancet 337:694–697

Mann JR, Lovell-Badge RH (1984) Inviability of parthenogenones is determined by pronuclei, not egg cytoplasm. Nature 310:66–67

Mann JR, Gadi I, Harbison ML, Abbondanzo SJ, Stewart CL (1990) Androgenetic mouse embryonic stem cells are pluripotent and cause skeletal defects in chimeras: implications for genetic imprinting. Cell 62:251–260

McGowan R, Campbell R, Peterson A, Sapienza C (1989) Cellular mosaicism in the methylation and expression of hemizygous loci in the mouse. Genes Dev 3: 1669–1676

McGrath J, Solter D (1984) Completion of mouse embryogenesis requires both the maternal and paternal genomes. Cell 37:179–183

McGrath J, Solter D (1986) Nucleocytoplasmic interactions in the mouse embryo. J Embryol Exp Morphol (Suppl):277–290

Monk M, Surani MA (1990) Genomic imprinting. Development (Suppl)

Monk M, Adams RLP, Rinaldi A (1991) Decrease in methylase activity during preimplantation development in the mouse. Development 112:189–192

Moore T, Haig D (1991) Genomic imprinting in mammalian development: a parental tug-of-war. Trends Genet 7:45–49

Nagy A, Sass M, Markkula M (1989) Systematic non-uniform distribution of parthenogenetic cells in adult mouse chimeras. Development 106:321–324

Nagy A, Gócza E, Diaz EM, Prideaux VR, Iványi E, Markkula M, Rossant J (1990) Embryonic stem cells alone are able to support fetal development in the mouse. Development 110:815–821

Nicholls RD, Knoll JHM, Butler MG, Karam S, Lalande M (1989) Genetic imprinting suggested by maternal heterodisomy in non-deletion Prader Willi syndrome. Nature 342:281–285

Nicholls RD, Rinchik EM, Driscoll DJ (1992) Genomic imprinting in mammalian development: Prader-Willi and Angelman syndromes as disease models. Sem Dev Biol 3:139–152

Palmer CG, Schwartz S, Hodes ME (1980) Transmission of a balanced homologous t (22q; 22q) translocation from mother to normal daughter. Clin Genet 17: 418–422

Paro R (1990) Imprinting a determined state into the chromation of Drosophila. Trends Genet 6:416–421

Reik W (1988) Genomic imprinting: a possible mechanism for the parental origin effect in Huntington's chorea. J Med Genet 25:805–808

Reik W (1989) Genomic imprinting and genetic disorders in man. Trends Genet 5:331–336

Reik W (1992) Genome imprinting. In: Grosveld F, Kollias G (eds) Transgenic animals. Academic Press: 99–126

Reik W, Collick A, Norris ML, Barton SC, Surani MAH (1987) Genomic imprinting determines methylation of parental alleles in transgenic mice. Nature 328: 248–251

Reik W, Howlett SK, Surani MA (1990) Imprinting by DNA methylation: from transgenes to endogenous gene sequences. Development (Suppl):99–106

Sapienza C (1989) Genome imprinting and dominance modification. Ann NY Acad Sci 564:24–38

Sapienza C (1990) Sex-linked dosage-sensitive modifiers as imprinting genes. Development (Suppl):107–114

Sapienza C, Tran TH, Paquette J, McGowan R, Peterson A (1987) Degree of methylation of transgene is dependent on gamete of origin. Nature 328:251–254

Sapienza C, Paquette J, Tran TH, Peterson A (1989) Epigenetic and genetic factors affect transgene methylation imprinting. Development 107:165–168

Sasaki H, Hamada T, Ueda T, Seki R, Higashinakagawa T, Sakaki Y (1991) Inherited type of allelic methylation variations in a mouse chromosome region where an integrated transgene shows methylation imprinting. Development 111:573–581

Searle AG, Beechey CV (1978) Complementation studies with mouse translocations. Cytogenet Cell Genet 20:282–303

Searle AG, Beechey CV (1990) Genetic imprinting phenomena on mouse chromosome 7. Genet Res 56:237–244

Singh PB, Miller JR, Pearce J, Kothary R, Burton RD, Paro R, James TC, Gaunt SJ (1991) A sequence motif found in a Drosophila heterochromatin protein is conserved in animals and plants. Nucleic Acids Res 19:789–794

Solter D (1988) Differential imprinting and expression of maternal and paternal genomes. Annu Rev Genet 22:127–146

Spence JE, Perciaccante RG, Greig GM, Willard HF, Ledbetter DH, Heijtmancik JF, Pollack MS, O'Brien WE, Beaudet AL (1988) Uniparental disomy as a mechanism for human genetic disease. Am J Hum Genet 42:217–226

Surani MA (1991) Genomic imprinting: developmental significance and molecular mechanism. Curr Op Genet Dev 1:241–246

Surani MAH, Barton SC, Norris ML (1984) Development of reconstituted mouse eggs suggests imprinting of the genome during gametogenesis. Nature 308:548–550

Surani MA, Barton SC, Norris ML (1986a) Nuclear transplantation in the mouse: heritable differences between parental genomes after activation of the embryonic genome. Cell 45:127–136

Surani MAH, Reik W, Norris ML, Barton SC (1986b) Influence of germline modifications of homologous chromosomes on mouse development. J Embryol Exp Morphol (Suppl):123–136

Surani MA, Barton SC, Howlett SK, Norris M (1988a) Influence of chromosomal determinants on development of androgenetic and parthenogenetic cells. Development 103:171–178

Surani MA, Reik W, Allen ND (1988b) Transgenes as molecular probes for genomic imprinting. Trends Genet 4:59–62

Surani MA, Allen ND, Barton SC, Fundele R, Howlett SK, Norris ML, Reik W (1990a) Developmental consequences of imprinting of parental chromosomes by DNA methylation. Philos Trans R Soc Lond B326:313–327

Surani MA, Kothary R, Allen ND, Singh PB, Fundele R, Ferguson-Smith AC, Barton SC (1990b) Genome imprinting and development in the mouse. Development (Suppl):89–98

Swain JL, Stewart TA, Leder P (1987) Parental legacy determines methylation and expression of an autosomal transgene: a molecular mechanism for parental imprinting. Cell 50:719–727

Temple IK, Cockwell A, Hassold T, Pettay D, Jacobs P (1991) Maternal uniparental disomy for chromosome 14. J Med Genet 28:511–514

Thomson JA, Solter D (1987) The developmental fate of androgenetic, parthenogenetic, and gynogenetic cells in chimeric gastrulating mouse embryos. Genes Dev 2:1344–1351

Thomson JA, Solter D (1988) Chimeras between parthenogenetic or androgenetic blastomeres and normal embryos: allocation to the inner cell mass and trophectoderm. Dev Biol 131:580–583

Voss R, Ben-Simon E, Avital A, Zlotogora Y, Dagan J, Godfrey S, Tikochinski Y, Hillel J (1989) Isodisomy of chromosome 7 in a patient with cystic fibrosis: could uniparental disomy be common in humans? Am J Hum Genet 45:373–380

Subject Index